On a Cushion of Air

by

Robin Paine & Roger Syms

On a Cushion of Air

by

Robin Paine and Roger Syms

ISBN: 978-0-9568978-1-7

This book is published by Future Reality Limited in conjunction with **WRITERSWORLD**, and is produced entirely in the UK. It is available to order from most bookshops in the United Kingdom, and is also globally available via UK based Internet book retailers.

Copy edited by Ian Large
Cover design Jag Lall

WRITERSWORLD
2 Bear Close Flats, Bear Close, Woodstock
Oxfordshire, OX20 1JX, England
☎ 01993 812500
☎ +44 1993 812500

www.writersworld.co.uk

The text pages of this book are produced via an independent certification process that ensures the trees from which the paper is produced come from well-managed sources that exclude the risk of using illegally logged timber while leaving options to use post-consumer recycled paper as well.

For all those who worked for Hoverlloyd
on both sides of the English Channel.

Foreword by
His Royal Highness The Duke of Edinburgh

BUCKINGHAM PALACE.

As with so many original ideas, when they are first proposed they appear to be so obvious that it seems odd that no-one has thought of them before. When Sir Christopher Cockerell suggested that heavy weights could be supported on a cushion of air, and hence made easy to move around, it seemed to be too good to be true – until the problems of converting the idea into practical reality came to be tackled.

This intriguing book traces the process of overcoming the mountain of practical problems which arose in the process of developing a working hovercraft, and then in converting that into a wholly new form of transport. Problem solving has always been the challenge for engineers, and this book makes it evident that the engineers involved in the development of a practical money-earning hovercraft faced more than their fair share of problems to be solved.

When I was invited by Sir Christopher to visit his establishment on the Isle of Wight, I jumped at the chance to see this strange new vehicle and to drive/fly his invention on the waters of the Solent. Just over ten years later I was invited to open Hoverlloyd's Pegwell Bay Hoverport and to go for a trip in the giant SR N4 '*Swift*'. Progress had indeed been swift, and this account of those hectic ten years, and of the ensuing thirty years of cross-Channel operations, tells a remarkable story of persistence, ingenuity, original thinking and grinding hard work.

Philip

About the Authors

Robin Paine

Robin Paine was born in the USA in 1942. He went to prep school in Tonbridge, before entering the Nautical College Pangbourne, following which he joined the British India Steam Navigation Company in 1959 as a cadet serving in cargo and passenger ships to India, Ceylon, the Persian Gulf, East Africa, Hong Kong, Malaysia, Australia and New Zealand. After nearly six years with the company he joined Cunard with a First Mate's Certificate as 2nd Officer on the trans-Atlantic cargo ships, one of which lost its rudder in a storm 600 miles from Newfoundland, resulting in an SOS and a 17 day tow back to Southampton. After obtaining his Master's Foreign Going Certificate in 1967, he decided to look to a life closer to home. He joined Townsend Ferries on their new roll-on roll-off ship, *Free Enterprise III*, on the Dover-Zeebrugge service and within a week found himself as chief officer.

But the lure of the hovercraft was too great and he managed to get in, more or less on the ground floor, with Hoverlloyd's new operation with the giant SR.N4 out of Ramsgate. A Private Pilot's licence, combined with his Master's Certificate may have helped to secure his position and he was also fortunate enough to have been given a few hours training in the smaller SR.N6 before formally joining the company on 1st January 1969 as 2nd Officer, helping to develop the techniques for high speed navigation on the larger hovercraft. Within a year he was promoted to captain and then Safety Captain in 1972. He remained with Hoverlloyd until 1979, before moving on to pursue his business interests, but has always maintained a keen interest in the concept of hover transport.

Roger Syms

After 11 years at sea, with BP Tankers and the Royal Fleet Auxiliary – the stores arm of the Royal Navy – carrying everything from cases of beer to nuclear devices, Roger came ashore as a Master Mariner to study for a degree in Nautical Science.

By 1969 when Hoverlloyd's cross-Channel service commenced, he was beginning to realise that he was studying for a qualification that rendered him suitable for the very thing he didn't want to do, most of which would involve working in a London office, just the sort of employment he had originally gone to sea to avoid.

An offer of a navigator's job with Hoverlloyd, just for the season, came as a life-saver. As a further bonus, halfway through the summer, he was offered permanency.

Roger was promoted to captain in 1972 and then went on to be Flight Manager in Hoverlloyd at the time of the expansion to four craft in 1977. At the time of the merger, and the creation of Hoverspeed, he took up the position of Flight Manager in the merged company. Unable to settle into the new company's culture Roger left in 1982.

He now lives in Launceston, Tasmania after a successful second career as an academic starting with the Australian Maritime College and then moving on to run his own training company in Australia and overseas, mainly in India and Pakistan.

He, like most other Hoverlloyders, looks back on his 13 years spent with the 'big machines' as the most exciting and rewarding time of his life.

Acknowledgements

His Royal Highness The Duke of Edinburgh, in his foreword to this book, has clearly revealed his keen interest in the hovercraft industry from its very inception. We would like to thank His Royal Highness for his continued interest so amply demonstrated by his readiness to provide the foreword.

Thanks also go to BAE Systems for their generosity and support, originally by providing a substantial donation to the failed memorial project, and then by very kindly transferring those funds towards the publishing expenses, which helped to make this book possible.

It was always our intention that this book should not just be our story but it should include the stories of as many of those involved as possible. We therefore asked our erstwhile colleagues to send us their recollections, impressions and anecdotes, to which they responded magnificently. So much so, that we have a very large number to thank, with the attendant risk that out of the huge number of e-mails, letters, phone calls and conversations, some may have been lost in the crowd. If anyone has been left out, please accept our apologies.

Something that was not originally intended was the subsequent broad range of the book, which spread during its writing, almost by a will of its own, to cover the whole of the hovercraft industry beyond our own company Hoverlloyd. We realised early on that solely trying to write the history of one company was akin to trying to describe a horse while only concentrating on one leg.

The consequence of this expanded view was the necessity to conduct in-depth interviews with many more people, particularly those involved before and during the early stages of hovercraft development, or who were in senior management, and were therefore able to provide previously unknown detail and insights into this fascinating history.

First and foremost we must thank David Wise, who is the only surviving senior manager of Hoverlloyd. As one of the company's 'originals' in 1966, moving on to become Associate Director of Operations, and finally retiring as Operations Manager in 1993 of the merged company Hoverspeed, David's inside knowledge of the cross-Channel operation, together with his strong links with the aircraft industry and the British Hovercraft Corporation in particular, provided us with unique and invaluable information without which the book would have been a whole lot poorer.

In the same light, two other pioneers, Bill Williamson and George Kennedy, who enviably can claim to have been there at the absolute beginning, joining the very first commercial hovercraft enterprise, Clyde

Hover Ferries in 1965, gave us a colourful picture concerning the illusions of both operators and manufacturers at that early stage. Both were also instrumental in the development of Hoverlloyd during the evolutionary period of SR.N6 trials and the first difficult years of the SR.N4 at Pegwell Bay; again priceless information that few others could provide.

Other major contributors from Hoverlloyd were June Cooper, deputy chief stewardess and her sister Betty Dowle, who supplied between them, both on paper and in interview, a mass of material on the all-important cabin staff.

Without Mike Castleton, 'our man in Calais', half of the chapter *Making a Profit* couldn't have been written. At the same time we must thank Laurette Wacogne and Manu Heatley for adding an essential French flavour to that part of the story.

It is time we exposed our 'sleeper' in the engineering department – Crew Chief Mike Fuller. His full and detailed account of the technical side of keeping the SR.N4s moving was invaluable and he was constantly on hand to answer our numerous queries. We have a lot for which to thank him.

Trying to write the history of the engineering side of the operation was made more difficult than usual by the fact that the whole technical management team are no longer with us. One stroke of luck, quite late in the day, was tracing Technical Director Emrys Jones' wife, Joan, who very kindly sent us a pile of his documents and papers, which helped to fill in some of the gaps.

Similarly, our thanks go to Jane Wilson, daughter of Les Colquhoun, Hoverlloyd's first managing director, who made available her father's collection of material.

Before we embark upon mentioning the many ex-Hoverlloyd employees, who kindly passed on their memories, photographs and anecdotes, it should be explained that, although everyone deserves our thanks for their contribution, it has not been possible to use all the material. In instances where they covered a similar topic, we picked out the piece which most fitted the flow of the narrative, or more fully expressed a point we were making, so we offer our apologies to those whose efforts did not make it into print.

Our special thanks to all who contributed: **Flight crew:** Ted Ruckert; Hugh Belasyse-Smith; Linton Heatley; David Ward; Dennis Ford; Mike Rowland-Hill; 'Big' John Lloyd; John 'Biggles' Lloyd; Mick May; Tony Quaife; Ken Mair; Rodney Lake; Peter Gray; Geoff Riches; Jon Morris; James Read; George Lang; Robert Case. **Car deck:** Ted Unsted;

Dennis Holness; Peter Bennett. **Cabin staff:** Maureen Langford; Katrina Patchett; Cathy Withnall; Val Hughes; Jacquie Case; Di Stewart; Lynn Gibbons; Madeleine Marsden; Janine Dawson; Sally Burn; Brenda King; Delice Purvis; Lorna Monje. **Engineering:** Mike Thompson; Mick Bacon; Russell Mees; Paul Quaile; Mick Prior; Peter James; Bernie Chilvers. **Traffic Department:** Francis Balloy; Laurette Wacogne; Benoit Rozan: Nick Smith; Pat Lawrence; Vivienne Babbington. **Reservations:** Mary Ford. **Administration:** Janet Thomas; Peta Colquhoun. **Customs:** 'Pedro' Humphreys.

It is worth noting that our photo archive alone numbers around 1,500 items in donated photographs, not counting hundreds of documents, press cuttings and articles, all meticulously catalogued by our untiring and helpful archivist Tony Quaife. Thank you, Tony.

Once we inexorably moved away from Hoverlloyd to fill in the broader industry background, we were indeed lucky to meet up with two of the remaining pioneer members of the British Hovercraft Corporation, Ray Wheeler, Chief Designer of the SR.N4, and Bob Strath, test pilot.

Ray, a friend of Sir Christopher and his wife Margaret, worked in the stress and design departments throughout the years of hovercraft development. He was heavily involved with work on the skirt structure, without which the industry would never have, literally, got off the ground. Bob, starting out as assistant to Peter 'Sheepy' Lamb, and latterly as a test pilot in his own right, was involved with hovercraft testing from the SR.N1 through to the SR.N4 Mk.3.

Considering we are looking back through some sixty years of history, it is not surprising that both these amazing gentlemen are well into their eighties, but thankfully both have clear memories of the period in question. Their detailed accounts of the BHC days are priceless and we count ourselves fortunate to have them. In addition, we are extremely grateful to Ray for allowing us to use material from two of his books *From River to Sea* and *The Beginning – The SR-N1 Hovercraft* in relation to Saunders-Roe and the early development of the hovercraft. There seemed little point in us trying to 'reinvent the wheel'. Our thanks go too to Bob's son-in-law, Tony Baker, for scanning Bob's documents and pictures and, where necessary, emailing them to the other side of the world.

We have not included an in-depth history of Sir Christopher, nor have we been able to gain access to his papers, as his daughter, Mrs. Frances Airey, plans to have a full biography written of her father's achievements. We would, however, like to thank her for granting an interview and for providing some information about her father and mother during a period in the development of hovercraft that was not always as rewarding as it might have been.

Also we should not forget John Chaplin, who moved on from BHC early on and eventually became involved in the exciting developments in military hovercraft in the USA. His input presented a different perspective on the commercial hovercraft sphere, which was very welcome.

Our thanks too go to Anthony Brindle, who in 1966 acted as midwife and wet-nurse to the nascent British Rail company Seaspeed. Without his entrepreneurial gifts and unbounded enthusiasm the company would not have advanced beyond its first year or two. He showed the same infectious enthusiasm when asked to contribute to this volume, for which we are very grateful.

We were also lucky to have the help of two of the very first Seaspeed captains, Peter Barr and John Syring. Peter saw the whole saga through, almost to the end, not quite making it to an innings 'not out' when he retired in 1996. His experience, from the very start, of the various stages of Seaspeed's development and the final days with Hoverspeed, were an asset to the overall story. John was most helpful with his early hovercraft experiences on both the SR.N6 and SR.N4 and his subsequent time with the Civil Aviation Authority (CAA) as their test pilot.

Thanks also go to another Seaspeed 'original', Brian Laverick-Smith, who very nearly 'carried his bat', retiring in 1996 due to ill health. His enthusiasm to help in all aspects of the project has been much appreciated.

Although strictly not part of the cross-Channel story, we felt that the history of Hovertravel could not be ignored, for no better reason than the inescapable fact of its longevity. The company was one of the first to start up a commercial hovercraft operation, within days of Clyde Hover Ferries in 1965, and still prospers to this day. Our grateful thanks go to Christopher Bland and Barrie Jehan for granting interviews to tell their fascinating stories about Hovertravel and its associated company Hoverwork.

We are indebted too to Peter Yerbury and Robin Wilkins for bringing us up to date with the Hoverspeed story, and in particular a clear account of the management buy-out events. Without the people directly involved the whole epic would not be complete.

Although not directly related to the book, we would like to take this opportunity to thank the committee of ex-Hoverlloyd employees, who made valiant efforts on behalf of us all to get the Pegwell Bay memorial off the ground. Unfortunately, owing to funding difficulties and hostility from local residents at Cliffsend, the project failed, replaced in its stead by this volume, which we hope will be regarded as an adequate substitute.

Thanks go firstly to Dennis Ford, who from his vantage point in the Isle of Wight, was well placed to handle negotiations with the Hovercraft Museum and GKN, the name that superseded Westland, in order to refurbish an SR.N4 propeller kindly donated by Warwick Jacobs, Curator of the Hovercraft Museum. It is a great shame that Dennis' hard labour came to naught. We are doubly grateful to Warwick for providing so many pictures for the book from the Hovercraft Museum library and for his great fund of knowledge relating to the history of hovercraft.

David Ward and Geoff Riches, and many other supporters, deserve our gratitude for steadfastly tackling the hostile environment in many of the Cliffsend resident meetings and for successfully obtaining planning consent for the memorial, only to have our hopes dashed by promised funding not materialising.

'Big' John Lloyd[1] also helped with the initial design of the plinth and put a great deal of work into investigating suitable materials.

As with all projects of this nature, which requires an inordinate amount of time, thought and dedication, the 'better half' usually bears the brunt of it. Fortunately we just about managed to save both marriages, and for that we are extremely grateful for the forbearance and support of our wives.

Living as we do in the bottom right hand corner of the world, and even then, fifteen hundred miles apart in New Zealand and Tasmania, without the wonders of modern IT, the task would have been almost impossible. So thanks to Microsoft e-mail and Skype VOIP for the means of instant access across the globe. One cannot imagine how we would have coped without it. Our particular thanks go out to Google for the chance to use Google Earth and Google Maps as a basis for our many diagrams.

Our grateful thanks finally go out to the Writersworld team, Graham Cook, Ian Large, Charles Leveroni and Jag Lall, whose experience, skill and dedication to this project were inestimable. As babies in the minefield of self-publishing we were well served and could not have done without them.

Roger Syms
Robin Paine

[1] This is by way of a posthumous thank you. Sadly John, a stalwart Hoverlloyd 'original', died aged 70 on 3rd December 2011.

Contents

Introduction

In the summer of 1979 the Pegwell Bay Hoverport at Ramsgate, the size of an average domestic airport, was in full swing. Each day, four large SR.N4 hovercraft carrying full loads of cars and passengers departed for Calais and returned, covering a busy 27-departure schedule from 6.00am to late in the evening, a movement on and off the landing area roughly every 15 minutes. The approach road was crammed with arriving and departing vehicles, the waiting car parks were full, as were the check-in area, the departure lounges and cafeterias. All was frenetic activity and bustle.

After an uncertain beginning in 1969, in a pioneering effort to make a new and untried concept work, the company had persevered through a series of incidents and near disasters to arrive a decade later as a highly motivated and efficient operation, which was at the very pinnacle of its success. By then, Hoverlloyd was moving a million and a quarter passengers and their cars annually, most of which were carried in the summer holiday months, June to September.

It is hard to accept now that, after such success, a little more than two years later, Hoverlloyd as a company ceased to exist.

Although Hoverlloyd faded into history in 1981, this was not the end of the cross-Channel hovercraft story. It carried on as a merged Hoverlloyd and Seaspeed company under the name of Hoverspeed, operating out of the port of Dover. Strictly speaking Hoverlloyd had not died, but simply changed its name. However, for us dedicated Hoverlloyders it was not just the end of a passionate affair with a much loved organisation, but also the fact that we had lost our home and, like the Israelites in the Bible, we had been driven into exile, with our version of the waters of Babylon being the waters of Dover Harbour some 19 miles down the road.

The Pegwell facility stayed operational for another few years just as a maintenance base, but in 1995 the hoverport buildings, after a long period of neglect and deterioration, were razed to the ground leaving only the concrete foundations. What remains today is an overgrown dirt-strewn pad with little indication of what it once was.

So well has this small particle of history been erased that when we approached a bank, within a short distance of the old hoverport, to open an account to deal with the expenses of the 40-year reunion in 2009, we were somewhat shocked that the young bank employee had never heard of Pegwell Bay Hoverport or the hovercraft. Perhaps not so surprising, assuming this particular young man to be in his early twenties, was that when the hoverport ceased to function he would still have been a toddler.

Hoverspeed continued on for almost another 20 years, but in 2000 eventually succumbed to the inevitable. The painful truth was that large hovercraft, as far as present technology is concerned, were uneconomical compared with conventional ship ferries and therefore simply could not compete. The very small profit margin cross-Channel hovercraft could achieve was totally nullified with the loss of the lucrative duty-free sales when the European Union removed tariffs between member nations. The final nail in the coffin was the Channel Tunnel, with which even the ferry companies are still finding it difficult to compete.

So one motivation for this volume is to tell the full story of a phenomenon that appeared and disappeared in a relatively short period of time, with Seaspeed's start of operations in Dover in 1968 and ending, a little over 30 years later, also in Dover. In fact, Dover featured even earlier in the narrative when in 1959 the first experimental hovercraft, the SR.N1, with its inventor Christopher Cockerell on board, roared up the harbour beach to complete the first crossing of the English Channel by hovercraft.

The beginning and end at Dover also points to what is in fact the main reason for relating this history. As Hoverlloyders we are strongly motivated to tell *our* story for the simple reason that in telling the Dover to Dover story the parallel history at Pegwell Bay has tended to be forgotten. Much has been written since the last hovercraft left Dover in 2000 but in many instances Hoverlloyd is either ignored altogether or is mentioned in an off-hand manner; and often dismissed in terms of, "Oh! And there was also another operation up the road at Ramsgate", forgetting that for most of the first decade when Hoverlloyd existed, it was by far the major cross-Channel hovercraft operator in terms of number of hovercraft and cars and passengers carried.

Finally, there is a third reason for the book. Someone had to put on record what an amazing and truly unique company Hoverlloyd was. Companies that engender such an amount of devotion and loyalty in their employees are very few and far between. There is not one ex-Hoverlloyd employee who does not look back on their brief time working there and not proclaim it to be the happiest and most fulfilling time of their lives. Such an achievement cannot be allowed to go by without at least some acknowledgement.

Roger Syms
Robin Paine

SR.N4 *Swift* arriving at Pegwell Bay on 14th January 1969.

Prologue

In the western world, 1969 heralded a new age of advanced aeronautical and aerospace technology: The Boeing 747 flew for the first time; the Hawker Siddeley Harrier Jump Jet went into service with the RAF; *Apollo 10* flew within 16km of the Moon's surface; Neil Armstrong and Edwin Aldrin then landed on the Moon in *Apollo 11* and became the first humans to walk on the Moon's surface; *Mariner 7* made its closest flyby of Mars; Concorde first flew and later that year broke the sound barrier for the first time; the first computer-to-computer link, the predecessor of the contemporary global Internet, was established by Arpanet; and Charles Conrad and Alan Bean in *Apollo 12* became the third and fourth human beings to walk on the Moon.

Although less heralded, and to a large extent now almost forgotten, 1969 saw a 'new technological kid on the block' making its entry as the first regular[1] car and passenger cross-Channel service between Pegwell Bay near Ramsgate in England and the French port of Calais. It was neither wholly aeronautical, nor marine, but a combination of both. It was a giant 'hovercraft'.

---

[1] Seaspeed ran the first SR.N4 cross-Channel service for a few weeks in the summer of 1968 before it was withdrawn for major modifications.

On Wednesday 2nd April 1969 at Pegwell Bay Hoverport, near Ramsgate, Captain Bill Williamson, Hoverlloyd's operations manager and senior captain, climbed the ladder to the flight deck, some 13 feet above the car deck. At 0900 precisely, sitting in the left-hand seat, he asked his 1st officer in the right-hand seat to open the throttles of the four Rolls Royce Marine Proteus gas turbine engines. The 130-foot long, 165-ton giant SR.N4 Mk.1 hovercraft, with its load of 254 passengers and 30 cars, lifted up on its 8-foot cushion of air. Bill manoeuvred *Swift* towards the hoverport's northern ramp and out across the Pegwell Bay mud, which, at low tide, extended for a mile and a quarter. With 13,600 shp (shaft horsepower) available, he increased the power on all four engines and set the four propeller pitch levers to give him a speed of 55 knots, (63mph). Each 19-foot diameter propeller, the largest of their kind in the world, turned at 600rpm as *Swift* sped across the Goodwin Sands (the graveyard of an estimated fifteen hundred ships) and out into the Straits of Dover, renowned as the busiest shipping lanes in the world, on the 31.5 nautical mile route to Calais. The Marine Proteus engines were consuming aviation kerosene (Avtur) at a rate of 1,000 gallons (4,546 litres) per hour.

Bill listened attentively to his navigator, the 2nd officer, who was peering into a Decca 629 radar in a north-up relative motion display on the six-mile range. Sitting behind the captain and 1st officer, and surrounded by curtains to keep out the glare on the screen, he called out the bearings and distances of the multitude of targets in the Channel that needed to be avoided – a new concept in two-dimensional high-speed collision avoidance and navigation over which, unlike on ships, the captain had no direct control.

Meanwhile, in the two passenger cabins either side of the car deck, six attractive young stewardesses were serving drinks and duty free goods, while on the car deck the five car-deck crewmen kept an eye on the vehicles they had lashed by their wheels to ring bolts on the car deck. Now they were awaiting the order from the flight deck to 'unlash', shortly before arrival at Calais.

Forty minutes later *Swift* arrived on the pad at the new specially built hoverport at Calais, just to the east of Calais Harbour.

In the late afternoon of Sunday 1st October 2000 at the merged Hoverlloyd and Seaspeed's hoverport at Dover Western Docks, Captain Nick Dunn ordered his 2nd officer to close the throttles of the four up-rated Marine Proteus engines on the 185-feet, 320-ton SR.N4 Mk.3, *The Princess Margaret*. The 321 passengers and 52 cars disembarked and so ended exactly 31 years and six months of continuous SR.N4 operations across the Channel – the era of the giant car and passenger carrying hovercraft was closed.

Although the hovercraft survives to this day, and flourishes as a utility and military vehicle and as a small passenger carrier in many parts of the world, it is unlikely that a giant car and passenger hovercraft of the size of the SR.N4 will ever again be seen plying to and fro across the English Channel at speeds, which, as far as records go, even the Channel Tunnel car and freight trains have been unable to match.

This is the story of how private foreign enterprise in the form of Swedish Lloyd and Swedish America Line, who formed a British company called *Hoverlloyd*, galvanised the British government into supporting this new concept in transport through the formation of a British Rail subsidiary called *Seaspeed*. It is a story, told by those who were there, of how young adventurous men and women, mostly in their twenties and early thirties, took on the exciting challenge of getting an operation, in which they all believed, off the ground. It tells of the difficulties and near disasters, through lack of experience, that nearly wrote off the industry in the early days; the clashes of cultures between the free enterprise and government operations before and after they merged; and why, after so much early promise, the great adventure with the giant car and passenger carrying hovercraft came to an end.

The hovercraft story begins with Saunders-Roe in 1870, the boat and flying boat builder, which eventually agreed to investigate, and subsequently develop, Christopher Cockerell's discovery in 1953 that heavy weights could be supported on a cushion of low pressure air and, more importantly, that the concept could be practically applied. Much has already been written about Christopher Cockerell, later Sir Christopher, and the development of the hovercraft and the industry to the present day. Those relevant parts are repeated in this book, together with new material that has come to light, to provide a comprehensive narrative of the history of the development of the hovercraft, together with the early machines and the companies that operated them, culminating in the giant SR.N4 cross-Channel operations.

Recently, Angus Watson, a highly regarded freelance journalist who contributes regularly to most of the British national newspapers, was asked by an editor to write an article listing 'Britain's Greatest Inventions that have enriched the lives of billions'. The published article listed: *The steam engine; the telephone; the television; the Brompton folding bicycle; the clockwork radio; the Worldwide Web; and the hovercraft.* While the hovercraft may not have made as big a contribution as the others for enriching the lives of billions, it certainly created a source of fascination for millions which remains to this day, for it was indeed one of 'Britain's Great Inventions'.

Introducing the SR.N4 Mk.1

278

FRIDAY, JANUARY 17, 1969

Here she comes . . .

Giant Hover 'purrs' into Pegwell

Swift, Hoverlloyd's first SR.N4 hovercraft arrives at Pegwell.
(Note the yet to be officially christened craft lacks name and registration number.)
East Kent Times 17th January 1969

The First Arrival

The *East Kent Times'* front-page spread of *Swift's* arrival at Pegwell on 15th January 1969 marked the beginning of Hoverlloyd's adventure as a fully-fledged cross-Channel fast ferry operator. The accompanying text went on to say:

> *Two hundred workmen building the Ramsgate Hoverport*

stopped work on Wednesday morning to watch the first Mountbatten hovercraft arrive. In no time at all the speck on the horizon off Deal had grown into a huge machine as it skimmed across Pegwell Bay and settled on the concrete pad in a mist of spray. Spectators waiting to greet the big hovercraft were pleasantly surprised at the low noise level as it purred across the mudflats to the hoverport.

For those in the know, the references to 'purring' will raise a wry smile or two – another local paper, the *Thanet Times*, went even further and called the new arrival the 'Whispering Giant'. The truth of the matter was that due to the previously strong objections to the likely noise, raised equally noisily by the local residents at Cliffsend, *Swift* had been deliberately driven in under very low power. It has to be admitted that it was a bit of low-down cheating.

Of course, the subterfuge could not last, and it wasn't long before complaints about the noise surfaced again, this time in earnest. In April, shortly after the start of the service, when captains were more confident at using higher power settings, a national newspaper produced a short piece entitled:

ENGLAND'S 'NOISIEST VILLAGE'

Residents at Cliffsend, near Ramsgate, Kent, where hovercraft started operating in January, say that their village is the noisiest in England.

For some houses on the cliff top, the arrival of a hovercraft in Pegwell Bay brings a thrumming which rattles the windows. Some residents have already sought rate rebates.

A local resident was quoted as saying:

Now this terminal is built and it is all spoilt. In June, when the full service starts, hovercraft will be arriving every hour from 8.00am to nearly 11.00pm.

How much noise it makes depends on the direction of the wind. On a bad day it drowns the television and makes outdoor conversation impossible.

The lady in question had the times slightly awry but was more or less right. What she would have thought in later years when four hovercraft were running a half-hourly service, which meant movements in and out of the hoverport five times an hour, is not recorded.

002 – *Swift* – being wheeled out at the British Hovercraft Corporation's Columbine Hangar at Cowes on 10th December 1968. Note the three tracks.

Just five weeks later on 14th January 1969, *Swift* sat on the new pad at Pegwell Bay Hoverport for the press to admire, with the unfinished terminal building in the background.

Pomp and Circumstance

Such historic events, inaugurating a brand new era in transport, could not go unheralded, and in this case no effort was spared to emphasise the importance of the occasion of the craft-naming ceremonies. The christening of the two new hovercraft and the opening of the new hoverport occurred in quick succession during the first half of 1969 and must have been hard work for Hoverlloyd's nascent PR department.

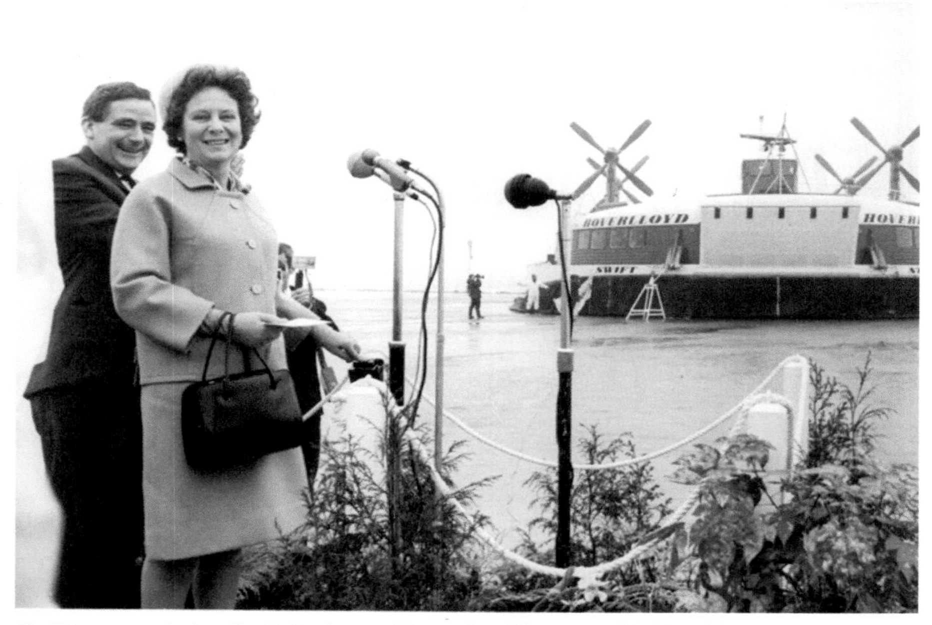

Swift is named. Leslie Colquhoun, Managing Director, with Mrs. Mary Wilson as she presses the button – Mike Fuller's 'smashing machine' works. (A much relieved Mike can be seen in his white boiler suit on the extreme left of the hovercraft.) 23rd January 1969.

The ceremony for *Swift* took place a little over a week after its arrival and was conducted by Mrs. Mary Wilson, wife of the Prime Minister Harold Wilson.

This was closely followed a month later when HRH The Duke of Edinburgh officially opened the Pegwell Bay Hoverport. He toured the not quite complete building and then accompanied Captain Bill Williamson and his crew on a short flight out to the Goodwin Sands where he was given the opportunity to pilot the machine.

The second craft, *Sure,* arrived in early April and went into service later that month. It was named on 3rd June by Mrs. Mary Soames, wife of the British Ambassador to Paris.

The Duke of Edinburgh boarding *Swift* with Captain Bill Williamson, Emrys Jones, Techical Director, and Leslie Colquhoun. Maggie Studt, Chief Stewardess is standing in the doorway. 2nd May 1969.

A Closely Guarded Secret

In line with the usual seafaring tradition, when each hovercraft was christened, a bottle of champagne was smashed over the bow. How this was to be achieved in safety with such an unusual machine required some innovative thinking. For instance, smashing a hard champagne bottle up against a 1mm duralumin hull would almost certainly result in the wrong item being smashed.

The successful solution devised by Mike Fuller was an early indication of how amazingly innovative and versatile the engineering department would turn out to be:

> *I was tasked by Emrys Jones, the chief engineer, with designing and building some sort of machine to smash a bottle of champagne on the front of an SR.N4 in readiness for the first launching ceremony.*
>
> *This I made at home in my shed and it had a timber frame with a hinged arm made from two SR.N6 plenum chamber struts. I*

fitted an Ever Ready doorbell battery underneath, which powered a rotating cam driven by a Meccano electric motor which released the arm. The bottle then fell forward and smashed on the anchor.

The Duke of Edinburgh climbs the ladder with Captain Bill Williamson to the flight deck.

That was the theory, but champagne bottles are tough and when I set it up at the hoverport against a cast iron drainpipe the bottle just bounced. This obviously is a real no-no in the world of launching ships so I pre-cut the bottle with a diamond glasscutter and it worked perfectly, much to my relief. Les Colquhoun supplied the empty bottles for me to test but promised a full one for the day.

That day, the 23rd January 1969, arrived and my machine was set up in front of Swift's *starboard anchor. Mary Wilson, the Prime Minister's wife, was the chosen dignitary and was pre-briefed to hold her finger on the bell push button until she saw the bottle move. I was alongside the machine with a screwdriver to manually trip the arm in case my Meccano motor failed but the thing worked as planned and the bottle fell forward and duly smashed on the anchor. Commander Peter Lamb, BHC's chief test pilot, started the four Proteus engines and* Swift *lifted off out into the bay. It was a great relief for me seeing that bottle break, and cover the anchor in R White's lemonade, as we had drunk the champagne the night before!*

The Duke of Edinburgh prepares to take over. George Kennedy is on the jump seat as 1st officer.

Sure was named on 3rd June 1969 at Calais Hoverport by Mrs. Mary Soames, wife of the British Ambassador to Paris.

Moving and Hovering

Unlike other aircraft, or any other vehicle for that matter, the hovercraft requires its engines to perform not one but two tasks. Energy is required to move the craft along, but also to provide air to lift the vehicle off the ground. Many designs use separate engines, one for propulsion and one for lift, but the clever trick is to design the transmission system so that one engine performs both tasks.

As can be seen from the following diagrams the 'one engine for both tasks' principle was adopted in the N4. Each of the four Rolls Royce 'Marine Proteus' gas turbines producing 3,400shp was linked to a 19-foot diameter propeller, the largest of their kind in the world, and a 12-foot diameter centrifugal lift fan. These were a developed version of the first troublesome Proteus engines, as used on the *Princess* flying boat.

To manage the link connecting the engine, variable pitch propeller and lift fan, two gearboxes were needed – the 'main' gearbox at the base of the pylon supporting the propeller and the 'propeller' gearbox in the pylon nacelle. As well as being able to change the drive direction, each gearbox was also designed to provide the different rpm output to the propeller and lift fan. The horizontal shaft from the forward engines was some 60 feet from the main gearbox they were driving.

One result of this dual system was that the torque (the load) on the propeller governed the whole performance. The harder the craft was driven, i.e. the more the propeller pitch put on, the slower the propeller, and consequently the slower the lift fan, reducing the lift. The whole system was designed to balance. In other words, it was not possible to drive the craft into the ground.

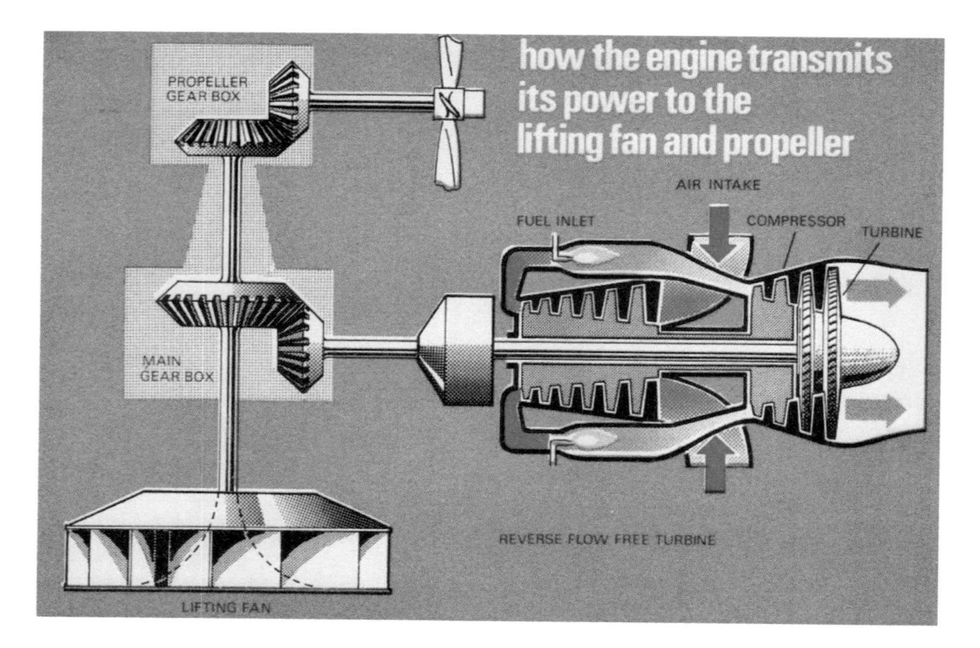

There was one very useful consequence: if the machine met an obstacle, an unexpected large wave for example[2], pulling pitch off all four propellers slowed the craft down to lessen the impact, but the reduction in torque speeded up the fan, so provided more lift to climb over it.

The pylons would swivel 30° either side of the centre line to give the hovercraft directional control. Two moving fins with coupled rudders at the rear of the craft provided additional control and stability. Electrical power was supplied by two 15/90 Rover gas turbines adjacent to the two engine rooms, one on each side of the craft.

The SR.N4 Mk.1 Mountbatten Class Hovercraft

The following pages are taken from the British Hovercraft Corporation's *Operating Manual* containing drawings of *Swift* and *Sure* as they appeared in 1969. The images on this and the previous page are from a brochure *How the Hovercraft Works,* which was handed to passengers in the early days on boarding.

---

[2] The problems with wakes of ferries will be covered later.

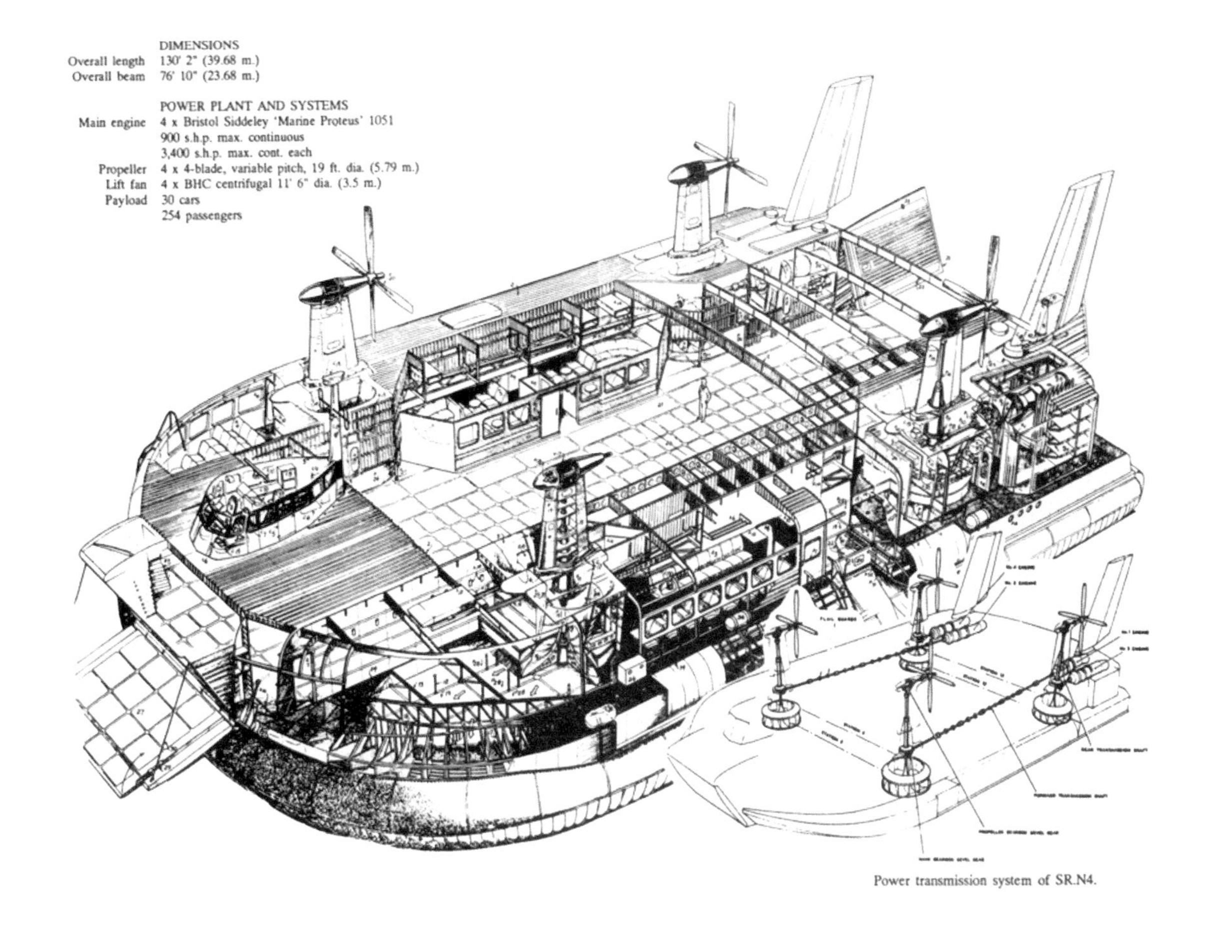

DIMENSIONS
Overall length 130' 2" (39.68 m.)
Overall beam 76' 10" (23.68 m.)
POWER PLANT AND SYSTEMS
Main engine 4 x Bristol Siddeley 'Marine Proteus' 1051
900 s.h.p. max. continuous
3,400 s.h.p. max. cont. each
Propeller 4 x 4-blade, variable pitch, 19 ft. dia. (5.79 m.)
Lift fan 4 x BHC centrifugal 11' 6" dia. (3.5 m.)
Payload 30 cars
254 passengers

Power transmission system of SR.N4.

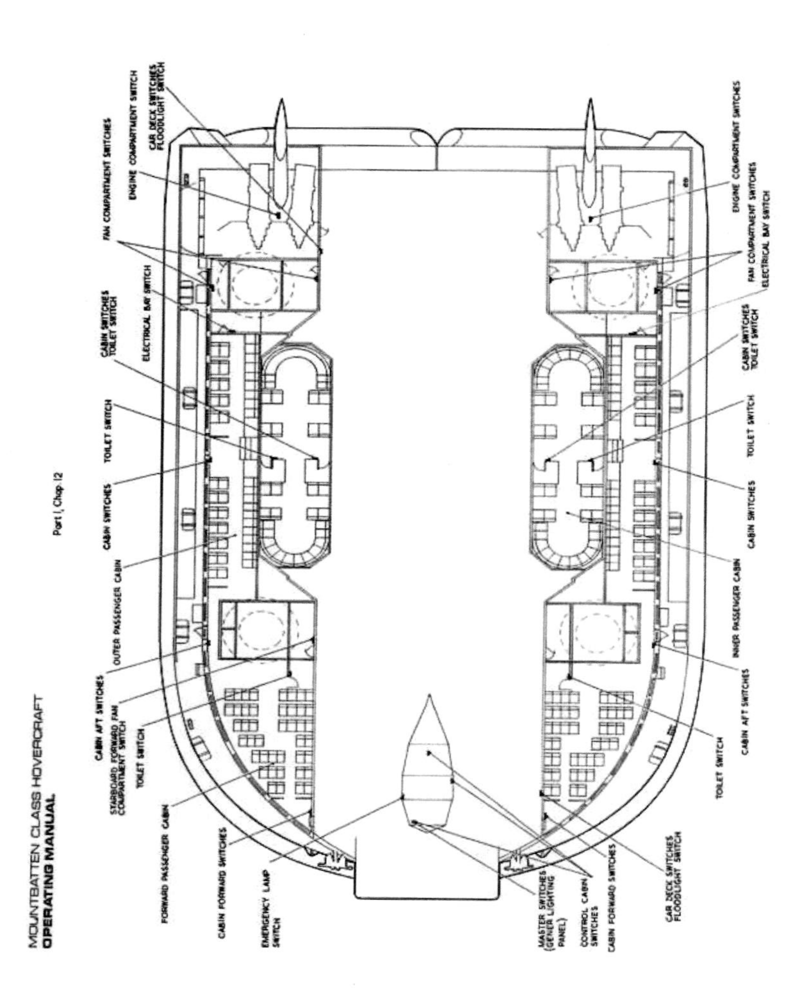

That Mysterious Skirt

All the general public ever saw was the outside of the skirt system; they had no idea how complex it was underneath. The following diagrams show what a work of art it was. Approximately nine tons of neoprene-reinforced rubber was formed into what was referred to as a 'bag and loop' system. The loops, more commonly known as 'fingers', attached to the bottom, had the ability to move up and down with the waves and therefore maintain a seal.

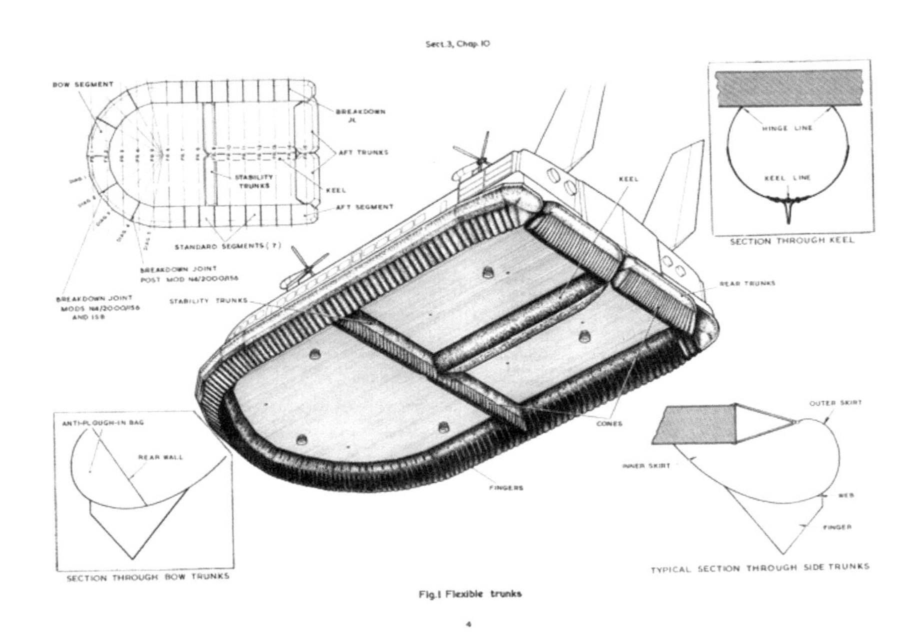

Fig.1 Flexible trunks

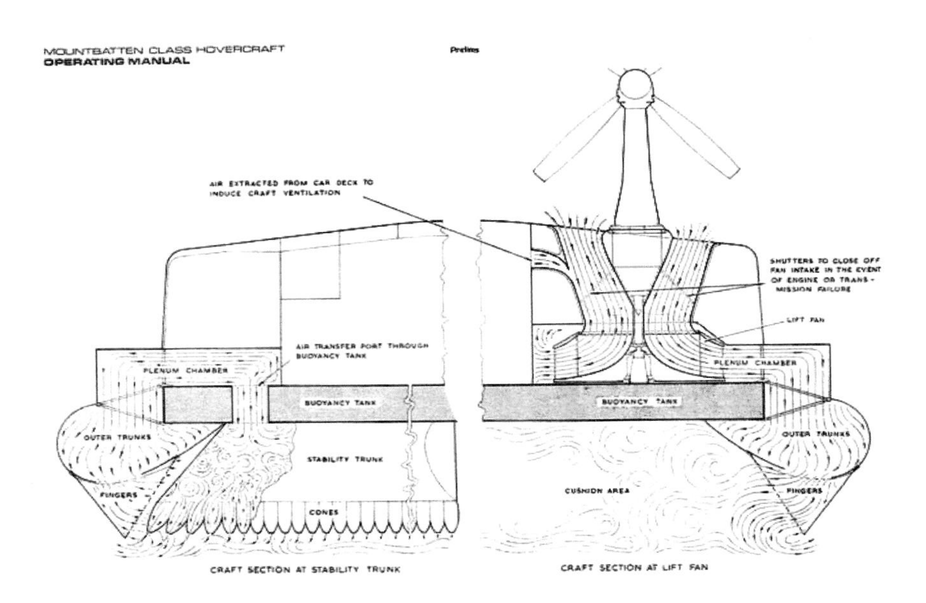

The cross-wise 'stability trunks' and 'keel bag' combined to keep an even pressure laterally and longitudinally. No, there were no wheels underneath, although many to this day think there were.

After the first teething troubles the system worked very well operationally, but the excessive wear on the fingers, requiring replacement of several on an almost daily basis, was a heavy impost from an economic point of view.

The second diagram on the previous page shows the airflow through the system, a masterpiece of design in itself.

Loading Vehicles

Cars were, after trial and error, eventually loaded up the forward ramp and discharged off the rear, which, unlike the front, didn't have its own built-in ramp system. Instead there was a portable ramp at the hoverports, which was driven up to lock into the craft once the rear doors were open.

The exceptions to the 'load-up-the-front rule' were coaches and other very heavy loads. The two centre lanes of the deck were reinforced to take heavier loads but only for two thirds of the length from the stern, so loading from the bow was not an option. Limitations of the area for coaches etc., coupled with the weight limitation, meant that only two could ever be carried at any one time.

A coach being loaded up the portable rear ramp. Coaches drove on and reversed off.

Passengers

In additional to 30 cars, accommodation for 254 passengers was split either side of the car deck. This was not ideal on the Mk.1, as, apart from the claustrophobic 'inner cabins', the sideways-facing seats in the main cabins couldn't have been all that comfortable either.

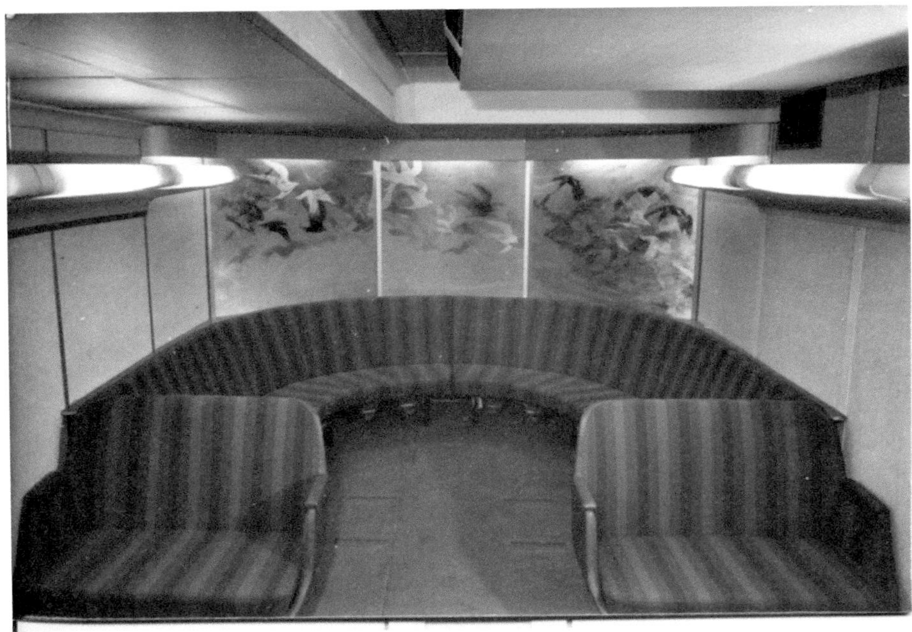

One of the two windowless inner cabins on the Mk.1, seating 40 passengers in each. An alleyway divided each cabin giving access from the car deck to the main cabin on each side. They were somewhat claustrophobic and not the most pleasant way to cross the Channel.

In the inner cabins the confined windowless space was only part of the problem – the design complement on each side was 40 passengers. This meant, on a busy day, squashing 12 people into the semi-circular seat at the end. Betty Dowle remembers it as not the easiest task for the cabin staff:

> *I think the worst thing, when we had a full load, was making people go down into those inner cabins, and you had to count the people that went on to that round semi-circle seat. I can't remember the number you had to get on, but they would keep spreading out and I would tell them we had to get more in there. They weren't popular cabins and we were glad when they went.*
>
> *The only time they were popular was when the craft wasn't full and a couple wanted to go down there on their own. Then you didn't look!*

Despite the British Hovercraft Corporation (BHC) being essentially an aircraft manufacturer, when it came to designing windows for this new venture at sea they diverged away from the usual pressure-resistant small windows of the average airliner and plumped for what can only be described as 'picture windows'.

A Seaspeed SR.N4 Mk.1 showing the main cabin seating.

No doubt this feature added to the attractive looks of the N4 and provided a very light and airy feel to the passenger accommodation, but there were drawbacks.

A particular concern for Hoverlloyd was the nature of the route from Pegwell Bay to Calais, traversing as it did large areas of mud flat in the bay and long stretches of sand across the Goodwins. The mixture of mud, sand and salt had an abrasive effect on the glass and that, coupled with the continuous cloud of spray, made visibility very limited at times.

The biggest problem was with the forward windows, which were consistently vulnerable to breakage in rough weather, so much so that the standard operating procedure was to clear passengers out of the forward end on bad days. Luckily, as the occurrence of rough seas was more likely to coincide with the winter months when passenger numbers were low, the forward areas could be cleared without difficulty.

The Long Journey

This may have been the beginning for Hoverlloyd's cross-Channel hovercraft service but, in terms of the history of a concept, it was merely the latest step in a long period of development stretching back to the first experimental SR.N1, which successfully crossed from Calais to Dover in 1959. It must also be conceded that it was not even the first N4 in service. That accolade goes to our rival company Seaspeed, the British Rail owned operation out of Dover, who had the dubious privilege of opening their service with the far from reliable prototype SR.N4 001 during the previous summer season in 1968.

Ann Deverson serving duty free in the port forward cabin on the Hoverlloyd SR.N4 Mk.1. *(Hoverlloyd publicity brochure.)*

Prior to the historic event of the N1 crossing, a similarly lengthy period stretches back to 1953 when Christopher Cockerell first conceived and patented his invention. In fact, it could be said to go much further back than that. Cockerell[3] was simply continuing a search for a means of lubricating a ship's hull to allow it to go faster through the water, a search that had been exercising the minds of naval architects for a considerable time before. The hovercraft takes its place alongside the hydrofoil, catamaran and the recent wing-in-ground (WIG) machine as a solution to the problem. In fact it could be said that Cockerell's was the best solution to date, as it could achieve faster speeds than the hydrofoil and the catamaran, and would have been probably more economical than the WIG.

So, if it was such an effective solution, why are we writing a book about a hovercraft operation that lasted just a little over 30 years? A good question, which perhaps this book may go some way to answer.

All that is left of *Swift* today is the control cabin, which is on display at the Hovercraft Museum at Lee-on-Solent and, even if it were complete, it would be an example of the Mk.2 version, not the Mk.1 as originally delivered. The first two craft were modified to Mk.2 level during 1972 and 1973.

---

[3] Christopher Cockerell was knighted in 1969 and received 33 honours and 98 patents. His patents reflect his work pattern: Mechanical engineering; electronic engineering; small craft for leisure; hovercraft; wave power; and intense personal research of everything unimaginable by most.

PART ONE

N1 to N4

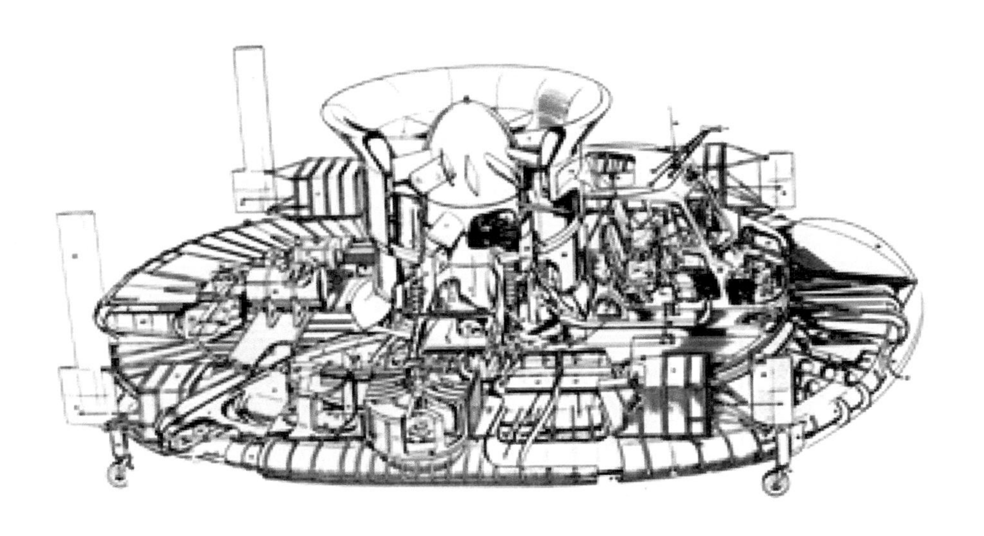

Chapter 1

The Inventor and his Invention

Christopher Cockerell

The name Christopher Cockerell[1], MA, is the best known in the hovercraft world and will remain so as the inventor of the hovercraft principle. Born in 1910, the son of Sir Sydney Cockerell, the Curator of the Fitzwilliam Museum at Cambridge, and his wife, Florence, who was well known as an illuminator of manuscripts, his antecedents were humanist and artistic rather than technical.

Christopher Cockerell, knighted in 1969 for his services to engineering.

However, his interest in things mechanical became evident at an early age when he used a toy steam engine to drive his mother's sewing machine, and constructed a wireless set for his preparatory school in the early days of radio. When, later, he went to work for Marconi Wireless Telegraph Co., his father promised him £10 for every invention or improvement of his that the firm took up, but soon had to put a stop to the arrangement as it was proving much too expensive.

He went to school at Gresham's School in Holt, Norfolk, and read engineering at Peterhouse, Cambridge. On leaving the university he effectively served an apprenticeship for two years in Bedford with W.H. Allen & Sons. After two further years spent at Cambridge, where he obtained his Master's Degree in Radio and

---

[1] Ray Wheeler's book *In the Beginning, The SR-N1 Hovercraft* records in great detail his own personal experience of the early days, including his knowledge of Christopher Cockerell, so with Ray's kind permission we have been able to reproduce parts of it here, in addition to adding Bob Strath's and John Chaplin's recollections of events.

Electronics in 1935, he joined the Marconi Wireless Telegraph Company at Chelmsford. Cockerell's talents were soon recognised when he became Head of Marconi's Aircraft Radio Development Establishment at the age of 27. His ability was such that, during his 15 years at Marconi, 36 important patents were taken out in his name. In 1939 he headed the team that developed the radio direction finder that went into every British bomber – known to the RAF as the *Drunken Men*, because it had two needles that swayed and crossed like two drunks helping each other home. In 1935 he married an art student, Margaret Elinor Belsham, who proved a tower of strength to him during the Second World War when he worked exceedingly long hours.

The war had taken its toll and, although he was offered new posts, he decided they involved too much administration, which he did not enjoy. He had started sleepwalking so he decided he was in need of a complete career change. He resigned his senior position from Marconi in 1951, sold his Regency country home and moved to a small caravan at Old Wherry Dyke at Somerleyton, near Lowestoft, which had been a quay for a Victorian brickworks that had stood on the site and was being transformed into a marina. Using an inheritance of Margaret's they set themselves up in a boat-hiring and caravan-building business.

He had no experience of naval architecture but, in designing his own small boats, his ideas were very much concentrated on how to make a boat go faster on less power. His training and background generated an attitude of mind that automatically questioned everything and wondered whether there is a better way of doing it. In order to find out why boats were designed the way they were and what were the limiting factors on their performance, he went to the local library and read every book he could find on the subject. In a biographical sketch, he noted:

> *Very soon in reading about boats one comes upon the problem of wave resistance. This, one notes, is a law of nature – that is something one cannot remove. All one can do (if other things permit), is to design it so that it shall be to a minimum. Then one comes to skin friction. And, of course, if you design for minimum wave resistance you are likely to get maximum skin friction and vice versa.*

By 1953, the year of Queen Elizabeth's coronation and the conquest of Everest, he was working on an idea of how to air lubricate a hull – a concept that he had had in the back of his mind for some time. His first attempt involved fitting a centrifugal fan in the bow of a punt, driven by the main engine, and then a 20-knot ex-naval launch called *Spray*, with the aim of blowing a thin layer of air under a slot in the bow, but it proved too difficult to lubricate the whole of the bottom. He then

The complete simple testing arrangement showing a vacuum cleaner electrical pump mounted above the tins blowing air onto kitchen scales.

realised that a thick cushion of air was required and started experimenting with hinged doors, water jets and air jets at the bow and stern of the boat and cushion-retaining extensions along the boat's sides. The problem was that the rigid hinged flaps at the bow and stern could not be expected to follow the shape of the wave tops and therefore the air could escape. He considered various ideas to fix the problem, but in the end decided that the best solution was to use air curtains at the bow and stern[2]. The next step was to replace the side walls with an air jet cushion, which also had the advantage of making the boat amphibious.

This resulted in Cockerell's famous experiment with the Kit-e-Kat tin inside a slightly larger Lyons coffee tin. When connected to a vacuum cleaner air blower over a set of kitchen scales, it showed that his rudimentary annular jet (having the form of a ring) could create a load on the scales within 10% of his calculations, namely, three times greater than could be generated with a straight jet. To demonstrate the principle he designed a 2.5-foot long, 2-foot wide pear-shaped model, which was powered by a model aeroplane engine. The balsa wood model, weighing just over 5 lbs, was built by A.D. Truman of Oulton Broad and its performance delighted Cockerell when the small model achieved a speed of 13 knots over water. Thus, in 1953 the hovercraft was born.

Selling the Idea

Cockerell's next challenge was to attract interest in his concept. With his experience in airborne electronics he took his idea to some of the executives in the aircraft world with whom he was acquainted. One

[2] Cockerell had basically invented the sidewall hovercraft, more about which is written later.

chief project engineer told him he saw hundreds of inventions, but the unusual thing about this one was that it seemed to work. However, he declined to take it up. Shipbuilders were unable to get their heads around it either. He wrote patiently:

> *This is not surprising. It is always difficult to get new ideas taken up and the extreme newness of the idea makes it doubly difficult.*

Cockerell was not deterred and felt he needed to press on with all the means at his disposal. He had written several reports about his invention and one of those reports produced in October 1956 stated:

Christopher Cockerell displaying his first complete pear-shaped model in 1955.

> *It can rise and hover and then move away and therefore needs neither harbours nor runways. It can travel over the sea, over a marsh, up the beach and on to the land.*

Cockerell took his model[3] to his friend and landlord, Lord Somerleyton, who, after seeing a demonstration on his lawn, was so impressed that he wrote to Lord Louis Mountbatten to enlist his support for the concept. Fortuitously, Mountbatten saw the potential and in 1956 arranged for a demonstration of the model in the basement of Cockerell's patent agent in the presence of Admiralty representatives, Ron Shaw (see page 45) and the patent agent. Cockerell recalled that:

> *The model 'whizzed' round the room, Shaw jumped for safety on to the furniture and the room was filled with noise and fumes.*

Despite having Cockerell's report as to the hovercraft's capabilities, the senior Admiralty representative was unimpressed claiming that there

---

[3] The Hovercraft Memorial, set as a column with a replica of his first complete hovercraft with the pear-shaped model on top, was unveiled on 4th June 2010 in the grounds of Somerleyton Hall by Lady Somerleyton, to commemorate the local invention of the hovercraft by Sir Christopher Cockerell on the 100th anniversary of his birth.

was nothing new in the concept, but Shaw, a keen supporter, played for time by asking for the idea to be placed on the Secret List for three months while the Ministry of Supply examined its potential. As Cockerell later, ruefully noted:

> *The Navy said it was a plane not a boat; the RAF said it was a boat not a plane; and the Army was 'plain' not interested.*

Ron Shaw was undoubtedly the unsung hero in bringing Cockerell's invention to fruition.

Christopher Cockerell filming his model in free flight. 1956.

He tried to interest Short Brothers[4] in the invention to no avail, and then he and Cockerell visited Saunders-Roe at Cowes, who agreed to examine the idea and write a report, provided they were given a contract. The three months period of classification was extended several times until, after eighteen months, when during a discussion on hydrofoil concepts with Carl Weiland, a Swiss inventor, who also claimed to have invented a means of creating an air cushion, there was an inadvertent disclosure of Cockerell's patent. This meant that the secret classification could no longer be applied to the patent. By the time Saunders-Roe Aircraft Company received their eight month classified contract from the Ministry of Supply on 26th August 1957, four frustrating years had ensued for Cockerell since he had first discovered the concept.

---

[4] It is unclear why Ron Shaw went to Short Brothers first, especially as Saunders-Roe had just lost the SR.177 contract resulting in 1,500 redundancies. Shorts, however, was founded in 1908 and had a long history in the manufacture of flying boats. They had moved to Belfast from the Isle of Sheppey in Kent in 1936.

The underside of Christopher Cockerell's first model showing the jets which provided stability and part of the thrust. 1956.

Nine years later Ron Shaw wrote:

One good measure of the real novelty of a new concept is the objections it raises and in this the hovercraft certainly ranks with the aeroplane. Both suffered the flat, disbelieving, uncommitted faces of committee and mercifully survived. Another measure of novelty, but also of real worth, is the capacity of the new concept to generate enthusiasm in its converts. This the hovercraft has certainly done, as anyone in the business knows, but the surprising thing is the number of converts that must have been made without their ever being personally involved at all. The idea itself, without sight or word of its success, grasps the imagination.

In some aspects, of course, hovercraft are not new – the seeds were there. Earlier inventors than Christopher Cockerell had had some of the idea, but the circumstances were not favourable for growth. A fairy godmother, by way of state support and a concentration of skills, such as provided by the team at Saunders-Roe, were needed to nurture the infant idea. It even needed the death of the SR.177 rocket interceptor to make an unusually strong team suddenly available. The step from the cloth caps in a City basement and the tiny wooden model roaring around on a string, to the clinical blueprints, the strange new shapes in tank and wind tunnel, and the first glossy brochures, was only eighteen months.

Ron Shaw - Hovercraft Godfather

Mr. Ronald Andrew Shaw, OBE, MA, CEng, FRAeS, was born in Liverpool in 1910. He gained a 1st Class Honours in Part 1 of the Maths Tripos Honours and 1st Class Honours in the Mechanical Sciences Tripos at Cambridge. In 1932 he was appointed Junior Staff Officer at the Royal Aircraft Establishment (RAE), Farnborough, and from 1932 to 1938 he worked on wind tunnels and on in-flight fuel jettisoning. In 1938 he was appointed Senior Scientific Officer at the Marine Aircraft Experimental Establishment, Felixstowe, for work on flying boats. He continued there throughout the war and was also made responsible for full-scale and model work on anti-submarine weapons. His next promotion was to Principal Scientific Officer. He became attached to the Council for Scientific and Industrial Research, and from 1945 to 1947 took charge of the Aerodynamics Section of the Aeronautical Laboratory, Fishermen's Bend, Melbourne. From 1947 to 1950, he was attached to the Aerodynamics Division at the National Physical Laboratory, Teddington, for work on supersonics. In 1950 he was posted to the Joint Services Mission in Washington, DC, with responsibility for liaison in aerodynamics. In 1953 he was posted to headquarters, London, as Assistant Director Aircraft Research with responsibility for research in aircraft and later for hovercraft.

In the business of aircraft manufacture, and as well as many of its other projects, Saunders-Roe had been contracted by the government, or companies, on a cost-plus basis to build and test prototypes, mostly military, very few of which ever went into production. Unlike Shorts, although it had built small commercial aircraft, the *Princess* flying boat was the only one designed for scheduled passenger services, and that never saw the light of day in that role. But the eventual success in Ron Shaw bringing together Christopher Cockerell's hovercraft invention with the highly experienced and well equipped Saunders-Roe was to change the life of Cockerell and the future direction of Saunders-Roe, as well as creating a unique form of transport that would fascinate the world.

The Annular Jet Principle

Before we go any further, we need to look at exactly what Cockerell had discovered. His hovercraft concept provided an engineering solution to the problem of air lubrication of the hull of a ship using a deep enough air cushion to minimise the wave contact with the under-surface of the craft in rough seas.

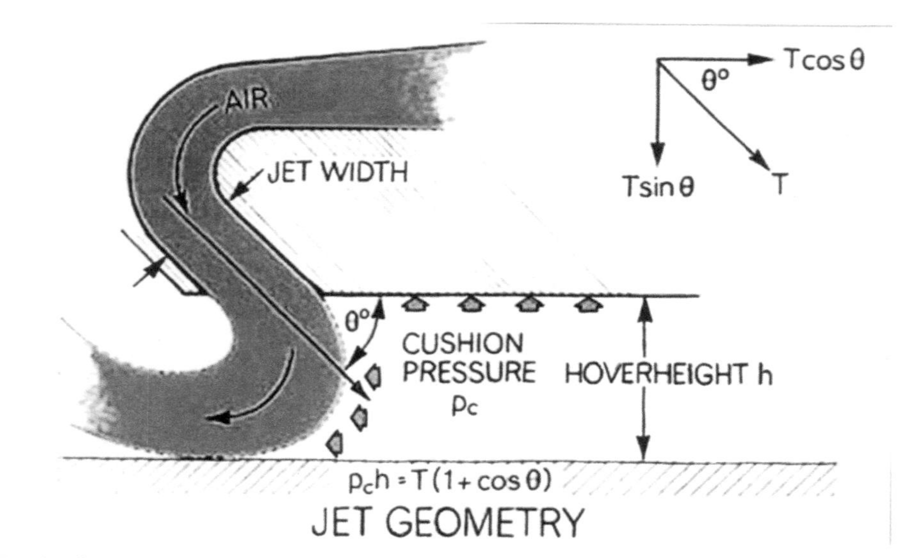

The basic annular peripheral jet principal.

The basic principle of the hovercraft was the generation of an air cushion having a pressure higher than ambient and bounded above by the under-surface of the craft and at the sides by a curtain system. Cockerell's invention used an air curtain ejected downward and inwardly from nozzles extending around the whole of the perimeter – hence the term annular. This inward turning of the jet required a region of high pressure underneath the craft to deflect the air curtain outward so that it had a circular profile touching the ground, or water surface, before exhausting to the atmosphere.

This unique annular jet hovercraft feature was that the craft does not contact the surface and thus both water and hard surface friction are eliminated. In fact, the hovercraft is more comparable to the planing craft using static displacement of the surface in the hover, and then a changing combination of displacement and downward particle acceleration as the speed increases. It also generates a wave drag which peaks during the transition from hover to planing speed in a similar manner to the planing craft; the classical 'hump drag'.

The success of the hovercraft as a transport system would depend on the basic trade-off between the power saved by the reduction in surface drag and the power required for the generation of the air cushion.

The Initial Studies

The contract awarded to Saunders-Roe Aircraft Company in August 1957 was to carry out a theoretical and experimental analysis of the Cockerell concept. The study was headed by the Special Project Engineer (Wilfred Egginton), reporting to the Saunders-Roe Chief Designer (Maurice Brennan), who sent the Chief of Aerodynamic Research (John Chaplin) to Cockerell's boatyard in Lowestoft to review his experimental work. This was a unique chance for Chaplin to experience the enthusiasm and inventiveness of Cockerell and the charm and hospitality of his wife, Margaret. Chaplin recalls:

> *My first reaction to it was that it was bloody amazing. It was something I hadn't seen before and as an engineer it was intriguing. I don't think at that moment I sat down and thought there would be SR.N4s or anything running across the Channel. That all came later when we started doing the engineering studies, but initially it was wonderment at something one hadn't thought about or seen before. My main emphasis at the time was making sure I did a good job of checking all the rigs and figures. I'd been sent there to verify it all, almost like a technical assessor. I had to report to Maurice Brennan and he was an interesting, fierce Scott character. When he sent me there he said, "You bloody well find out whether this is true or not because if it isn't we're going to waste a lot of money." So I had to report back to him and then he started thinking about it.*

After two days of running the test rigs, reviewing Cockerell's test reduction and some friendly arguments on the correct axis to use in the aerodynamic data plots, Chaplin reported to the chief designer that the experimental methods and data reduction used by Cockerell were sound. About Ron Shaw, Chaplin remembers:

Ron Shaw came down at that time and he and Maurice Brennan were certainly talking about great things in the future for a new form of transportation. Whenever I met him you could talk technicalities with him and he seemed to understand it, but he was a sort of sponsor type chap in the MOD who got enthusiastic about things.

Wind tunnel, tow tank and free flight tests with Cockerell's original model were carried out and the data from his research analysed. A two-dimensional test rig of the annular jet principle was constructed to provide further data and validation of the basic principle.

The results of this exploratory contract were delivered by Saunders-Roe in two reports in May 1958 – *Assessment of the Hovercraft Principle* and *The Potential Application of the Hovercraft Principle.* It was reported that Cockerell's theories and experimental data were valid and that lift within 10% of the theoretical value could be obtained over land and possibly a slightly higher efficiency over water. The conclusion was that the hovercraft principle showed considerable promise, justifying a further programme of research and development.

Research and Development

In July 1958, at the request of the Ministry of Supply, Saunders-Roe proposed the second stage for the development of Cockerell's hovercraft concept. Unfortunately, at the same time it was realised that since there was no defined military need for hovercraft (a not unreasonable situation since it had just been invented) it was impossible by law for the Ministry of Defence to authorise further funding. The programme was therefore cancelled.

But, once again, keen supporter of the hovercraft, Ron Shaw, came to the fore and, with his encouragement, a proposal for further research and development was submitted by Saunders-Roe to the National Research Development Corporation (NRDC).

NRDC was an independent public body set up in 1948 by Act of Parliament, in association with industry, to encourage and sponsor promising ideas and inventions that were in the national interest. It was not a government department, but was required to have Board of Trade approval for its major ventures and could, with the board's approval, borrow money from the Treasury. NRDC would issue licences under patents assigned by the inventors and in return receive a royalty on sales to recover its investment.

Cockerell was referred to NRDC by Lord Caldecote of the English Electric Company and the first meeting took place with the Secretary of the Corporation, Mr R.A.E. Walker on April 16th, 1958. Mr Walker wrote to Sir Owen Wansbrough-Jones, then Chief Scientist of the Ministry of Supply, who was also at that time a member of the corporation's board:

> *Our first acquaintance with this project was made at a meeting with the inventor, Mr C.S. Cockerell, on the afternoon of Wednesday last. While the Ministry of Supply had placed a preliminary study contract with Saunders-Roe, the latter's report would not be available for some weeks. His present difficulties were concerned with the finance of overseas patent applications. To secure Convention priority (desirable because of the activities of an intervening Swiss inventor) these applications had to be on file overseas by the 2nd May. He had been unable to make previous arrangements for these because the secrecy provisions of Section 18 of the Patents Act had been imposed. This imposition was, however, on the point of being withdrawn.*

Lord Halsbury, Managing Director of NRDC[5], decided on the spot to take crash action and finance the foreign filings. Instructions were accordingly given to the inventor's patent agents within a few hours of first seeing the inventor:

> *We shall be reporting this to the Board tomorrow, and shall also seek their preliminary views as to the suitability of the project for the Corporation's sponsorship should we be asked to support further development – e.g. if the Saunders-Roe report is favourable, but the Ministry of Supply decide not to take any further action.*

The next day the NRDC Board, under the chairmanship of Sir William Black, confirmed the decision to support the hovercraft project and quickly realised that it was likely to become the largest project in terms of finance on which the corporation had yet embarked, and for this reason they went on to set up a subsidiary company, Hovercraft Development Ltd. (HDL)[6]. Dennis Hennessey, who had been a principal patents officer at the Ministry of Supply before he joined NRDC

---

[5] Lord Halsbury was succeeded in 1959 by John Duckworth, who, after taking a 1st Class Honours in Physics at Oxford in 1938, played a leading part in radar research and development during the war.

[6] HDL was incorporated on January 8th 1959, and made provision for shareholdings by Christopher Cockerell, some of his initial private backers and by the corporation. The company was thereafter employed as the corporation's instrument for promoting the development of hovercraft.

in 1950 as patents manager, became its first managing director and chairman, with Cockerell as technical director. The NRDC officials were more perceptive than others in seeing the potential of Cockerell's revolutionary principle and, having agreed to provide financial support, authorised for the contract to Saunders-Roe to go ahead.

Saunders-Roe considered that, in addition to more theoretical and experimental studies, it would be essential to carry out a programme of tests on a large-scale radio-controlled free-flight model on the River Medina and Solent to provide the data necessary to move on to the next stage. On 4th September 1958, Saunders-Roe submitted a proposal to NRDC.

Chapter 2

Saunders-Roe – Boats, Planes, Rockets and Hovercraft

Saunders-Roe 1870 – 1918

To understand why Saunders-Roe was so uniquely qualified to be entrusted with the development of the hovercraft, it is worth looking at the background of the company in more detail[1]. This potted history, as related here, barely scratches the surface of the projects in which Saunders-Roe was involved, but nevertheless illustrates the point that it would take a company with great depth and experience in the field of science and engineering to bring the hovercraft to practical reality.

Sam Saunders (left) escorting King George V at Columbine Yard in 1913.

The history of Saunders-Roe, with its origins dating back to 1870, had, by the time the hovercraft came along, extensive expertise in developing high-speed craft both at sea level and in the air, in addition to a sophisticated electronics division, experience in bonding metal alloys and a water-tank testing department.

It all began at the Swan Inn at Streatley in Berkshire, where Moses

[1] The full history of Saunders-Roe is published in Ray Wheeler's books *From River to Sea* and *From Sea to Air*.

Saunders was not only the publican, but was also responsible for the construction of the weirs and locks on the River Thames. As there was no bridge between Goring and Streatley, he took on the additional duty as ferryman across the Thames, and it is extremely likely that with the skills he possessed he built his own ferry boats and dinghies for his work on the river. His son, Cornelius, produced seven children with his wife Eleanor and, of the three sons who showed an interest in boat building, it was Sam who stood out. Saunders Boat Builders of Streatley was established in 1870, probably by Cornelius, as Sam was only 13 at the time, and went on to build steam launches for use on the Thames. Cornelius died in 1880 at the age of 43. In the same year, with the advent of the German Daimler combustion engine, Sam Saunders, who had by now established the company's reputation for high quality workmanship, saw another opportunity to expand the business. He claimed to be the first person to produce power-driven boats and his success in building them partially came from his lightweight construction techniques, which he initially developed by importing canoes made by the North American Indians from Canada and copying their methods.

The Ravaud Aero-Hydroplane of 1911.

The demands for cash to expand the successful business were such that in 1900 Sam incorporated 'Saunders Patent Launch Builders Syndicate' to attract finance. Because of the size and speed of the Saunders launches, the Thames was considered to be too constraining and so the syndicate moved to Cowes on the Isle of Wight. By 1906 the syndicate had built 150 boats each of up to 75 feet in length. Despite its success, and with the initial agreement coming to an end, Sam decided to leave the syndicate.

He then leased the extensive, but somewhat derelict Liquid Fuel Engineering Company Works in East Cowes, called Columbine Yard, which was still one of the properties of Westland Aerospace Limited in the days of hovercraft manufacturing. Sam soon had the yard in working order and within two years had gained a reputation as a modern marine engineer and launch builder.

By 1909 Sam had taken an interest in the new science and engineering of aviation and had formed a department to design and build 'everything required for aero navigation'. One of the early aircraft-related vehicles was the strange-looking Aero-Hydroplane made at Cowes to the design of a Frenchman, called M. Roger Ravaud who was introduced by M. Henri Fabre, the first man to fly off water on 28th March 1910. This Ravaud craft, which was intended to skim the surface of the water, but not fly (although it incorporated a number of aircraft features) was scheduled to enter the aeroplane and motor boat contests to be held in Monaco in the spring of 1911. No records are available as to what happened to the craft, but as it did not appear at Monaco, it must be assumed that the project was unsuccessful.

The Sopwith *Bat Boat* of 1912.

In 1911 an order was received for two gondolas for the Naval Airship *Mayfly*, which were assembled in Barrow-in-Furness. A second order in 1912 was from *Maple Leaf IV's* pilot, Tom Sopwith, for the hull of the first amphibian flying boat in Europe. Flown by Australian Harry Hawker in July 1913, this aircraft, called the *Bat Boat*, won the Mortimer Singer prize of £500 for seven successive take off and landings on water. This led to several more flying boat orders. Thus, by the start of the First World War, Sam had now achieved outstanding success with fast boats and some success with aviation.

Sam would have expected to concentrate on the production of boats for the Royal Navy, but the government wanted him instead to turn his hand to the production of aircraft, with the result that his contracts for coastal craft were given to other companies with the exception of lifeboats. This did not please him as he realised that at the end of the conflict it would then be difficult for him to prosper in the fast boat and launch business. Even so, he embraced the war effort and in the first four years of the First World War the company produced over a thousand aircraft – 669 biplanes and 358 seaplanes – justifying the government's faith in him as a competent manager and rewarding him with a well-earned OBE at the end for services to the nation.

1918 – 1939

The rapid expansion of the company had again stretched Sam's resources and in 1918, at a meeting with the board of Vickers, a new agreement was drawn up in which Sam effectively lost control of S.E. Saunders with Vickers owning 51% of the company. They also purchased a valuable property portfolio built up by Sam. The reason behind the acquisition by Vickers was their interest in flying boats and Saunders' international reputation for the design and construction of the hulls of very fast motor boats.

Sam was allowed a fair degree of autonomy, but his company still needed a large amount of capital to continue with its projects, one of which was to turn the Folly Works, which had produced aircraft, into a plywood factory at a cost of £38,000. The Vickers Finance Committee was unimpressed and refused his request. There had been other clashes too. In 1918, the government's Air Board had ordered three N3 *Valentia* flying boats at Vickers and, after some forthright board meetings, it was decided that the hulls should be built at Cowes and that all flying boats should be known as Vickers-Saunders machines. There were management and culture clashes between the two companies and, with the starvation of finance by Vickers for Sam's projects as the final straw, he bought back S.E. Saunders from Vickers[2], which included all the original properties, for just £20,000! Soon after the acquisition, Sam decided he had no option but to close the factory temporarily in the light of continuous interference by certain trade union officials, following the national strike of shipyard workers, in order to reorganise it as a non-union shop. It is not certain when the unions eventually returned to Saunders-Roe – they were certainly there in 1945 – but they did not pose any problems for the company.

[2] Westland bought Saunders-Roe in 1959 and then in 1966, with Vickers, became the major shareholders of the British Hovercraft Corporation.

Saunders struggled on, always in need of finance, until 1928 when it was decided that a radical financial restructuring was required to save the company. Edwin Alliott Verdon Roe[3] and John Lord acquired the company and Sam resigned his directorship, but remained president on a salary of £250 for life. And so Saunders-Roe came into being. It was at the end of 1933 that Sam passed away after an illness at the age of 76.

Edwin Alliott Verdon Roe in 1953.

In addition to designing and building flying boats for commercial and military use, the company turned its attention back to fast motor boats. One of particular note was the 36-foot *Miss England II*, which was commissioned in 1930 to be one of the most powerful racing boats ever made. It did in fact reach speeds of over 100mph, but failed to win the ultimate accolade of the World Water Speed Record.

True to the culture of the company, also in 1930, a new venture was commenced, which combined the marine and aeronautical skills of the staff. Launched in 1931 and known as the *Hydroglider*, it was a 28-foot, seven-seater craft, flat bottomed from midships to the stern, with a 6-foot propeller driven by a 140hp engine mounted at the aft end. The first and only craft, which achieved a speed of 25 knots on half power, was bought by the Solent Hydroglider Company.

Orders for boats and aircraft were patchy and efforts were made to persuade the Royal National Lifeboat Institution (RNLI) to allow Saunders-Roe to once more build their lifeboats after their exclusive contract had been terminated in 1919. Again, in a precarious financial position, the company launched a major effort in 1932 to sell the concept of 35 to 52-feet coastal cruising boats.

---

[3] Sir Edwin Alliot Verdon Roe (1857-1958) was the first Englishman to make a powered flight in 1908 at Brooklands and was the founder of Avro, which had numerous landmark designs such as the Avro 504 trainer in the First World War, the Avro Lancaster, one of the pre-eminent bombers of the Second World War, and the delta wing Avro Vulcan, a stalwart of the Cold War.

The *Hydroglider* and its four-bladed pusher propeller, which was purchased by the Solent Hydroglider Company of Portsmouth. 1931.

At this time it was fortunate that the aircraft side of the business was to receive its biggest order in its history for a flying boat of its own design. In 1931 the Air Ministry had circulated a requirement for a twin-engine open-sea flying boat to a certain specification, which Saunders-Roe won with Henry Knowler's design, designated the *A27*, later known as the *London*. It was 56 feet 9 inches in length with an 80-foot wing span and an all-up weight of ten tons. It flew for the first time in 1934. With a speed of 155 knots, the crew of five were provided with three Lewis machine guns and 2,000 lbs of bombs or depth charges. Altogether, 31 of these flying boats were built, providing much needed work for the factory, with the last one coming off the production line in 1938.

The board had sanctioned 'various extensions to aircraft buildings', one of which was the new aircraft shop at East Cowes, known as the Columbine Hangar, to cope with the production order for the *Londons*. It was completed in 1936, but, when the final costs were known in 1937, they had risen from £32,343 to £48,816. Even so, it was a tribute to the Board of Directors and the architects that this shop proved large enough to accommodate the construction of the mighty *Princess* flying boats ten years later and subsequently the giant SR.N4 hovercraft.

In addition to the *London*, another larger monoplane flying boat, with an all-up weight of 18.5 tons and with a speed of 200 knots, called the *Saro A33*, was ordered by the Air Ministry. The prototype flew in 1938 but it had design flaws, which resulted in it bouncing three times on takeoff, reaching 50 feet and crashing back down on the water, causing irreparable damage. Fortunately, no one was killed but no further aircraft were ordered.

The Saro *London* general purpose reconnaissance flying boat. 1934.

1937 was a bumper year for the boatyard and *Yachting Monthly* magazine reported that if you wanted a boat built for the season you had to book early to avoid disappointment. The situation was to change within two years. There was no mention of commercial boat building in the records in 1939, but the company said it was extremely busy carrying out work for the government.

In 1938/39 Saro Laminated, a sub-division of Saunders-Roe, had introduced a method of pre-forming wooden aircraft components and were in quantity production in wartime for the Albemarle and Mosquito aircraft. Following the disaster of the *Saro A33* and the cancellation of the associated contracts, the company produced a twin-engine deep-hull monoplane flying boat designated the *Saro A36* and called it the *Lerwick*. With a wingspan of 81 feet and a 64-foot long hull, the *Lerwick* had a top speed of 215mph. She was launched in 1938 and trials began during which it was found she had severe 'porpoising' problems – rising up and then diving towards the water again in a similar motion to a porpoise in the water – during takeoff and landing. One of the problems was that it is always more difficult to get things right when marrying a prototype engine with a prototype airframe.

The issues were eventually resolved in 1939 and a total of 21 were built for RAF Coastal Command and the Royal Canadian Air Force, the last being delivered in 1941. These aircraft were replaced by the *Sunderlands* and *Catalinas* by 1943.

Supermarine *Walrus* single-engined Bristol Pegasus general purpose amphibian flying boat, of which 491 were built.

1939 – 1945

As in the First World War, with the advent of the Second World War Saunders-Roe was to receive a significant production order from the government. Supermarine (the company that designed and built the famous *Spitfire*) found that during the re-armament period in 1939 it took all its capacity to keep up with their *Spitfire* production and was therefore obliged to subcontract the build of their *Walrus* amphibian. Saunders-Roe received the contract in that year and ultimately manufactured 491 of these flying boats as well as 290 of its development of the *Sea Otter*.

With the rapidly expanding business due to the war came the need for additional buildings in which to carry out the work. Government contracts included the construction of hundreds of pontoons of various types, the majority of which were made from wood on metal frames, and towards the end of the war the company became a pioneer in the introduction of aluminium alloy for pontoons and bridges.

Supermarine *Sea Otter* single Bristol Mercury engined general purpose amphibian flying boat, of which 290 were built.

At Saro Laminated Wood Products large quantities of plywood were being made and it was estimated that the company supplied 40% of the total plywood requirements for the British aircraft industry during the war. The company also made components for motor launches, motor torpedo boats, air sea rescue craft and folding boats.

In 1942 bombing raids destroyed a considerable proportion of the factory floor at Cowes and East Cowes. The new shipyard works at Cornubia, part of the building programme for the war effort, and the Solent Works at West Cowes were also completely wrecked as was the repair shed at Somerton. Fortunately, dispersal proved its worth by minimising the disruption to output.

In 1943 Arthur Gouge, the famous chief designer of Short Brothers, joined Saunders-Roe and was immediately appointed vice chairman and chief executive of the company. His decision to join Saunders-Roe may have been a combination of disaffection with Shorts and an interest in a new large flying boat weighing 60 tons, with a top speed of 220mph and a range of 1,500 miles. Both Saunders-Roe and Shorts submitted designs, but the government responded by insisting on a joint design, which was finally submitted as the Shorts-Saunders-Roe R14/40. Shorts was awarded the contract with an instruction regarding the involvement of Saunders-Roe. The aircraft became the *Short Shetland,* but Saunders-Roe built the wings, flaps, nacelles and ailerons and was responsible for the hydrodynamic design of the hull. The flying boat first flew in December 1944 and performed well in respect of design on both the water and in the air. The end of the war, combined with a disastrous fire at her moorings at Felixstowe, persuaded the Ministry of Supply not to proceed. The second prototype was redesigned as a civil transport plane but did not fly until September 1947, by which time the *Saro Princess* (*Princess* flying boat) was regarded as the civil flying boat of the future.

The *Saro/Short Shetland,* four Bristol Centaurus engined, very long range reconnaissance flying boat. 1944-45.

1945 – 1959

After the war the design and development teams, who, in 1940, had been transferred to Beaumaris on the Isle of Anglesey, returned to the Isle of Wight. The facilities at Beaumaris were taken over by Saunders Shipbuilding Limited, who went on to build a prototype airborne lifeboat for evaluation by the Air Ministry, a batch of pram dinghies, a motor tug and a 75-foot, 42-knot motor torpedo boat (MTB) for the Royal Navy, as well as 300 double-decker buses for London Transport.

While Anglesey was beginning to prosper, very exciting progress was being made at Cowes, with the first flying boat jet fighter, designated the SR.A1, making its first flight in 1946. Although the trials were successful with a top speed of 525mph, the company failed to convince the Air Staff of its value and the project was eventually abandoned.

In 1943 the company had been investigating the feasibility of a flying boat to fly non-stop across the Atlantic. The design was completed in 1947 for the SR.45, later to be known as the *Princess.* She was required to cruise at 350mph, at a height of 36,000 feet from London to New York against an 89mph headwind, with a payload of 20,000 lbs.

The 148-foot long flying boat had a wing span of 219 feet 6 inches and was pressurised from the bow to a bulkhead at the base of the fin, in which the crew and 100 passengers would be accommodated on two decks with their facilities. She was powered by ten 3,200hp Bristol Proteus gas turbines. The two inboard engine bays in the wings on each side carried coupled engines driving 16-foot 6-inch diameter contra-rotating de Havilland propellers, while the outboard engine bay on each side contained a single engine, driving a four-bladed single propeller. Many advanced ideas were built into the *Princess*, the most advanced of which was the fully power-operated flight control system.

The *Saro SR.A1*, the world's only jet engine flying boat fighter aircraft. Three of these were built at East Cowes. The first flight was July 1946.

The Princess flying boat was launched on 20th August 1952 and two days later took off for a 35-minute flight around the Isle of Wight. BOAC[4] had placed an order for the *Princess*, but cancelled it in 1951 as its needs had changed.

Bob Strath was one of two master observers on the *Princess:*

---

[4] BOAC became British Airways in 1974 with the merger of the two government-owned airlines, BOAC (British Overseas Airways Corporation) and BEA (British European Airways).

Geoff Charmers, the senior observer, and I shared the master observer duties and, of the 100 hours flown on the only Princess *flying boat that flew, we did about 50 hours each, although sometimes we were both on it. The flight crew consisted of captain, co-pilot, two flight engineers, and the Bristol Engine Representative, Doug Kemball*[5]*, plus Geoff and me. Doug, who had been seconded to Saunders-Roe and had installed all ten Proteus engines, which proved extremely unreliable in the early days, reckoned that more turbine blades were scattered over the French coast than there was flak during the Second World War. We rarely took off with ten engines working. On one occasion at the Farnborough Air Show we took off on nine and came back on seven.*

The magnificent Saunders-Roe *Princess* flying boat. With ten Bristol Proteus engines, she first flew on 22nd August 1952.

Construction continued on the basis that they would be used as RAF transports, but this never materialised either. Only the first of the three ever flew, accumulating a total of 100 hours and, the other two, although completed, were mothballed at Calshot until they were scrapped in the late '60s.

The unreliable Proteus engines took many years of development before

[5] Doug Kemball was subsequently involved in the historic Channel crossing of SR.N1 with Bob Strath in 1959.

they were successfully installed in the *Britannia* aircraft and, subsequently, the marine version in the SR.N4 hovercraft.

The flight deck on the *Princess* flying boat.

The need to provide instrumentation for the comprehensive test programme required on the *Princess* resulted in the formation of a small electronics group within the research and development department. As a result, Saunders-Roe developed and took out patents on several inventions including the design and manufacture of miniature radio controls, telemetry equipment and servos. By 1952 they had constructed their first analogue computer, which led on to both large and small computers and they were able to apply their expertise to basic flight simulation. By 1954, having taken over Technograph Printed Circuits and the Cierva Autogyro Company, the group became the Electronics Division of Saunders-Roe Limited to solely concentrate on the modern technology.

The division's expertise led it into the supply of ground test equipment for the de Havilland *Firestreak*, English Electric *Thunderbird* and other guided missiles, and subsequently the production of Saunders-Roe's *Black Knight* high altitude re-entry research vehicle, and the *Black Arrow* satellite launcher, among others.

In 1952 the business in Anglesey was renamed Saunders-Roe

(Anglesey) Limited to make it clear that it was an integral part of the Saunders-Roe Group.

Due to hydrodynamic problems with the design of speedboats and flying boats, the first of three test tanks was constructed in 1947. The tank was 618-feet long and 8-feet wide and could accommodate models up to 6-feet long and 1.2-feet in beam, with a speed range from 0.05-feet per second to 43-feet per second. The carriage to which the model was attached could also support four observers. The second tank, opened in 1950, was a 400-foot circumference open-air test facility with a large gravity-operated catapult used to launch models at any required speed, angle of incidence and rate of descent back into the tank in either head or beam seas. The third tank, opened in 1956, was 250-feet long, 12-feet wide and 6-feet deep and was equipped with an unmanned carriage drawn along a monorail at speeds of up to 50-feet per second. At his station the operator not only controlled the model, but had all the data presented to him immediately on dials and pen recorders.

The manned carriage of the No. 1 tank in operation in 1949.

The company engineers had also developed the skills required to manufacture free-flight self-propelled radio-controlled models so that all characteristics of the craft could be checked, including controllability and manoeuvrability both on the sea and in the air. The research and development facilities, which were established in what was originally the old stables of Osborne House to the east of Cowes, included extensive equipment for structural static testing, vibration and fatigue testing, pressure testing, hydraulic and electrical testing and material and metallurgical testing. All disciplines of the research, development and experimental team at Osborne carried out work on a commercial basis for companies worldwide as well as those in the Saunders-Roe Group.

A large open water tank for ditching and free test flight. Waves are generated from two directions at right angles. 1950.

Model under test in the No. 2 tank suspended from an unmanned carriage. 1957.

It is worth mentioning here that following the end of the Second World War, an extremely important establishment was created at Osborne works site for the training of company apprentices. At the peak of the company's staff requirements in the 1950s, some 300 apprentices would be receiving their training at any given time. Apprentices, of whom Ray Wheeler was one, were given every encouragement to obtain an engineering degree. The full facilities included a residential centre with its own chaplain.

Henry Knowler, Chief Designer from 1926 to 1952 and then technical director until his retirement in 1956.

With the first boat jet fighter SR.A1 flying in 1946 and the detail design of the *Princess* proceeding, the research and development team turned their attention to the next stage of fighter design with the capability of taking off and landing on water. The fundamental difficulty in designing a waterborne aircraft capable of transonic or supersonic speeds is that the craft density becomes so great that insufficient buoyancy is available. Therefore, some other means of lifting the craft from the water had to be found. The team investigated the use of hydrofoils, hydroskis and a combination of the two. In the end, what became known as Saro Project P121, the aircraft was fitted with retractable skis, which were claimed to be lighter than the undercarriage of a conventional landplane fighter. In 1951, however, the concept was rejected by both services and the Ministry of Supply, and the company decided to proceed with the design of the land-based rocket-powered landplane fighter, which, after discussion with the Royal Navy and RAF, was modified to a mixed-power fighter with both a rocket and a jet engine.

In February 1952 the government issued a specification for a similar aircraft proposed by Saunders-Roe, which, not surprisingly they won against the competition. Thus the SR.53 was born and they were given the order to proceed in October 1952. The success of this mission led to the promotion of Henry Knowler to technical director and Maurice Brennan, a fiery Scotsman, to chief designer. The SR.53's fuselage was 41-feet long and the wing was a 25-foot low-aspect ratio-delta platform. It had to outperform a Mach 1.5 bomber at 60,000 feet, which meant a capability approaching Mach 2 and have a climb rate of

35,000-feet per minute. It was powered by an Armstrong Siddeley *Viper*[6] turbo jet with 1,640 lbs. of thrust and a de Havilland *Spectre* rocket of 8,000lbs thrust.

Maurice J. Brennan, Chief Designer of SR.53, SR.177 and SR.N1. Joined the company in 1936, appointed chief designer in 1952 and resigned in 1959.

Bob Strath was taken off flight testing on the *Princess* before the end and sent to Boscombe Down to study the latest techniques in instrumentation:

> *It wasn't possible to instrumentate the SR.53 the way we had the* Princess. *I was brought up to speed with telemetry, recorders and miniature recording devices. At the time Saunders-Roe was building the successor to the SR.53, the SR.177 prototype. I returned to Cowes by which time the SR.53 was coming up to first flight date. There was no way it could be flown from the Isle of Wight, so we went to Hurn Airport near Bournemouth and took over a wartime hangar and associated brick buildings in which we set up a complete site test unit, including offices and showers, to do all the trials for the SR.53. As time went on either Hurn got smaller or the*

---

[6] The *Viper* jet was subsequently used in the SR.N1 as the main means of propulsion.

SR.53 takeoff weight went up, because it became obvious that the runway wasn't going to be long enough, so the SR.53 was taken to Boscombe Down.

There was now nothing going on at Hurn, but somehow Saunders-Roe got involved in sub-contract flight testing for other companies. OR946 (Operational Requirement) was a new flight presentation system for all the new up and coming aeroplanes such as the Lightning, Buccaneer, Bristol 188, *and* Sea Vixen. *Kelvin Hughs and S.G. Brown, who were producing the master reference gyro based on the German Anschutz gyro that stabilised the V2 World War Two German rocket, had no means of flight testing it, so Saunders-Roe got the job. I was sent to Hurn to run the place sometime after 1954 until April 1959. We were testing instrumentation and in-flight presentation systems. My last active days as a flight test engineer were in the back seat of a Meteor night fighter. We were testing the true airspeed units and vertical tape presentations etc. – it seemed I was set for life.*

The Saunders-Roe SR.53 mixed power unit fighter jet. The aircraft was powered by an Armstrong Siddeley *Viper* jet and a de Havilland *Spectre* rocket.

The first flight of the SR.53 had taken place in May 1957 at Boscombe Down by the company's chief test pilot, Squadron Leader John Booth. A second aircraft was completed, but the first aircraft crashed for reasons unknown, killing Booth. The test flights, however, showed it to be successful in every way and the government decided to proceed with a larger version, designated the SR.177, 20 per cent bigger than the SR.53, which could carry the latest air-to-air radar, new homing devices and other associated equipment. It would be armed with two Red Top guided missiles, all at an increased speed of Mach 2.35. In September 1956 the contract was increased to cover the supply of 27 aircraft.

New staff were engaged in all departments and many design and production sub-contracts were placed. With 50 per cent of the aircraft complete and 91 per cent of the design, the company was devastated by a government white paper on defence produced by Duncan Sandys, Minister of Defence, in the then Conservative government, stating, *"Fighter aircraft will in due course be replaced with a ground-to-air guided missile system".* On 24th December 1957 the contract was cancelled, resulting in 1,500 redundancies at Saunders-Roe and the end of the design and build of fighter aircraft by the company, although test flights still continued with the SR.53 until 1960.

The Saunders Roe SR177 developed version of the SR53. Cancelled before any aircraft were completed in 1958.

As a result of their work on a mixed power unit fighter and their close amicable collaboration with the Royal Aircraft Establishment at Farnborough, in 1955 the government awarded Saunders-Roe a contract in support of the British medium-range ballistic missile *Bluestreak* programme, to develop a liquid-fuelled ballistic research rocket called *Black Knight* and the *Black Arrow* satellite launcher. The contract was for the complete package of design, development, manufacture and test of their flight control systems[7] and

[7] The SR.N4 hovercraft control system was based on the technology from the *Black Knight* and *Black Arrow*.

instrumentation, together with the ground firing sites in UK and Australia. The 35-feet long, 3-feet diameter *Black Knight*, which, when fully fuelled with high test peroxide and kerosene (as in the SR.53's *Sceptre* rocket) weighed 12,800 lbs, could reach a height of 500 to 600 miles and achieve a re-entry speed of 12,000 feet per second. First tested in 1957 at High Down near the Needles on the Isle of Wight, it was taken to Woomera in Australia and successfully launched in September 1958. Altogether, 22 successful launches were made without a single failure – a remarkable record without parallel in ballistics rocket development.

Black Knight single stage rocket. Twenty two were successfully launched at Woomera, Australia.

In 1952 a contract was placed with Saunders-Roe by the British Admiralty, in conjunction with the Defence Research Board of Canada, to test a model of an existing Canadian hydrofoil called *Massawippi*, employing the Bell Baldwin hydrofoil system. The system of ladder hydrofoils was invented in 1919 by the Canadian, Casey Baldwin, and Alexander Graham Bell. A hydrofoil craft built by them reached a speed of 70.86mph in the same year on Lake Bras d'Or in Nova Scotia, making it the fastest boat in the world at the time. The contract required Saunders-Roe to develop this system so that a more efficient craft could be designed. After completion of these tests, which involved extensive tank testing, an order was placed with the company in 1954 for the design, development and manufacture of a full-scale experimental craft. The design was carried out at Cowes in close collaboration with Beaumaris, who were to build the craft. The 59-foot long high-speed craft was supported by three 9-feet high ladder-hydrofoil units of similar size; one on each side forward and one at the stern. Two 1,750hp Rolls Royce Griffon IV aero engines provided the power, which was transmitted by a complicated series of bevel gearboxes to a pod at the bottom of a central strut, resembling a keel with a propeller at each end. Launched in 1957 it underwent successful trials with only minor problems, achieving a speed of 36 knots instead of the intended 50 knots. The boat was accepted by the Research Board and shipped to Canada where further extensive trials took place on the ladder foil system. Nothing much happened after that and the craft is now preserved in the Museum of Science and Technology just outside Ottawa.

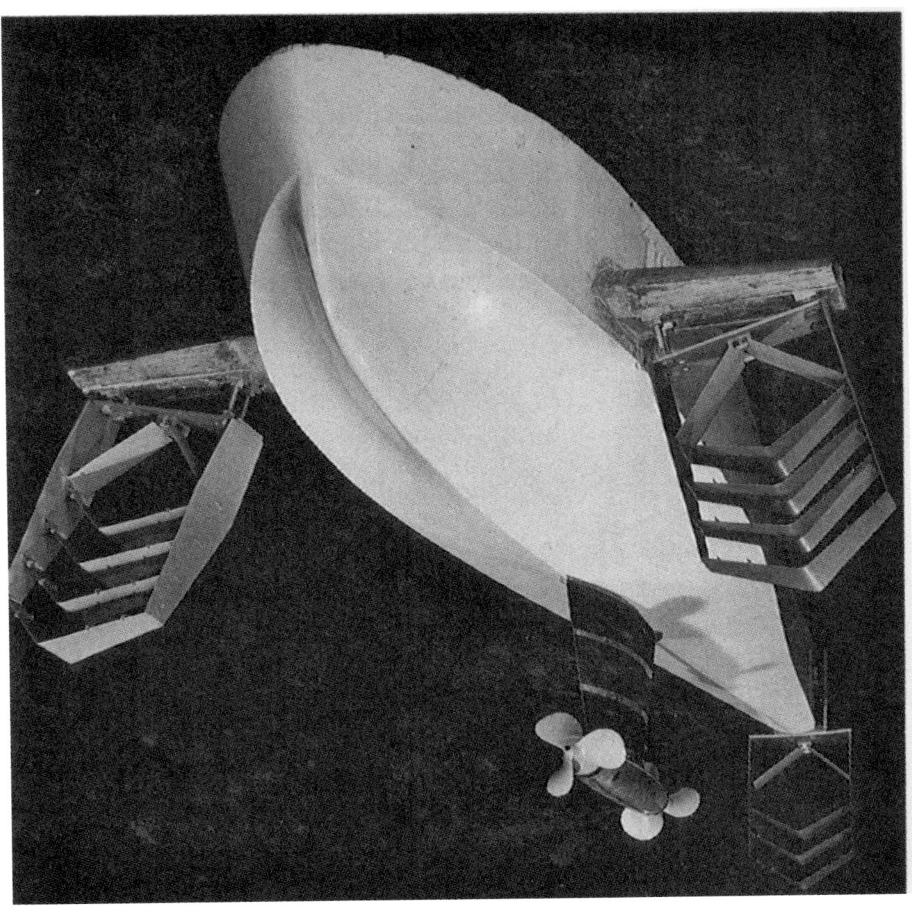

A scale test tank model of the *Bras D'Or* hydrofoil. 1957.

During the 1950s there seemed to be no end to the diversification of the Saunders-Roe Group. Their extensive high-tech facilities and skilled staff attracted a variety of projects. The Beaumaris experience in the building of bus bodies combined with the Cowes' need for support for the handling of high test peroxide for the SR.53, led to a number of contracts with the Ministry of Supply. Beaumaris manufactured a wide range of equipment for this purpose, from road tankers with a total capacity of 2,700 gallons to 2 gallon non-spillable containers.

Bras D'Or in the Menai Strait in the spring of 1957.

In the post-war era, one of the mainstays of profitability for Saunders-Roe was sub-contract work for other aircraft companies. In 1951 an agreement with Vickers saw the company produce 400 sets of wings at Eastleigh, in Hampshire, for the Vickers *Viscount* airliner until 1959 when the contract ended. In 1954 the company accepted an order for the design and lofting of the *Viscount* fuselage. At Cowes, between 1951 and 1956, 105 sets of pressurised crew cabins, bomb doors, elevators, rudders and ailerons were made for the *Valiant* bomber.

De Havilland was another large sub-contract customer for Saunders-Roe. Some 300 sets of wings, tailplanes and elevators were manufactured for the *Vampire* jet fighter, the second jet to enter service with the RAF after the *Gloucester Meteor*. Various spares were manufactured for the ill-fated *Comet I*, and subsequently the *Comet II* and *IV*, involving metal-to-metal bonding, an activity which was to become a major future activity of the company. All the hovercraft subsequently built made substantial use of this method of construction. In 1956 de Havilland's parent company, de Havilland Holding Limited, took a major shareholding in Saunders-Roe.

By 1958, Saunders-Roe was desperately searching for work to fill the void left by the cancellation of the SR.177, the combined supersonic rocket and jet fighter. As a result, a new company, Saunders-Roe Structures Limited, was set up to build alloy structures in the field of civil engineering, based on their experience of aluminium alloy in aircraft. The largest, and most interesting of the projects, was for a 345-feet long and 43-feet wide cylindrical segment stressed-skin roof for the North Thames Gas Board's oxide plant at Beckton. Other projects followed, but this entrepreneurial company ceased operations in 1960 when a disastrous fire destroyed the Folly Works based on the East Bank of the River Medina, Cowes.

Test tank model of a 1:96 scale of a nuclear submarine cargo vessel. 1958.

In the same year Saunders-Roe Nuclear Enterprises was incorporated. It was already familiar with the principles of nuclear engineering through a contract with the United States Navy to investigate fitting a nuclear plant to the *Princess* flying boat. For this purpose the senior engineers at Cowes had already undertaken a comprehensive course in this new branch of engineering. The initial project investigated was for the whole plant to be used for the pasteurisation of fruits and vegetables, and the sterilisation of surgical dressing by irradiation from a nuclear source. What became one of the main products was low-level nuclear lighting called *Beta Lights*, which were first manufactured at Osborne.

An interesting project was commissioned in 1958 and 1959 by Mitchell Engineering Limited. The comprehensive study was to consider nuclear-powered submarines specifically designed for carrying iron ore throughout the year from the Diana Bay region of Northern Quebec, Canada, to the United Kingdom with a capability of descending to 400 feet. The study concluded that a 600-feet long, 72-feet diameter submarine of 50,000 tons displacement, with a 50,000hp engine, giving a speed of 25.5 knots, was both technically feasible and an economic proposition on that route.

Another project was an 18-foot experimental amphibious beach survey craft known as the *WALRUS*, which stood for Water and Land Reconnaissance Survey Unit, whose purpose was to survey beaches and shallows on approaches to airfields. It was driven by two 12-inch diameter contra-rotating helices, similar to a threaded screw, but of the same diameter from end to end. They were mounted below the hull, one on either side, and were almost as long as the craft. The craft was capable of nearly 9mph over water and 2mph over land. It was duly delivered and the customer stated their intention was to fit a television link, sounding gear and beach survey instruments.

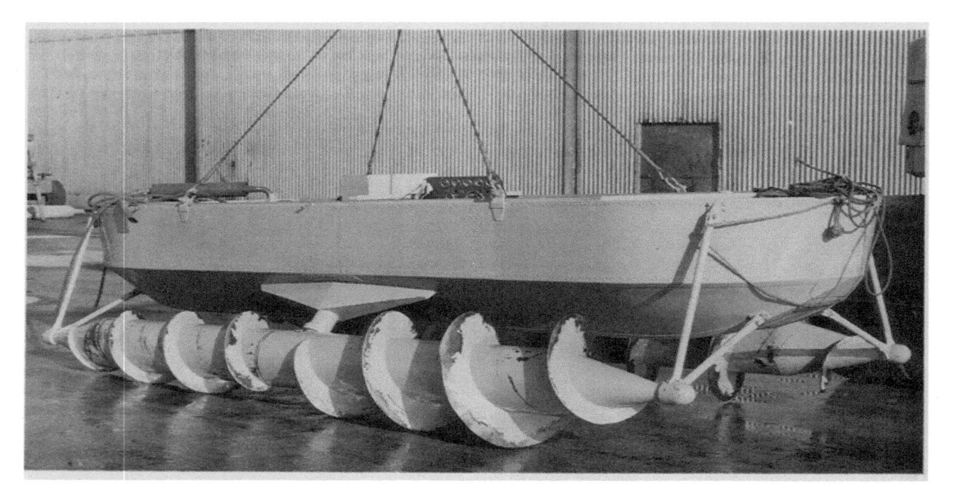

First tethered trials of *WALRUS* off Columbine slipway in 1960.

But events were moving on. As Bob Strath recalls:

> *After John Booth had been killed in the SR.53 in 1957 we operated for a whole year with a chap called Jock Elliot, Chief Test Pilot of the Airspeed Division of de Havilland, who, by that time, was also flight testing the Sea Vixen. All the planes operating from Hurn were called CA Fleet (Controller Air Fleet). When it was decided that Saunders-Roe would continue flight testing the SR.53, Sheepy Lamb left the Royal Navy in 1958 to become the company's chief pilot. Sheepy successfully flew the second SR.53 at Boscombe Down and also came down to Hurn when he had to carry out flight testing on the Meteor night fighter. We had all the breadboard models plastered all over the place and that's how we got to know one another. In fact, people used to think I was in the Royal Navy because Sheepy kept referring to me as 'My Observer'.*
>
> *One day the phone rang and Maurice Brennan's (Chief Designer) secretary said, "We're lucky to get you in – you're*

never there. No cricket matches on then?" "No, dear, the cricket season hasn't started yet." "The boss wants to see you rather urgently." I came back to Cowes and marched into the great man's presence – he was a fiery little Scotsman. He said, "What's the situation over there, what are you doing?" I gave him a report and he told me to wrap it up. I asked what would happen to me and he told me I was coming back to work on a secret project. I said, "What's that?" at which he growled at me, "You'll be told in good time, buggerlugs!" – that was his favourite expression for me – "There's no need for you to know at this stage."

I spent an anxious weekend and then on Monday morning I took the first Yarmouth to Lymington ferry to get back to Hurn. It was one of the old boats on the run known as 'the old coughing lung specialist'. I staggered on there with the place full of smoke and there was a chap sitting in this fug, which you could cut with a knife, with a copy of the Daily Mail. As I walked by I saw a photograph of Sheepy Lamb. So I peered over him and the article said something to the effect that Sheepy Lamb would be Britain's first 'hovernaut' and would be testing the 'flying saucer'. That was the first inkling I had – it was March 1959. When I eventually got to Hurn Airfield I picked up the phone again and said, "This bloody secret thing I'm supposed to be involved in – is it the hovercraft?" Of course there was all hell to pay because the Daily Mail reporter had blown it. I was coming back to join Sheepy and the pair of us were going to do the trials. I was the only flight test observer left by that time, but that was what I was there to do.

Nothing came of the *WALRUS* amphibian craft, or the concept, even though it was still being tested in 1960. It was eclipsed, as were some of the other projects, by a new and exciting form of amphibian transport, the hovercraft, which would not only have a substantial influence on the affairs of Saunders-Roe, but also eventually on the way in which passengers and cars were carried across the English Channel at speeds the like of which had never been experienced before and are unlikely ever to be so again. From the start the project was to prove a huge challenge fraught with difficulties.

Chapter 3

SR.N1 – The First Manned Model

The Saunders-Roe Nautical One (SR.N1)

The second stage of the contract, which was awarded in October 1958, involved more advanced research into the basic principle and development of the air cushion as proposed by Cockerell in 1956: optimum cushion platform and intake design; directional stability and control of the hovercraft; and design studies of 70-ton, 250-500-ton, 1,000-2,000-ton and 15,000-ton hovercraft. For this purpose, two manned models with elliptical plan forms were proposed, and after review with NRDC and Cockerell, *Model A* was selected.

The dimensions were to be 30.3-feet long, 17.2-feet wide, 12-feet high, all-up weight 4,000 lbs. The hover height would be 2.25-feet at zero speed and 1.5-feet at 30 knots.

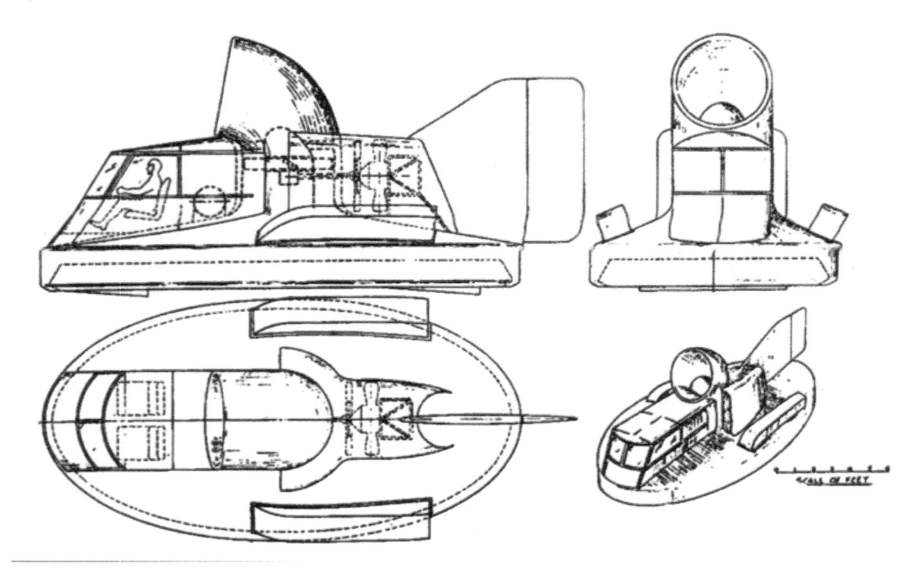

Manned Model Proposal Design 'A' developed from the original preliminary design using a vertically-mounted Alvis Leonides 521/1 450hp radial air-cooled helicopter engine.

There was, however, concern expressed with the design that the single annular peripheral jet would not provide adequate pitch and roll stability and that diagonal stability jets, as on the original Cockerell model, had not been incorporated. Even though the aerodynamics office had managed to convince the chief designer, Maurice Brennan, that the single peripheral jet, with no diagonal jets, was a good design, continuing doubts resulted in the construction of a three-dimensional model, not included in the contract, in the wind tunnel.

The sceptics proved their point as the model was extremely unstable, and when it was shown to Maurice Brennan it was realised that the craft would be too difficult to control even with the best piloting skills. An immediate 'stop design' was ordered on the cushion system. Further tests proved conclusively that the pitch and roll of the existing design with single peripheral jets was still inadequate. The use of transverse jets across the centre of the machine, as on the original Cockerell model, was not a viable solution as the buoyancy tank, which formed the platform for the cushion of air under the craft, for what was now designated *Saunders-Roe Nautical 1* (SR.N1), was already under construction. It was therefore decided to extend the plan form of the craft and put an extra peripheral jet all the way around the outside, so that there would be an inner and outer peripheral jet. A relatively simple analysis showed that there had to be a reasonable distance between the two jets, plus a bit more, to ensure an adequate margin of stability.

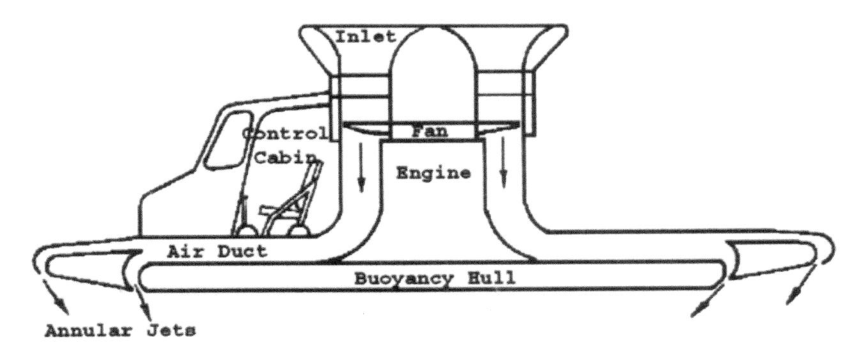

Final lift air ducting and peripheral annular jet configuration.

To check the stability of the double jet system, a one-third scale model was built and put through its paces. Still the stability problems persisted, but the eventual solution was to incline the bottom surface between the two jets at a shallow angle of 6 degrees so that the inner jet contacted the surface before the outer jet. By the time the solution to the stability problem was discovered, the outer segments of the SR.N1 were being built, but it was possible to quickly modify them in order to provide the necessary 6 degrees between the inner and outer jet.

Several models were used in the wind tunnel and tow tank testing to support the design development. A 1:6 scale free-flight radio-controlled model was tested in secrecy in the final stages in a remote bay on the Isle of Wight.

Engineers testing the radio-controlled model at Osborne Bay in the summer of 1959. Bob Strath is closest to the model in the water.

The final SR.N1 configuration had changed considerably from its original conception. It was now a twin-jet system with a total cushion area of 535 square feet with an estimated increase in weight by a third from 4,000 lbs. to 6,600 lbs. Two options were considered for bleeding off the propulsion air, but unfortunately the wrong one was chosen, which led to some instability problems that were not discovered until the initial SR.N1 hover tests.

The design and build of the SR.N1 proceeded through to the end of 1958 and into 1959 under Saunders-Roe's chief designer (Maurice Brennan) with the project engineer (John Chaplin) reporting to the chief of aerodynamics, Richard Stanton-Jones. The structural design and analysis was assigned to Ray Wheeler, reporting to the chief of stress, George Thomson.

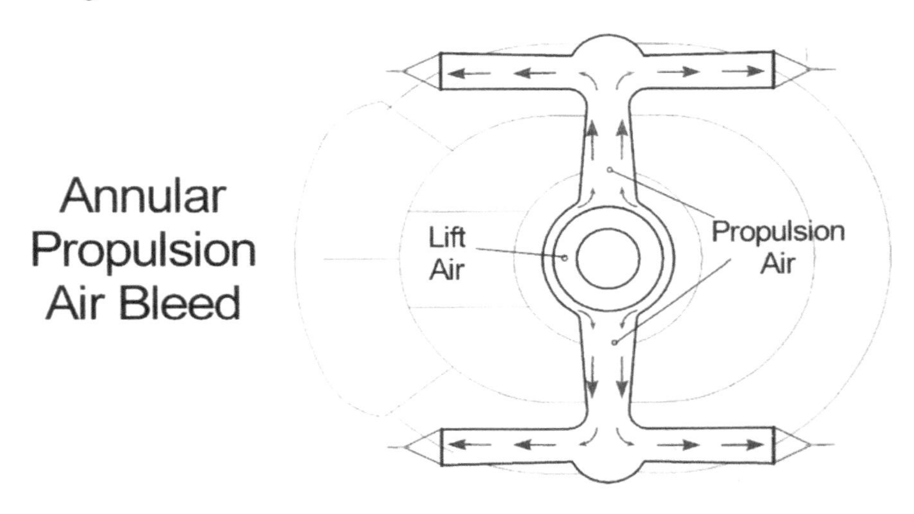

The craft was eventually built on a simple thin-riveted aircraft-grade aluminium alloy sheet buoyancy tank, 26-feet long, 20-feet wide and 10 inches deep and finished with a thin coating of pure aluminium for corrosion protection.

The outer jet, and what became known as the 'plenum chamber' (a box containing the air before it disperses through the jets under the craft) was mounted on closely-spaced 2-foot long ribs radiating from the buoyancy tank. The outer jet finished four inches above the inner jet, giving an inclined surface between the jets, which was plated and reinforced against wave impact or inadvertent ditching.

Longitudinal ducts 20-feet long and 2-feet square were mounted on the deck on both sides of the craft, supplied by air from what became known as the 'cotton reel', which was the cylindrical centre piece containing the engine and horizontally mounted fan, to give both forward and reverse thrust.

At both the forward and aft ends of the ducts, simple rotatable aerofoils were mounted to give both side and vertical control forces. At the aft end the vertical aerofoil was extended upwards to line up with the top of the cotton reel duct to form two conventional aerodynamic rudders. Finally, the control cabin was mounted on the deck immediately in front of the cotton reel. Unfortunately, the pilot's head was almost directly in line with the plane of the axial fan.

The Aerodynamics Department requested a very small clearance between the tip of the fan and the wall of the cotton reel on the grounds of aerodynamic efficiency. This could not be achieved without introducing into a section of the cotton reel wall a rectangular section beam or shroud, strut mounted, directly off the engine mountings. Ray Wheeler opposed this idea on the grounds that engine vibrations could cause the tip of the fan tip to hit the beam, resulting in a catastrophic failure in the intake. Alec Prickett, who jointly led the detailed design team, compromised but insisted on incorporating the idea and leaving the actual measurements to be made early in the trials. Ray then suggested that, in that case, Alec should be in the control cabin, which was adjacent to the cotton reel, and be the first to operate the engine controls – Alec smiled!

Interior bulkheads of the buoyancy tank forming watertight compartments.

When Sheepy Lamb, Saunders-Roe's chief test pilot, also expressed concern about the control cabin being positioned adjacent to the cotton reel containing the engine and fan blade, the mechanical design engineer replied with dry humour:

> *Don't worry – your seat is stressed to 10g horizontal, the engine mount to 12g; you have a 2g lead on the engine!*

But he did direct that the exterior of the cotton reel was to be heavily reinforced with thick plate to protect the crew.

Deck reinforcing added to the top of the plenum chamber and the four 'cotton reel' support struts in position.

The Alvis Leonides piston engine installed above the buoyancy chamber complete with the circular ring adjacent to the axial fan. The ring was removed after the first engine runs.

Who's in Charge?

During the early stage of the design process the coordination with Christopher Cockerell was on an informal basis and it soon became apparent that a more formal procedure was required to provide him with a way to continue to be involved in the design of the SR.N1, which was compatible with the organisational procedures at Saunders-Roe headed by the chief designer. John Chaplin explains:

> *Christopher Cockerell was like a lot of inventors in that he just wanted to get into everything. He always had a solution for anything. He had a brilliant mind – there's no doubt about it. Maurice Brennan, who was this fiery Scot, told me that I was going to be the project engineer and I was to handle it. We'd have a meeting and Maurice Brennan would say, "OK, I approve of this and we'll use that as a main spar", or whatever it was, and then Dougie Matthews, who was doing the design, would walk round the drawing board the next day and there was something entirely different on the drawing. We'd say, "What the bloody hell's that doing there?" Dougie would say, "Oh, Mr. Cockerell came in and said he wanted it this way." Maurice would just go into orbit. So I was left with the task of trying to placate him. At the time Cockerell was just walking in every day and wandering around the design department, looking over calculations or model tests or anything like that. He was very nice – just bubbling with enthusiasm – and saying, "No, no we should do this". He was the customer – he was representing the NRDC who was paying us, so we had to say, "Yes sir, no, sir". Maurice was just going ape. It wasn't a question of being right or wrong. There were two ways to do something – Maurice's way and Cockerell's way. In front of Cockerell, Maurice controlled himself and realised he was the customer, so all this fiery blowing up that I got, Cockerell never got. In the end it was agreed that we would ask Cockerell not to come in when he wanted to, but we would have weekly or bi-weekly design reviews where we would present to him what we were doing and he could make his comments then and discuss it, after which we would make a decision. It worked pretty well.*

Hovercraft Development Limited (HDL) was established in 1959 as a subsidiary of the NRDC, of which Christopher Cockerell was appointed technical director to handle the SR.N1 contract. But its primary task was to build a substantial patent portfolio based on the early Cockerell hovercraft patents, which, at the time, included the annular jet cushion, the plenum chamber cushion, compartmentation jets for stability, powered and non-powered recirculation and various configurations of flexible skirts.

Shortly before HDL's formation, Cockerell started to assemble a small technical team in October 1958, two of which included a craftsman and a model maker from Saunders-Roe. In January 1959 HDL purchased a cottage called 'The White Cottage' in East Cowes, conveniently situated near the Saunders-Roe Osborne office, in which Cockerell and his family took up residence. An ex-Northwood Army Camp shed was installed at the bottom of the garden to provide office space and accommodation for the equipment for model making and test rig fabrication. Early work performed by the staff consisted of the development of Cockerell's theory and analysis, and static experiments into advanced forms of the annular jet concept, in addition to the key activity of the documentation and patent application of new inventions. HDL subsequently moved to a new facility on the west side of the Solent at Hythe. The number of engineers would grow to over 40 and more challenging hydrodynamic testing would be accomplished, which also included the development of the HD-1 and HD-2[1].

Richard R. Stanton-Jones MA, FEng, FRAES,MRINA. Chief Designer, Saunders-Roe from March 1959 to 1966.

---

[1] The HD-1, built as a full-scaled test bed for skirts and control, was of mainly wooden construction and was described as the first non-amphibious craft with air curtains at both the bow and stern, before being transformed into an amphibious craft. The HD-2 was produced as a full-sized test vehicle to research hovercraft controls. In October 1967, the development and research aspects of HDL were transferred to the National Physical Laboratory to form the NPL Hovercraft Group. HDL continued with the licensing of the hovercraft patents.

As the SR.N1 progressed, confidential board meetings were being held with Westland Aircraft Limited, who was primarily interested in acquiring the Saunders-Roe Helicopter Division facilities and the *Skeeter* light aircraft programme to add to a new rotary wing monopoly. The future of the hovercraft business, and whether it would be included in the acquisition, was not clear at this time, but the SR.N1 team was unaware of these developments.

In March 1959, Maurice Brennan, probably anticipating that something was up, resigned as chief designer to join the Folland Division of Hawker Siddeley and Richard (Dick) Stanton-Jones was promoted to chief designer of Saunders-Roe.

Trials

SR.N1 hovering. The corner safety guards and landing pads were removed before the English Channel crossing.

The SR.N1 was completed and the first engine run made on 29th May 1959 with Sheepy Lamb and Bob Strath on board. The concern by the aerodynamic engineers that there was too much clearance between the fan blade tips and the cotton reel casing, thereby causing a significant loss in performance, was initially resolved by placing a section of duct wall in the plane of the fan, mounted on struts connected to the engine so that it would move synchronous with the fan and allow a small clearance between it and the fan tips. Ray Wheeler had recommended accelerometers be mounted on the shroud during the initial engine runs. The data from the first run of the engine was sent by the project engineer to the Stress Office and, shortly after, the crew were startled to see Ray sprinting across the apron clutching the data and shouting, "Stop!" The data showed that the stress levels were such that it would only have been a matter of minutes before the assembly failed with disastrous consequences. It was then decided that structural integrity took precedence over aerodynamic efficiency with the result that, amongst other things, the integral shroud was removed, giving greater clearance for the fan blade tips.

SR.N1 controls. Left stick controls the elevons in the hovercraft's propulsion ducts. Yaw fins in the same ducts are actuated by the rudder bars. The right stick moves forward or backward to give directional propulsion thrust. Engine output is controlled by twisting the grip.

On 30th May, with the modifications complete, the first full power engine tests were carried out. The craft was tethered on the slipway under a camouflage net to hide it from any prying press cameras. With Chris Gear, the mechanical design engineer responsible for the tests in the pilot's seat and Chief Test Pilot, Sheepy Lamb, alongside him in the observer's seat, Gear slowly increased the engine revs and checked the engine instruments, while the mechanics watched the engine and systems. The engine control was a twist grip, mounted on the fore and aft thrust control lever in the pilot's right hand, similar to the engine control on the collective of a helicopter. At about 50% power it appeared that Gear started to cycle the thrust control lever backwards and forwards and the SR.N1 began to surge accordingly like an animal fighting its tethers. An angry Sheepy, renowned for his forthright way in dealing with situations, ordered the craft to be shut down and in no uncertain terms made it clear to a miffed Chris that he was only responsible for the engine operation and that he, Lamb, would carry out all the flight checks. Chris replied that he fully understood the test protocol and that it was the control that was moving his arm and not the reverse. Further engine runs with Sheepy at the controls showed the thrust control system valves were unstable and that even his strong arm could not hold the lever.

Having identified the problem, requiring modification to the propulsion gate valve system, and with the scheduled press day only 11 days away, the design and production department worked round the clock to design and install the necessary modifications. Having settled the issues for the aerodynamic controls system, the final pilot controls were: a right-hand lever to control the thrust valves, with fore and aft movement combined with a twist grip engine throttle; aircraft rudder pedal to control the yaw; a central joystick connected to the duct valves and elevons to control pitch and roll. It was something akin to patting one's head with one hand and rubbing one's stomach with the other.

Bob Strath was to become more involved in the testing and launch a new career as a hovercraft test pilot:

> *Sheepy still had the SR.53 to fly, but, in addition, he was travelling up to Blackburn to fly the* Buccaneer *prototype. The SR.53 was being modified quite a bit. This put a huge demand on Sheepy's time as he had to keep up to date with his hands-on experience, by which time we got into the initial trials of the SR.N1. Then one day he said to me, "I think I'd like you directly on my staff." At that time I was still with the design office employed under Bill Worner, who was the chief flight test engineer. Sheepy said, "Would you be happy to come and join me?" to which I said I would be delighted, as we had become great buddies.*
>
> *He wrote to the insurers and the company explaining that the pressure of work on him was such that he wanted to be able to leave the development test flying of the SR.N1 to me. He detailed my past experience with the company and pointed out that I had acted as flight observer to him from day one. That's how I became a pilot.*
>
> *Bearing in mind that the first time we ever hovered we were there together, scratching our heads trying to figure out as to how it might work, so we learnt together. I don't know how many hours I had piloting before I went solo, but it wasn't many. Sheepy and I just grew with this machine together.*

Peter 'Sheepy' Lamb

Before joining Saunders-Roe Limited in 1958, Lieutenant Commander Peter Melville Lamb had a distinguished career in the Fleet Air Arm. During the later years of the Second World War he flew with 807 and 808 Squadrons in the Mediterranean and with 800 Squadron in the Far East. Afterwards he completed advanced courses at the School of Naval Air Warfare and Empire Air Navigation School to qualify as an instructor.

In 1950 he flew as a senior pilot with 800 Squadron in Korea and was awarded the DSC. He then completed No. 10 Course at ETPS and became Senior Pilot of the Naval Test Squadron at Boscombe Down from 1952 to 1954, during which time he received the AFC. From 1955 to 1957 he commanded 810 Squadron and received a bar to his DSC after the Suez operations. He subsequently took over command of 700 (Test and Trials) Squadron before joining Saunders-Roe as a test pilot. In 1958 he became Chief Test Pilot and continued to work in this capacity after the takeover by Westland. He then became involved exclusively with hovercraft and, on the formation of the British Hovercraft Corporation, assumed responsibilities for the testing of all BHC development and production craft as Chief Hovercraft Commander.

There were no tests between 1st and 6th June, but on 7th June all the system checks required by the Design Office Test Schedule were successfully completed. The SR.N1 was then tested for the first time in an untethered 'free' hover on the East Cowes Columbine slipway, but it was discovered that the craft had a pitch instability over a small range so that it could not be operated over that range of pitch trim. Later tests showed that had the annular propulsion bleed system, which was the alternative to that chosen, been installed, the problem would have been solved. The elevons and modified duct flow control valves proved to be ineffective in controlling the roll or pitch trim. With three days to go before press day, the elevons were removed and an all-out effort was made to resolve the other issues, some without success.

On 11th June, as scheduled, the SR.N1 was presented to an enthusiastic press, who immediately dubbed the new machine 'The Flying Saucer'. The plan was to limit the operation to static hovering and low speed manoeuvring over the slipway. Cockerell was not invited. According to Ray Wheeler and John Chaplin there was a complete clash of personalities between Christopher Cockerell and Dick Stanton-Jones, to the extent that Dick was determined not to have Cockerell there. Why this decision was not overridden by the directors is unclear, but whatever the reasons, Ray and John think it was vindictive, unnecessary and a huge mistake. A very determined Cockerell, however, persuaded a reluctant policeman to admit him on the day, following which the directors allowed him to stay. He did not steal the limelight, although he made his presence felt.

Ray and John are adamant that Cockerell was never disliked, but he may have made a nuisance of himself with the senior management. Whatever the reason for his exclusion from the official invitation list, it is sad to record that Cockerell was later to receive the same discourtesy for both the rolling out and launch of the SR.N4 some ten years later.

Christopher and Margaret Cockerell on Press Day.

The SR.N1 land-based demonstration received an enthusiastic reception from the assembled reporters, but they refused to accept that the demonstration was over and clamoured for the demonstration to continue at sea. Saunders-Roe management would have preferred to carry out the first sea-going trials in private, but after discussions with the design team and Chief Test Pilot Sheepy Lamb, who expressed confidence that the SR.N1 would operate satisfactorily over water, they reluctantly, with a silent prayer, gave the order to proceed.

The SR.N1 passing the Cunard liner Queen Mary during the first sea operation on Press Day. 11th June 1959.

Instead of charging off down the slipway in the hover, the craft was launched by winching it down into the water, the standard procedure Saunders-Roe had used for many years for launching flying boats. She was towed out of the harbour by the company launch and then, in a cloud of spray, accelerated through the hump speed and set off down the Solent at a rate of knots. With amazing good luck the SR.N1 was photographed passing the Cunard Liner *Queen Mary* on her way to Southampton. Rave notices appeared in the national press the next day and thus the success in demonstrating this new concept in transportation to the world was hailed as another triumph for British inventiveness and marked the dawn of an industry.

An unexpected factor in the waterborne demonstration was the amount of spray generated by the cushion jets, which blew back over the craft and the crew in the open-sided control cabin, as a result of which the glamorous white flying suits were quickly discarded in favour of waterproof marine jackets and trousers. The windshield wipers were replaced with the Kent Marine 'Clear View' rotating screen, which worked well over water only, but was subsequently found to be wanting when operating over muddy water or sand.

Press demonstration of the SR.N1 on the Saunders-Roe East Cowes Columbine apron 11th June 1959. Note the photographer, bottom left, lying down to make sure nothing is touching the ground.

Sheepy Lamb at the controls with Bob Strath in the observer's seat.

The SR.N1, having been lowered into the water on the old flying boat dolly wheels, is being towed out of Cowes Harbour for its first sea trial on Press Day.

The second sea trial was on 13th June and included full power runs and emergency ditching trials. During the operation, the United States Lines *SS America* entered the Solent and a decision was made to try to get some more PR photographs. Unfortunately, as the craft sped past the liner it encountered the ship's bow wave which pooped over the round bow and brought the craft to a sudden stop. Having established there was no structural damage, the SR.N1 returned to base and, after discussions with the project engineer and chief designer, it was decided to incorporate a hydrodynamic planing bow to reduce the tendency to dig in to waves before any more trials.

Trials continued with a sortie to Osborne Bay with the first transition from land to water, to test its true amphibian capabilities. Up until then the hovercraft had either hovered on land or on water but had never made the transition between the two, which was probably the most important characteristic of the hovercraft. Bob Strath explains:

> *It was the gorgeous summer of 1959. When we packed up for lunch, for some silly reason we moored the thing in about three feet of water. Osborne Bay was old Queen Victoria's private bathing beach. It was a gorgeous day – we were both in swimming gear and we swam ashore. It's difficult to explain, but it was all so new. Everybody's smart now, but in those*

days every day brought something fresh. It seemed silly to moor as we had been going up and down the beach at Osborne and from land to water and vice versa all morning and then for some reason or other we moored it. Why? For goodness sake we could have parked it on the beach and stepped ashore.

So here we are on the deserted beach with beer and sandwiches that had been delivered by the company van, it's a gorgeous day, and the sea's blue. We had our lunch, chatted about the resort and all sorts of things and then we both nodded off. I can remember Sheepy, bless his heart, waking up in a panic and saying, "Christ! The tides gone out – we're high and dry!" But it didn't matter because we could hover on land and water and go from one to the other. It was just all so new.

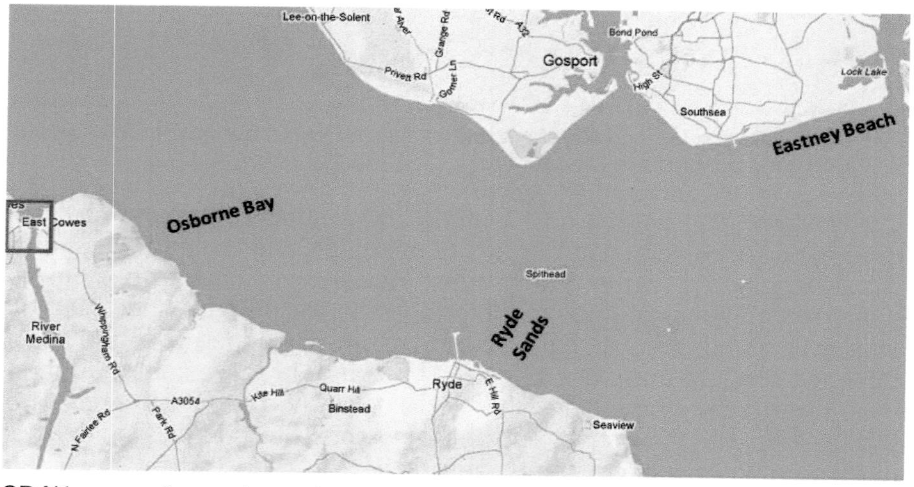

SR.N1 range of operations prior to the English Channel crossing.

After lunch Sheepy and Bob were joined by Dick Stanton-Jones and John Chaplin for the return from Osborne. While travelling over the Shrape Mud Shoal, where the water was only a few feet deep, the SR.N1 appeared to be lower in the water. At first the crew thought the hull buoyancy compartment had been holed by stones hitting the underside on the rough pebbled beach, and taking on water. The initial reaction of Sheepy Lamb was to issue the traditional naval command to lighten ship. As Stanton-Jones and Chaplin, realising they were the only jettisonable cargo, debated as to whether they should depart over the side, the craft returned to a more normal operating condition and safely returned to the slipway. On reviewing the situation with the coxswain of the chase boat, it was realised that for the first time hovercraft shallow water drag had been experienced. Had the senior executives attempted to dive overboard it might have been vaguely reminiscent of a slapstick Keystone movie.

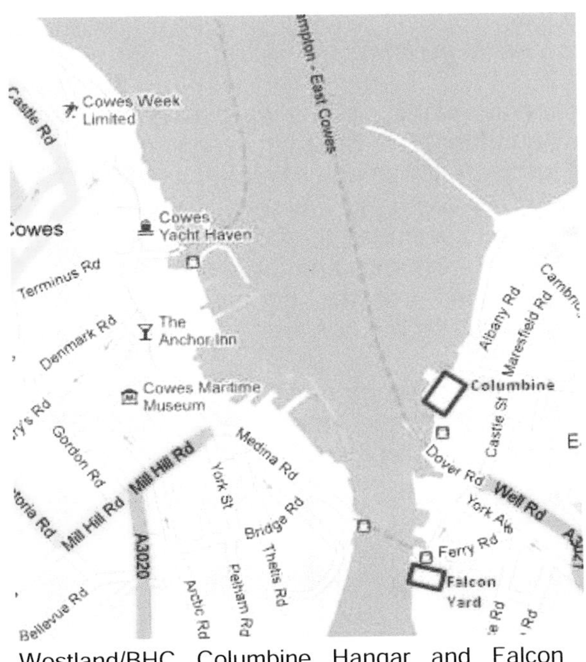

Westland/BHC Columbine Hangar and Falcon Yard at East Cowes

Only 11 days after the first launch, the SR.N1 made its first 'operational' sortie on Eastney Beach, near Portsmouth, for the Royal Marines exercise called 'Operation Run Aground'. It was so successful that the craft won commendations from the services, who had been so sceptical just over a year before.

Trials continued over sands at low tides, measurements of hover height were taken, towing drag tests carried out and VIP demonstrations given. The sea operations confirmed the basic performance predictions, but also showed the minimum control characteristics that were typical of most of the early hovercraft. Turning of the SR.N1 was achieved by yawing to a large sideslip angle and then using the thrust to provide a centrifugal force and turn, much in the same way as a racing car driver controls a four wheel drift into a corner. Although the heading had changed, it took some time before the direction in which the craft was travelling changed. The laws of motion prevented the hovercraft from eliminating this characteristic – it could only be reduced – and, as we will learn later, it lead to many anxious moments for the captains and navigators of the giant SR.N4 when operating across a heavily congested English Channel in thick fog.

During one of the early SR.N1 debriefings it was stated that if the craft was on a collision course with a large ship in the Solent, it would be difficult to avoid a collision, but the pilot could choose any part of the SR.N1 with which to hit the ship!

Overland, without the motion attenuation of wave drag, the control was even more difficult, with the result that considerable pilot skill and anticipation was required to counter the effects of ground slope and cross winds. Sheepy, in one of his characteristic flight reports, noted:

> *Driving the SR.N1 overland is a bit like driving an elephant, in as much as you can get an elephant to go anywhere you want provided the elephant wants to go there.*

On 23rd July, more fodder for the world's press was provided when it was announced that the Saunders-Roe Board had agreed that the company was to be sold to both Westland Aircraft Limited and de Havilland Holdings Limited[2]. Westland Aircraft took over the aeronautics part of the group, which included the hovercraft division and the main plant at East Cowes and Osborne, and the Helicopter Division at Eastleigh. The operational side of the company became the Saunders-Roe Division of Westland Aircraft Limited.

This was just one day before a momentous event in the history of the hovercraft took place, making front page headlines in most of the national newspapers and the lead story on the BBC News.

SR.N1 operating in Osborne Bay transiting from land to water. A Beken photograph.

[2] De Havilland took over Saunders-Roe (Anglesey) Limited and Saro Laminated Wood Products Limited.

Chapter 4

The First Crossing of the English Channel

While the early SR.N1 trials and demonstrations were going on in June, the NRDC was looking for publicity opportunities, which was often a prerequisite for attracting further government funding. It was Cockerell who realised that 24th July 1959 was the 50th anniversary of Bleriot's crossing of the English Channel and that France had designated it a day of celebration.

The SR.N1 on board the Royal Fleet Lighter *RN 54* heading for Dover. Picture taken from the company *Rapide* aircraft.

Cockerell's idea was that the SR.N1 should participate in the celebrations by making the crossing from Dover to Calais. Key proponents of the plan were Lt. Cdr. Ashmeade, from NRDC, and Ron Shaw, but the reaction of the Saunders-Roe project and operational personnel was negative on the grounds that the craft had only operated in the protected waters of the Solent. The craft barely had 39 engine hours on the clock, 14 sorties totalling 24 hours at sea, a maximum water speed of fewer than 30 knots in calm seas, with the longest sortie being six miles, also in calm conditions. To consider a cross-Channel attempt at this early stage in the SR.N1's life, with the vagaries of the wind and sea over such a long route length was a high risk and to some foolhardy. The rewards for success, on the other hand, would be huge in terms of publicity.

The *HMRT Warden* provided by the Royal Navy as escort for the SR.N1 English Channel crossing.

After several days of top management meetings with NRDC and Saunders-Roe, a last-minute decision was made to take a punt and the operational team were instructed to proceed with the planning and implementation of a cross-Channel attempt. To transport the craft to Dover, the services of the Royal Navy (RN) were enlisted to provide a self-propelled lighter, *RN 54*, accompanied by the tug, *HMRT Warden*. It was then realised that there was no suitable lifting gear to transfer the craft on to the lighter. On the Saturday morning, six days before departure, Ray Wheeler and a colleague, Doug Matthews, were instructed to see to the design and manufacture of suitable lifting gear and have it ready by Monday morning. They found the materials and Ray produced a sketch on a beer mat. Doug took charge of having a lifting beam manufactured, together with the slings and bolts, and it was tested with a huge crane on the Sunday, whereupon the necessary test certificate was issued[1].

Chief Test Pilot Sheepy Lamb had decided that he would take Bob Strath to handle the engine and craft systems and John Chaplin, Head of Aerodynamic Research, for the craft trimming and performance optimisation.

On 21st July the crew took the company's de Havilland *Rapide* aircraft to survey the route and look for possible beaches in the Dover area from

---

[1] Several months later the chief stressman, George Thompson, asked Ray, "Where are the calculations?" When Ray told him that he had lost the beer mat, but all that mattered was the test certificate, George had sense of humour failure.

which to launch the SR.N1. The final decision was the beach inside Dover Harbour and the SR.N1 was loaded on to the lighter.

Captain Barlow supervising the unloading of the SR.N1 at Calais. The SR.N1 twin jet arrangement is clearly shown.

Change of Plan

On 24th July[2], the crew and support team arrived in Dover shortly after the lighter only to discover that the weather forecast for the next few days was for strong north-easterly winds, which would make it difficult, if not impossible, for the SR.N1 to complete a crossing.

The only feasible alternative was for an attempt to be made from Calais to Dover. After approval for the change was obtained, the lighter was all set to sail for Calais when the hand of bureaucracy delayed the departure on the grounds that Saunders-Roe did not have an export licence for the SR.N1. Fortunately, Saunders-Roe managed to persuade higher authorities that they had no intention of parting with their beloved creation and that, with any luck, it would be returning to Dover imminently.

---

[2] The day after, Saunders-Roe became part of Westland Aircraft Limited.

So in the late afternoon the lighter with its precious cargo, together with *HMRT Warden*, with the Saunders-Roe crew and support team on board, and a Royal Air Force (RAF) Air Sea Rescue launch supplied by the MOD, they arrived at Avant Dock in Calais. The weather on the crossing – a nominally Force 4 north-easterly, producing white caps on the 4-foot waves, plus a moderate swell was well beyond the capabilities of the SR.N1. They had sat on the *Warden* in mid-Channel for part of the afternoon studying the conditions and, after reviewing the situation, Sheepy decided to postpone the cross-Channel attempt until the next day.

The SR.N1 was offloaded from the lighter into the dock and moored alongside with a growing number of people gathering to view this strange machine. It was low tide and the SR.N1 looked a long way down. The publicity for the crossing had preceded the SR.N1 and, as it was a national holiday in France, the crowds, including the press, continued to grow. Bob Strath continues the story:

> *Was I on the crossing? Well, yes and no! I had crossed over to Calais in the RAF high speed launch and, with the others who had arrived in* Warden*, we decided to repair in a British Rail café on the dockside for some sustenance. Unbeknown to us Christopher Cockerell had taken it upon himself to tell a sole gendarme that under no circumstances was anybody to be allowed on the hovercraft.*
>
> *In the café, Christopher Cockerell, Dick Stanton-Jones, Sheepy, Ron Shaw, Lt. Cdr. Ashmeade (NRDC), George Saunders and John Chaplin were all sitting round a table, while Doug Kemball, the Bristol engine chief inspector and I sat up at the bar. Dick kept saying, "Get us another round of drinks and some sandwiches, Bob." He would give me a FF10,000 note and I was getting FF8,000 change each time this happened. This went on and on, "Bob, get us another round", and I was accumulating a pocket full of French francs, which was going to save me the next morning. In the centre of all this was Christopher having the time of his life.*
>
> *Meanwhile, the crowd of interested spectators, who were trying to get a glimpse of the little hovercraft, had grown from a few tens, when we had set out for the café, to hundreds and so it was decided that we would put on a demonstration for them and the press. There is a picture of a young lady French reporter and me taken the evening we did the demonstration. Sheepy sent me back with Doug Kemball to get things ready and start the engine. We left the café and had to push our way*

through the mass of people, who were crammed into a small area on the quay above the SR.N1 – it was unbelievable. I eventually pushed my way through, but Doug was some way behind me.

I had almost reached the ladder to start the long climb down when the sole gendarme trying to control the crowds, put his hand up to stop me. In my fractured French I told him I was part of the crew, but he said, "Non, Monsieur Moustache said no go on." – Monsieur Moustache was Christopher Cockerell. I said, "Well, Monsieur Moustache cannot stop me as I am the second pilot", and he let me go. I climbed over the wall and started down the ladder. I got just a few steps down when I heard this anguished cry coming from Doug, "Bob, he's going to shoot me!" There was this one uniformed policeman surrounded by hundreds of people and he'd actually pulled his revolver out and stuck it in dear old Doug Kemball's stomach. The story Doug told up to his dying day was that, "If it hadn't been for Bob Strath, Monsieur Moustache would have had me shot!" I couldn't make it up.

Bob Strath with the young French journalist, who went with him to Dover the next day. At Calais Docks prior to the evening demonstration.

At the request of Christopher Cockerell, Sheepy Lamb agreed to take him along for the demonstration and also as a passenger on the Channel crossing. Just after 7.30pm at low water, with the harbour clear of shipping, the SR.N1 departed. Sheepy's report stated:

> *In view of the fact that there was so little time to spread the word, it was remarkable to see the very large numbers which had gathered. The starting of the hovercraft engine was rather like the tune of the* Pied Piper. *Children appeared from everywhere to throng the harbour wall.*

The SR.N1 at Calais docks with Cockerell preparing for a late afternoon demonstration. 24th July 1959.

After the initial demonstration in the calm waters of the harbour, to show the manoeuvrability, acceleration and turning performance of the craft, the SR.N1 went outside the harbour and operated for almost an hour over the exposed sands to the south-west of the harbour, and in the lee of the breakwater. Sheepy continued:

> *By the time the hovercraft approached the sands a large number of excited children and equally interested adults were running hither and thither over the beach to obtain a good view*

at close quarters. The odd suicidal enthusiast decided to lie down in the path of the SR.N1 to ascertain there were no wheels underneath. This colourful, but somewhat over enthusiastic welcome by an excited crowd necessitated a restricted performance, especially as regards speed over the beach.

There was, however, one less than impressed Frenchman, who was digging for lugworms and never even looked up as the SR.N1 passed him by to within 50 feet.

SR.N1 returning to Calais Harbour after the demonstration on the beach the evening before the Channel crossing.

As the craft re-entered Calais Harbour it encountered some rough water, which stopped it dead in its tracks. At Sheepy's request, John Chaplin went aft to ask Bob Strath, who had been monitoring the fuel, to move up to the bow with him to help get the bow down in order to accelerate through the hump speed, but Bob had fallen when the craft had come to the sudden stop, receiving an incapacitating blow just below the knee from a hull stringer. On receiving assurance that Bob was otherwise all right, John went forward, but Christopher Cockerell, who was not aware of the situation, got out of his seat and went aft to Bob to tell him to get forward. When he saw Bob prone on the deck he went back to the control cabin, rapped Sheepy on the helmet and shouted, "Bob, Bob!" Sheepy cut the throttle to idle, got out of his seat and went to see what the emergency was with Bob. The SR.N1 was now drifting towards the breakwater and collided with its inside slope, which was almost the same angle as the craft's bow. It took some time, even with full lift, reverse thrust and pushing off with hands to free the SR.N1 and return to the lighter.

After the debriefing, Sheepy had a private word with Cockerell, explaining that the Channel crossing would be a much more hazardous operation. Sheepy fully understood the importance of Cockerell wanting to be part of this historic event and acknowledged his great contribution and enthusiasm to the development of the hovercraft, but requested that as a passenger he would not, unless asked, get involved in the operations. Cockerell agreed. How diplomatically Sheepy might have conducted this meeting, given his record for the use of forthright naval jargon, is a matter of conjecture, but what is certain is that Cockerell, given his great character and broad shoulders, would not have taken offence.

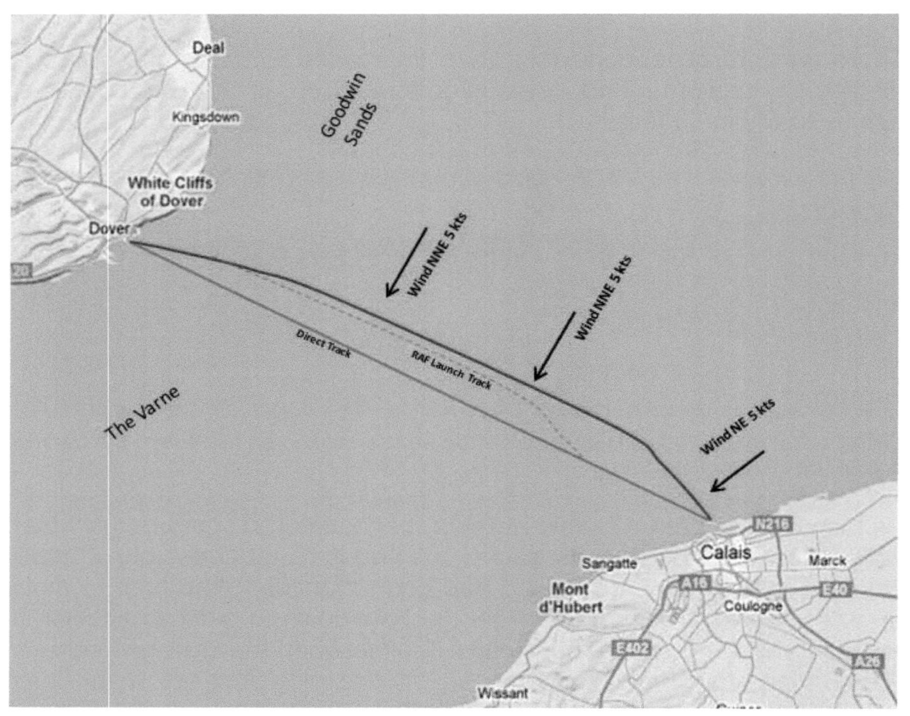

The SR.N1 and RAF chase launch tracks during the English Channel crossing.

Late in the evening, with weather reports still looking marginal for an attempted crossing, it was decided to stand down until 6.00am the following morning and the press were advised accordingly. NRDC had booked some rooms in a local hotel in Calais town prior to arrival when it was thought the crossing was going to be from Dover to Calais. Chief Designer Dick Stanton-Jones took most of the team to a late dinner. Bob Strath was the only one who went to the hotel after the demonstration for a shower and brush up before dinner and decided he would spend the night there afterwards. Sheepy, Chaplin and Cockerell, however, decided to accept the invitation of the captains of the *Warden* and the RAF launch to use berths on board.

Just after 3.00am the next morning, Sheepy woke to find that it was flat calm and immediately asked for the launch to take him outside the harbour to confirm the conditions. A radio message was sent from the launch to the ground crew in the harbour to fill the fuel tanks and fire up the SR.N1. To provide sufficient fuel for the crossing a second 50-gallon tank, in addition to the regular 25-gallon tank, had been installed on the SR.N1, but, as a safety measure, it was decided to carry a further 16 gallons in 8 two-gallon cans. The plan to have *Warden* stationed in mid-Channel was abandoned.

Attempts to locate Bob Strath, blissfully asleep in a hotel somewhere in Calais, were unsuccessful, and so the craft departed with Sheepy and John Chaplin as the crew, accompanied by Christopher Cockerell as the passenger. It cleared the harbour entrance at 4.55am, just ahead of the RAF launch and headed out for Dover in a light 5-knot north-easterly wind at about 18 to 20 knots. The true course was 294° but with the anticipated increase in wind after sunrise, Sheepy decided to track as far to the north as possible, so that he would be able to put the wind on his starboard quarter as it increased. The SR.N1's compass was unreliable, but Sheepy was able to use the loom of the South Foreland light on top of the white cliffs just east of Dover, which was still visible, but fast disappearing with the approach of dawn. Sheepy's report recorded:

> *For the next five miles navigation was made by dead reckoning, with the RAF launch, which had previously agreed to maintain the true track from Calais to Dover, well away on the port side. At approximately 5.30am the white cliffs of Dover, tinged red in the morning sunrise, were first visible*[3]*. Up to this time the SR.N1 appeared to have made extremely good progress at a constant setting of 2,700rpm. However, a slight swell was now apparent, which retarded the progress of the craft and on several occasions dropped her back below the hump speed.*

Meanwhile, back at Calais, Bob Strath had arranged a call for 5.30am:

> *In those days Calais was the major port for the holiday crowd and you couldn't get a room in Calais for love nor money, but NRDC had managed to book 10 or 11 rooms in an hotel. I was the only silly sod to use one. When I came down in the morning the foyer of the hotel was one mass of sleeping bodies – there were no rooms for them – English travellers coming back from their holiday awaiting their morning ferry*[4]*. As I went*

---

[3] The height of eye on the SR.N1 would have been 2 metres at the most, giving a sea horizon of about 5 miles.

[4] There was a very restricted cross-Channel service in those days with British Rail as the main carrier. Townsend was still using a converted frigate in 1959, and it was not until 1962 that their first purpose-built car ferry, *Free Enterprise I* came into service.

to leave, Madame suddenly stopped me and said, "Who's going to pay for all these rooms you haven't used?" I ended up paying FF22,000, which I was able to do as a result of keeping all Dick Stanton-Jones's change in the bar.

SR.N1 making heavy going in the swell. Chaplin checking the fuel gauge with Lamb at the controls and Cockerell at the bow for ballast.

I was a bit bleary eyed from the night before and as I was approaching the quayside, dear old Edgar Goodenough, who was the works superintendent, sitting on a bollard said, "She's gone." I said, "What do you mean 'gone'?" "Commander Lamb suddenly decided to go."

About half an hour later the press corps started to arrive, including that young reporter I had spoken to the previous evening, and all hell broke loose because the press and sightseers had turned up to witness the departure of the SR.N1 on its historic flight across the Channel only to find that it had left without warning. I was there and she remembered me from the day before. I was on the quay wondering what to do. On the bridge of Warden *I heard a voice calling out, "Single up to a line and a spring", and a little matelot up in the bow called out, "You coming with us, sir, or are you staying?" I said, "No, I'm coming with you", and slung my bag up to him. He caught it and I was just about to jump on board when the young reporter grabbed me and said, "I um coming wiz you." She pulled up her skirts and leapt aboard. I shall never ever forget it as long as I live. She was an attractive little thing, but she was not letting me go and just clung to me until we eventually arrived at a berth in Dover.*

From mid-Channel, the wind and sea increased and when the SR.N1 fell below the 12-knot hump speed, it was necessary to increase engine revs and head in a more westerly direction to get her going again. She was tending to slide down the swell, resulting in the port bow and propulsion duct nosing into the sea. To counteract this, Cockerell and Chaplin were working like a dinghy racing crew leaning out over the starboard bow while Sheepy, with great skill, tried to anticipate the nose-in, increase engine revs and turn on to a track more amenable to the swell.

Sheepy Lamb at the controls of the SR.N1. The open sided cabin gave little protection from the cushion spray.

At this time, a motor yacht of about 50 feet appeared on a collision course, over which the SR.N1 had right of way. The yacht's crew was either not keeping a proper lookout, or they were possibly misled by the SR.N1's large drift angle, whereby the craft was pointing in one direction, but was actually tracking in another. Sheepy, sensing that a collision could be imminent, altered course to starboard but the yacht, now bewildered by the track of the SR.N1 and probably wondering what this strange craft that looked like a flying saucer UFO was doing in the middle of the Channel, turned to port. There was nothing for it, but for Sheepy to cut the power and the SR.N1 came to a dead stop off the cushion. The yacht came right up to the craft's bow before coming to a stop. As soon as the yacht manoeuvred clear, Sheepy, in fine naval tradition, gave his explicit opinion of amateur mariners.

Just after this photograph was taken the craft hit the wake of a ship and Cockerell disappeared under the water over the bow, much to the consternation of Sheepy and John Chaplin.

After this incident the SR.N1, due to the difficult sea state, took an appreciable time to get above the hump again. Both Cockerell and Chaplin shifted positions to no avail. After proceeding for two miles below the hump speed, the craft encountered the 4-foot wake of a large ship proceeding north through the Straits of Dover. As the bow dug in, green water was shipped over the deck, totally engulfing Cockerell, who was still on the bow as ballast, which nearly swept him overboard. Until he re-emerged, still on board, Sheepy and Chaplin admitted later that they had both, for a moment, wondered how they would explain to the powers that be that they had lost one of Britain's most important inventors at sea. John Chaplin recalls:

> *The back of that N1 was a slippery place when trying to keep a foothold. Bob had been at the back when he fell the day before. Then I look back at the Channel crossing and think, Christ, I was running back there and pumping fuel etc. in the dark and if I'd slipped off and gone overboard, Sheepy would never have seen it and with Cockerell laid down on the bow, the buggers would have gone on and left me behind. It never struck me at the time, but now I get older I think, Christ, what the hell was I doing? The only thing was that the flight sergeant on the RAF launch gave me one of their survival suits which you zip up and inflate, so I probably would have floated for a while.*

Sheepy and Chaplin were becoming increasingly worried about Cockerell, who seemed to be suffering from exposure. Chaplin had his RAF immersion suit under his marine jacket, but Cockerell had chosen to wear a conventional waterproof jacket with a white towel at his neck, which unfortunately, in the cloud of misty spray, acted as a wick and carried the cold sea down inside his jacket[5].

SR.N1 approaching Dover. Chaplin and Cockerell standing by the fuel tanks.

As the SR.N1 continued on its journey to Dover, making less progress than at the beginning, Chaplin, who had taken over the assignments of the missing Bob Strath, continued to monitor the fuel consumption and periodically pump fuel from the secondary tank to the primary tank using a *Swicky* wobble pump (well known to all World War Two aircraft ground crews). The fuel level gauge was mounted directly on the main fuel tank and, since there was no fuel indicator in the control cabin, Chaplin had to continuously report the fuel situation to Sheepy. About two miles from Dover all the fuel from the secondary tank had been transferred and the fuel remaining in the primary tank was down to 2 gallons, leaving only a few minutes of engine running time. It was now imperative that the extra two-gallon containers, put on board at the last minute, be pressed into service. In order to save time, the designers at Cowes had replaced the

---

[5] His daughter, Frances, believes to this day that Saunders-Roe was negligent in not providing her father with the proper wet weather gear.

filler cap of the primary tank with the supply pipe from the secondary tank. The procedure was therefore to pour the fuel into the secondary tank and then pump it into the primary tank, which involved a three man operation – one to hold the funnel, another to pour the fuel and the third to pump, but with no Bob Strath the assignments had to be Chaplin to hold the funnel, Sheepy to pour and Cockerell to pump.

SR.N1 on the first landing at Dover. 25th July 1959.

With the SR.N1 set in a slow circle, and nobody at the controls, the refuelling began. Just as Sheepy went to pour in the first can the craft lurched as it hit a swell. The fuel missed the funnel and went straight into Chaplin's face. Without hesitation Sheepy grabbed him by the neck and pushed his head into the sea. This prompt action probably saved Chaplin from considerable future eye problems, but, for a brief moment, being grabbed by a big strong hand and held under water was, as Chaplin later admitted, rather alarming.

The refuelling now took on a greater sense of urgency because of the Leonides engine's reputation for being difficult to start when hot, and so Sheepy and Chaplin rushed to get the fuel in the secondary tank. After his gallant, exposed ride on the bow, Cockerell, cold, wet and fatigued and unfamiliar with the procedure, was waiting for a signal to start operating the *Swicky* pump. Sheepy, a man of little patience in tense situations, and not one to mince his words saw the inaction on the part of Cockerell. Reverting to type and with all the vocal force of a true naval officer, shouted, "Pump, you silly old ******, pump!" This outburst shows just how critical the situation was.

Cockerell responded admirably and the refuelling of the 16 gallons was completed with the Leonides still running smoothly. After landing, Sheepy apologised to Cockerell, who was magnanimous and agreed that the explicit direction was just what was needed at the time to save the operation.

SR.N1 arrives at Dover. First on board is HM Customs.

Sheepy's report ended:

> *After refuelling, the craft was clear of the swell in the lee of St. Margaret's Bay and it proceeded at the best speed it had obtained throughout the whole passage and up to and through the entrance to Dover Harbour. It is estimated that the craft probably achieved 30 knots in the last mad dash. The craft*

proceeded up the length of the harbour and beached adjacent to the Clock Tower, some 2 hours and 3 minutes after getting underway at Calais. The first personnel to board the craft were the customs officers with the request, "Anything to declare?" The obvious reply, "That it was good to be in England", appeared to satisfy one and all.

What Sheepy did not know was that Cockerell had stowed an expensive magnum of champagne in the forward cockpit locker, but fortunately the customs officers did not carry out a search.

The actual running time from departing Calais Harbour to arriving at Dover was 1 hour 48 minutes, taking time off for the refuelling, which was significantly more than Bleriot's 36 minutes, but still a reasonable time bearing in mind the low performance of the SR.N1.

As a relieved John Chaplin subsequently commented:

In the history of hovercraft, Sheepy Lamb was the right man at the right time. As an experienced test pilot, naval officer, seaman and ocean yacht sailor, he combined all the skills and maritime knowledge needed to lead the development of piloting hovercraft. On the Channel crossing I never had the slightest doubt, no matter what the adversities, Sheepy would get us to Dover.

As more press arrived and the crowds grew, it was decided to put on another demonstration, so, after refuelling, the SR.N1 was fired up and repeated the entry into Dover Harbour. At about 7.30am the crew was transported to the lawn of Dover Castle above the cliffs, where the BBC had sited a radio transmission trailer. After Big Ben struck eight, BBC's Christopher Serpell started the news with the words, "They have arrived", and went on to describe the Channel crossing, with a few comments from Sheepy. Cockerell wanted to be included, since he was the inventor, but the time ran out and the producer charged out of the trailer, drawing his finger across his throat in the classic signal to cut, just as Cockerell was about to get going. However, John Chaplin recalls 'a Cockerell moment' with the press:

We had arrived at Dover and after all the shenanigans with the refuelling etc. a press guy comes up to Cockerell and says, "What was it like?" and Cockerell replies, "A piece of cake, old chap!" Sheepy and I looked at each other and said, "Christ!"

From a public relations point of view that was surely the right answer, even though John and Sheepy were probably still getting to grips with what had been a huge ordeal, with much riding on its success.

After the broadcast had finished, Sheepy related an interesting coincidence. Fifty years earlier, another Lamb, Sheepy's uncle, as a young newspaper reporter, had interviewed Bleriot after he had landed on the White Cliffs of Dover.

Large crowd on Dover beach watching the re-enactment of the arrival.

The crew then went to a hotel in Dover for a well-earned shower and breakfast. Bob Strath eventually arrived at a berth in Dover on *HMRT Warden*:

> *When we arrived there was a message from dear old Sheepy saying that as soon as I could I was to join them in the White Cliffs Hotel on the waterfront for a celebration breakfast. The press were all there with Sheepy seated in splendid glory on the top table.*
>
> *When he saw me he said, "Bob, my dear chap, I'm sorry I had to leave you etc. etc." and then the French journalist appeared from behind me. She made a bee-line for Sheepy, who enquired, "Who's your friend?" thinking that she was a bit of crumpet I'd just picked up. Anyway, Sheepy gave her the exclusive interview she wanted. When the interview was finished, Sheepy said, "How did she get here?" I said, "She hoiked up her skirts and jumped aboard the* Warden *with me – I couldn't get rid of her."*

It was then we realised that she didn't have a passport and the question now was how we get her back to France without a fuss. It was said about Sheepy in those days that he could charm the birds off the trees. So the three of us went back to Warden *where Sheepy talked like a Dutch uncle to the captain and persuaded him to take her back to France, which he did, courtesy of the British taxpayer. The final piece of the story, as far as I was concerned, was when my wife saw a headline in the evening paper, 'Fourth Man Misses Historic Trip', at which she said to herself, "I bet that's Bob".*

June Cooper, later to become Hoverlloyd's Assistant Chief Stewardess, saw the SR.N1 arriving, probably for the second time:

My introduction to the hovercraft was on Dover beach that July day in 1959. A jarring engine sound, rather like a motorcycle from out at sea, getting closer and closer, preceded the arrival of a strange-looking craft, which did not stop at the water's edge, but glided straight up the beach before gently coming to rest on the pebbles quite close to us. Photographers and pressmen and quite a crowd gathered. I believe now that this was the SR.N1 and we had just witnessed the first crossing of the Channel by a hovercraft. Little then did I realise that some ten years later I would find myself involved in the operation of passenger services across the Channel with Hoverlloyd, which would give me probably the happiest 12 years of my working life.

But the final comment on the historic Channel crossing belonged to the old salt, Captain Barlow of the *RN 54* lighter. He turned to the Saunders-Roe crew chief as they were preparing to load the SR.N1 for the return to the Isle of Wight and said:

Edgar, I think we will stow your scooter a little further aft for the return voyage!

Chapter 5

The Dawn of an Industry

The on-going trials with SR. N1

The historic crossing of the English Channel in July 1959 by the SR.N1 did more than just attract publicity for the hovercraft. Westland and Hovercraft Development Limited (HDL) now agreed to proceed on a joint fifty-fifty basis with the design, development and construction of a 27-ton passenger-carrying hovercraft designated the SR.N2. Although the SR.N1 trials had been successful overall, they did illustrate that spray and dust would cause serious operational problems and the control and propulsion system was inadequate. So any new design would have to take account of these issues.

SR.N1 carrying 20 marines at the Farnborough Air show. September 1959.

While the design of the SR.N2 was underway, trials with the SR.N1 continued, which was all important to the development of the SR.N2. From August to September 1959 it had been detached to the Royal Naval Air Station at *HMS Daedalus*[1] where covered trials on the runways were carried out to prepare for the Society of British Aircraft Constructers' annual display at Farnborough. At the Farnborough Air Show the SR.N1 performed daily, carrying 20 fully equipped marines on each sortie. Fortunately the winds were light, enabling the craft to be manoeuvred on the runway without too much difficulty.

---

[1] This is the site of the Hovercraft Museum in Lee-on-Solent, Hampshire.

SR.N 1 Mk.3 with *Viper* engine making the transition from the Solent and across the road from the sea to the naval base *HMS Daedalus* at Lee-on-Solent. 1961.

On the return from Farnborough the SR.N1, which had been taken apart in order to transport it by road to Lee-on-Solent, was reassembled. On 21st September a demonstration was laid on for Lord Louis Mountbatten. Two days later another 'first' for the SR.N1 took place on the return to Cowes when the craft ascended the Columbine ramp under its own power. John Chaplin remembers:

> *There had been some concern that with the low thrust level it would be necessary to approach with sufficient speed to have kinetic energy to augment the thrust to climb the ramp. This would require judgment by the pilot since there was a limited depth of apron to stop before piling into the hangar doors. After the management, insurance and legal concerns were settled, the decision as to when the test occurred was left to the judgment of the chief test pilot.*

Bob Strath recalls the day:

> *It was 23rd September 1959. We were much more confident at handling the craft by that stage. Up until then we had always used the dolly legs to launch and recover the craft from the Columbine slipway. On this occasion, as we got back into Cowes Harbour, Sheepy looked at me and I looked at Sheepy*

– there were just the two of us and he said, "We've got to give it a go." We were out at the breakwater and Sheepy opened up the taps and gave it a long balls-out run at the slipway and to everyone's surprise we shot up the slipway on to the hard standing and from that time onwards no dolly legs.

On 16th December The Duke of Edinburgh, perhaps encouraged by his uncle, Lord Louis Mountbatten, paid a visit to Cowes to inspect this wonderful new invention for himself. Being a pilot he couldn't resist having a go at the controls. In a strong westerly wind, he sped past Osborne Bay at 35 knots, and was apparently very impressed. On return to the base 40 minutes later it was noticed that the plating of the bow had been 'dished'. This minor damage was never allowed to be repaired and it was known by all the staff as 'The Royal Dent'. In January the SR.N1 was exhibited at the Earls Court Boat Show in London.

The Duke of Edinburgh, accompanied by Sheepy Lamb and Eric Mensforth, President of Westland, at Osborne Bay in December 1959.

By the end of April 1960 a Blackburn *Marboré* jet of 800-lbs static thrust was installed behind the cotton reel as extra propulsion, purely to increase the speed from 25 to 45 knots. The craft now became the SR.N1 Mk.2. The *Leonides* still provided the lift and thrust as before. The *Marboré* air intake was a curious aft-facing horn to prevent the ingestion of as much salt spray as possible. Prevention of salt spray into the jet engines on all hovercraft was to be a problem that took many years to resolve satisfactorily.

In May the SR.N1 was taken by the same lighter that ferried it to France, *RN 54*, to Battersea Power Station where it was launched over the side into the then filthy River Thames. The object of the exercise was to put on a display outside the Houses of Parliament in front of Commonwealth Prime Ministers, but it would seem that the point was partially lost in that there was no demonstration of its amphibious capabilities. Bob Strath recalls:

> *Sheepy was piloting the demonstration and I was the flight observer. What I do remember was that there was an Appeal Court hearing going on at the time in the House of Lords, which had to be abandoned because of the noise from the* Marboré *jet engine – it didn't go down at all well.*

SR.N1 Mk. 2 passing the Houses of Parliament in May 1960.

SR.N1 Mk. 2 with *Marboré* engine. The spray filter is clearly shown. A Beken photograph.

The First Skirt

But a most important development took place in June 1960 when the first flexible extension was fitted to the periphery of the craft. It was a simple piece of 1-foot reinforced rubber that was held in place by a series of straps under the craft. This first attempt ended in disappointment when it was discovered that the hover height was only increased by 30% and the skirt was in tatters at the end of the first sortie.

Early 2-foot long rudimentary flexible extensions fitted to the SR.N1 in 1961 following the first 1-foot extensions fitted the previous year.

It should be pointed out that the idea of a flexible extension to the craft was not an entirely new idea. A certain Cecil Latimer-Needham, an aircraft designer, claims to have come up with the idea and was indeed granted a patent, which he eventually sold to Westland in 1961. Ray Wheeler comments:

> *I didn't think we needed to buy this patent at all. Latimer-Needham was of no consequence to us in the design department. He had a very general patent on fitting flexible skirts and it was so general it didn't say anything. It just said 'flexible extensions', but it was one of the Westland board who decided to buy it. We never even looked at it. We were told by Hayworth, who was the Group Patent Engineer that it was of no consequence.*
>
> *Christopher Cockerell should have fought him because Cockerell came up with the idea before Needham in 1957.*

By October 1960, development resulted in the 1-foot flexible extensions being fitted to both the inner and outer edges of the peripheral jet exits, held together with straps at intervals, which gave a 100% increase in the hover height of hard structure from the concrete apron when the skirts were new. Three months of trials with these skirts followed and, with all the modifications, the all-up weight had been increased to approximately 11,100 lbs, a 50% increase on the initial operating weight, but demonstrations were also made with an all-up weight of 16,500 lbs – a 100% increase.

SR.N1 Mk.3 with *Marboré* engine replaced with the *Viper* turbine, increasing speed to 64 knots. A sophisticated spray separation chamber enclosed the engine.

Development in skirt material was also progressing and in January 1961 the fabric was changed from an unsatisfactory off-the-shelf material to a custom-made rubber one reinforced with nylon and terylene. After this considerably improved material was fitted to the craft, the SR.N1 made a non-stop sortie on 23rd March around the Isle of Wight.

Meanwhile the design of the SR.N2 was nearing completion and the construction was well advanced when the decision was made to substantially modify the SR.N1 Mk.2 to become the Mk.3, in order to simulate as closely as possible the characteristics of the SR.N2. To this end the 800-lb thrust *Marboré* engine was replaced with a 1,500-lb thrust Armstrong Siddeley *Viper* turbo-jet engine, complete with a sophisticated engine housing to prevent salt spray ingestion. The speed was increased to 64 knots and the weight to 13,750 lbs.

Bob Strath remembers one of the unknown consequences of these improvements:

> *One day after we had the* Viper *engine fitted, Sheepy and I were belting down the Solent on a dead calm day – we had a water speed indicator by that time as well as the air speed indicator and we were up to 50 knots. All of a sudden there was a bang, a wall of water came up over us and we piled up against the windscreen. We shook ourselves, looked at one another and Sheepy said, "What the bloody hell happened?" I said, "I don't know – we just sort of ploughed-in." So that's how the term 'plough-in' came about. This had obviously happened as a result of the high speed we were doing and was a phenomenon with which we were not familiar. When we got in, dear old Gordon Elesley, who I think was our chief air dynamicist by this stage, blinded us with science by quoting all sorts of equations and then said, "You just slid off the cushion."*

The SR.N2

It was decided that the SR.N2 would be designed as a civil and military operational hovercraft rather than a further research vehicle, but engineered in such a way that the fan, transmission and control system could be used on a craft directly four times the size of the SR.N2. Further trials on the SR.N1 showed that a cushion pressure of 16-lbs per square foot was too low. It needed to be increased to 70-80-lbs per square foot for the new craft. The SR.N2's essentially ogival plan form would be 64 feet 6 inches long of 2 to 1 length to beam ratio and it was anticipated that this combination would give adequate lateral stability, in addition to which the natural bow shape would assist high-speed rough water operation.

The SR.N2 had more or less conventional ship's bows. Ray Wheeler explains the thinking at the time:

> *We simply thought it should look like a ship. We were doing a lot of tank testing on these things to establish the relationship between the centre of gravity and the centre of cushion. We initially got it wrong with the N4.*
>
> *After the SR.N3 we went to a rounded bow so you were always approaching the wave at the same angle. It also gave more support in the bow quarter.*

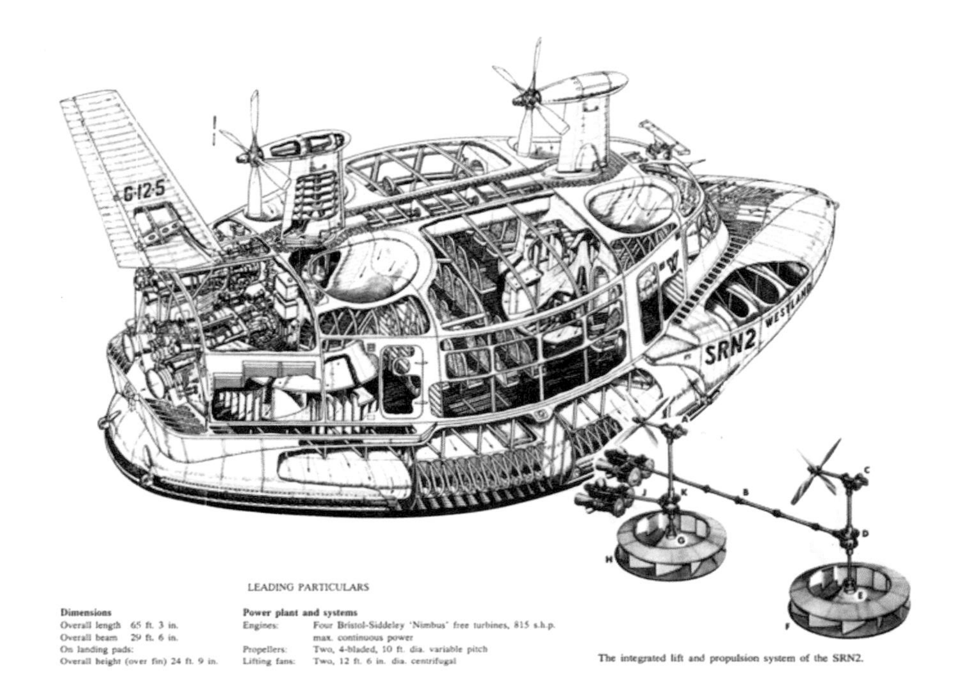

The integrated lift and propulsion system of the SRN2.

The basic structural buoyancy tank of SR.N2 in Columbine hangar in 1961.

Flow diffuser vanes surrounding the two centrifugal fans in position on top of the buoyancy chamber. These vanes were not used on future craft.

SR.N2 nearing completion in late 1961.

The SR.N2 would have two combined propulsion and lift systems, each system driving a 12-foot 6-inch centrifugal lift fan and a 10-foot diameter Dowty Rotol variable pitch propeller, each system being powered by two Blackburn *Nimbus* gas turbines, giving a total of 3,600 maximum continuous horsepower. A complicated series of shafts and gearboxes connected each propeller pylon, which could turn 30° either side of the centre line for directional control, together with the fixed-blade centrifugal lift fan, to each pair of engines. Because one third of the power was required for lift, and wave-making drag was proportional to the square of the cushion pressure, every effort was made to make the craft as light as possible. To this end the structure was constructed primarily of high-strength aluminium alloys suitably corrosion-protected as a result of the company's experience with flying boats. Designed to carry up to 76 passengers or eight tons of freight in a roomy cabin between the two lift and propulsion systems, it was anticipated that the hover height would be 2 feet 6 inches and the maximum calm water speed 75 knots. It has to be borne in mind that at this stage no skirt had been envisaged for the hovercraft. It had been calculated that the larger the hovercraft in terms of cushion area the higher the hover height that could be attained, so 2 feet 6 inches was a considerable improvement on the SR.N1's hover height of 1-foot 6 inches.

The SR.N1 changes shape to become the Mk.4

SR.N1 Mk. 4 fitted with pointed bow and stern to simulate SR.N2. October 1961.

In September 1961 it was decided to change the platform of the SR.N1 to approximate to the ogival shape of the SR.N2 by adding a pointed bow and stern and increasing the length of the flexible extensions at each end to 1-foot 5 inches. The weight of this penultimate Mk.4 version of the SR.N1 was now increased to 14,250 lbs, but the lower cushion pressure enabled a slightly higher speed of 66 knots to be achieved. The Mk.4 version was primarily used for the development of the flexible skirts. Apart from the control problems associated with the SR.N1 and the need to significantly improve the skirts, the modifications were regarded as successful and boding well for the SR.N2, but the big breakthrough in skirt design didn't come about until September 1962, some three years after the first Channel crossing.

Bob Strath recalls about this time:

> *We kept advancing into all sorts of unknown areas. I remember the first night trials we carried out and the shock we had when we were enveloped in spray as a result of which we couldn't see a bloody thing due to the reflected light. Old man Harley of the Harley Landing Lamp Company then got in on the act. Everybody wanted to get in on the act in those days, like Decca Radar. They got in on the ground floor.*

Thoughts turn to Navigation

It was about this time that Sheepy considered some thought should be applied to the navigation of the hovercraft to enable operation in all conditions, whether by day or by night. He appointed Bob Strath as the 'Navigation Specialist' to look into the aspects of all the requirements:

> *John Beatty at Decca got in on the ground floor both with the radar and the Decca Navigator[2], together with the flight log, the latter of which we had tested with him in the* Princess *flying boat trials. I remember phoning him at Decca and saying, "John, we've got to fit radar into the SR.N2 because we are going into night operations."*
>
> *He was always at Cowes and he said to Sheepy one day in my presence, "It would be a great thing if you were to get Bob qualified as a radar operator. I don't mean one of these 'fly by night' certificates – I mean the proper job at Warsash[3]."*

---

[2] The Decca Navigator was a piece of equipment that, for want of a simple explanation, converted signals from radio beacons located around Britain and Europe into numbered red, green and purple crossing lines which corresponded to those printed on a special Decca chart. You plotted your position on the chart by noting the numbers on the Decca Navigator and where they all met was where you were. The Decca Flight log was a vertical roll-up chart with a moving horizontal pen that automatically followed the numbers from the Decca Navigator to give an instant position, which, because of the speed, was the only feasible method of plotting in an aircraft or hovercraft.

[3] Warsash was one of the approved nautical colleges near Southampton that catered for both pre-sea training and courses for Marine Certificates of Competency.

Sheepy pointed a finger and said, "Go", and I went. But Sheepy also wanted me to take a condensed Master's Certificate.

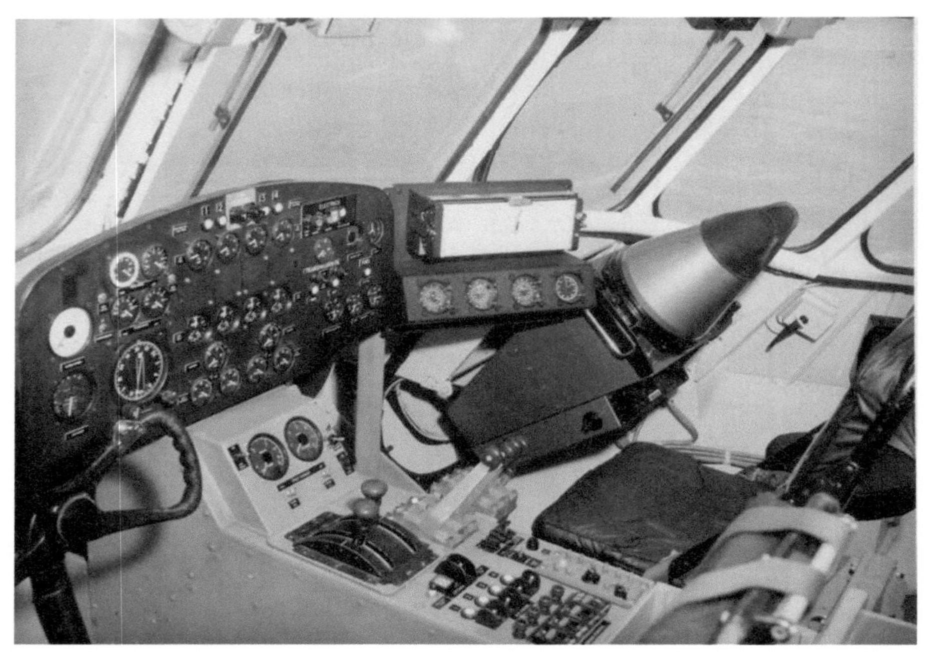

SR.N2 cockpit with Decca Radar and Decca Navigator with flight log. There are no rudder pedals.

I went to Warsash to do my radar course, but the actual condensed Master's was self study. The Board of Trade bent over backwards to help with this condensed exam, but I had to study for bloody weeks and weeks. I used to sit up in bed at night with my wife with cards on which there were flags and their meanings. She would hold a card up and say, "What's that?" I'd say, "C", and then she'd say, "What's it mean?" and I'd say, "I'm in a bad temper – steer clear." I didn't like that side of it much.

Then the big day came for the examination. I had to go for the Board of Trade exam in Dock Street, East End of London. There was a mixture of written papers, an oral and signals exam, but I failed signals first time on Morse.

For signals I sat in a large room in a corner seat and there was a nice guy, Captain somebody, sitting in a glass cubicle on the other side with this light in the corner of the wall, which I hadn't seen. I sat there, and I sat there and presently he said, "Is there something wrong?" I said, "No". He said, "Well, I've been signalling to you for the last ten minutes, but you don't seem to be doing anything." I said, "Where?" And there was this little light stuck in the corner. I said, "Christ, I can't bloody well see that." He said, "Do you wear specs?" I said, "Well I've just started, but I haven't brought them with me." So he failed me. When I got back Sheepy asked how I got on and I told him I'd failed. "Cor, what a useless bugger!" was his reply, but I had passed the writtens and orals.

There was another difficult bit, which was the First Aid. By this time I was quite well-known in the local community as one of the 'hovercraft men', as a result of which we were in demand to speak to the likes of Rotary. Sheepy would say, "You can do that, Bob, I'm too busy." Our GP was a member of Rotary and I said to him one evening, "I've got to get a First Aid certificate to get my Master's Certificate." "Umm," he said. He gave me a book and told me to mug up on it and then he'd test me. At the end of it he gave me a certificate and said, "God help any patient that comes into your hands." They were exciting days.

And so it was that the two prime pieces of equipment installed in the hovercraft for high speed navigation were Decca radars and the Decca Navigator with the aircraft flight log attached.

The SR.N2 is launched and a trial passenger service is operated

SR.N2 and SR.N1 Mk. 4. The 4-foot deep flexible skirts were not fitted at this stage.

The last three months of 1961 saw three significant events; the roll out of the SR.N2, together with the start of hovering trials; the formation of the Inter-Service Hovercraft Trials Unit (IHTU) by all three services at *HMS Aerial* at Lee-on-Solent; and the Ministry of Technology awarding a contract for the design of a 37-ton military hovercraft designated the SR.N3, which was a military version of the yet to be built, stretched SR.N2 Mk.2, capable of carrying 150 passengers.

Sea trials of the SR.N2 started in January 1962 with the primitive 1-foot 5-inch skirt as tested on the SR.N1 Mk.4, but the trials were dogged by problems with the Blackburn (soon to become Bristol Siddeley) *Nimbus* engines. By the middle of the year, however, she had achieved a speed of 73 knots (84mph) and a circuit of the Isle of Wight was completed in just over an hour.

In August, the SR.N2 was used on an experimental passenger service, operated by Southdown Motor Services Ltd., in conjunction with Westland, from Ryde on the Isle of Wight to Eastney on unprepared beaches, with a caravan on site for ticket sales. The areas of beach used for landing were roped off for safety reasons. It operated for two hours a day for ten days in the peak of the holiday season, carrying a total of 1,500 passengers. The year ended for the SR.N2 with it being used for trials at Portland by the IHTU.

SR.N2 wheeled out of Columbine Hangar on its beaching chassis in October 1961.

The pilot's and the navigator's seats in the SR.N2. The flight engineer's position and controls were behind and below.

SR.N2 landing at Appleby Beach, Ryde, Isle of Wight, when operating a cross-Solent service for Southdown Motors. August 1962.

SR.N2 leaving Appleby Beach. August 1962.

A breakthrough in Skirt Technology with the SR.N1 Mk.5

But 1962 also heralded a remarkable jump in skirt technology on the SR.N1 with the design of 4-feet 'U'-shaped deep flexible peripheral skirts, which encompassed the two original peripheral metal jets with a single jet.

Ray Wheeler recalls:

> *When Dick Stanton-Jones (Chief Designer), suddenly decided we'd have a 4-foot skirt on the N1, with a bag encompassing the two jets, we all thought he had gone crackers. We felt it would wobble all over the place.*

The stability issue was covered by dividing the cushion into two equal compartments by a longitudinal sausage-like bag inflated from its aft end. Drainage and a measure of an air jet were provided by a large number of holes along the base of the bag. The skirt was fitted in September to what had now become the final version of this prototype hovercraft – the SR.N1 Mk.5 – and, with its all-up weight increased to 15,000 lbs it was able to achieve a calm water speed of 47 knots, which was some 20 knots slower than before. The speed loss, however, was the result of a much larger frontal area and the fall in thrust from the *Viper* engine. The flip side was that the SR.N1 now had a vastly improved sea-keeping and obstacle clearance performance. In seas of 3 to 4-feet in height, speeds of 25 to 30 knots were achieved. It could negotiate gullies of 3 to 4-feet deep and 3 to 20-feet wide and clear rock walls of 3-feet 6 inches. Problems still occurred with pitch and roll stability, which was cured by fitting a wider 'keel' trunk and dividing the cushion with athwartship flexible trunks.

SR.N1 Mk.5 during later research with flexible annular trunks and pointed bow and stern simulating SR.N2.

Bob Srath recalls:

> *I did most of the 4-foot skirt development on the SR.N1, by which time Harry Phillips (an ex-Fleet Air Arm rotary wing test pilot) had joined us and I have photographs of us operating over the Beaulieu marshes. Once, when Lord Montague was away, I phoned up his manager and asked him if we could operate over the marshes. The guy at the time was quite reasonable about it, but there was a time when we had been threatened with all sorts of legal action if we went near the place.*

There was hostility from other quarters too. We used to go over to Lee-on-Solent on the lee shore to carry out high-speed runs. In the early days everybody used to wave to us, but I can remember this particular day as Sheepy tore down a line of yachties he said to me, "For goodness sake, Bob, wave back to the yachtsmen." I said, "Sheepy, I don't think they're waving at us – they're clenching their fists!" At one time it was all very fascinating, but it didn't take long before the public turned on 'that bloody noisy thing'.

SR.N1 Mk.5 trials over the Saltings at Beaulieu with Bob Strath and Harry Phillips.

By the end of 1962, and after three years of progress far beyond the wildest dreams of the original conception, the use of the SR.N1 for development had virtually come to an end. Although there was still much to be done in skirt technology, the success of the 4-foot skirt on the SR.N1 Mk.5[4] was the breakthrough that altered the thinking of the hovercraft design team. Bob Strath recalls:

One of the classic stories of the SR.N1 was that its design

---

[4] During 1963, the SR.N1 Mk.5 was confined to refining the 4-foot flexible skirt design and experimenting with new reinforced rubber materials. She participated in the Royal Engineers display at Chatham in June 1963 and was loaned to the IHTU from April to September 1964, after which she was sent, fully restored, still as a Mk.5, to the Science Museum Transport collection at Wroughton Airfield near Swindon in Wiltshire.

envelope was for a maximum of 25 operating hours or six months – that was the design requirement. Doug Matthews was the design engineer who was given the project and we wound up with something in excess of 500[5] *operating hours in all its developments over a period of five years. Dick Stanton-Jones sent for Doug one day and Doug, who was really chuffed with his achievement, thought he was going to get a gold watch, or medal, or at least a slap on the back, instead of which he got a bollocking for over cooking it. Dick said, "Your brief was to produce something for 6 months or 25 hours, but look, look, look, look, at this. What went wrong?!"*

The new Skirt changes the direction of Hovercraft Development

In the August of 1962, the original SR.N4, which had been conceived at the design stage of the SR.N2, was abandoned. That plan envisaged what was, in effect, two lengthened SR.N2s side by side, 135-feet 6 inches long with a beam of 53 feet and a maximum displacement of 125 tons, powered by four pairs of Blackburn A129 engines. Now the meeting decided the SR.N4 would be 125-feet long, with a beam of 65 feet and a maximum displacement of 168 tons, powered by three pairs of Rolls Royce Marine Proteus engines. Ray Wheeler explains:

The N4 was positioned in the sequence because initially we thought we needed a large craft to get a high cushion without a skirt. At that stage we went to the N4 because we thought we had to – the flexible skirt was still to come.

The 4-foot flexible skirt would indeed change the company's strategy. It was now considered possible to design a viable craft at under 10 tons all-up weight, which resulted in the SR.N4 being put on hold and the design of the SR.N5 commencing early in 1963.

The design of the SR.N3 had been finalised by December 1962. Westland then took the bold decision to proceed with construction to avoid a delay in the programme with the fall-back position, based on bitter experience with previous governments and their history of pulling out of contracts, of completing it as an SR.N2 Mk.2 in the event that a military order did not materialise. During a visit in March 1963, however, Julian Amery, Minister of Aviation in the then Conservative government, was so impressed that the development contract was awarded in May, just months before completion.

---

[5] The recorded engine hours for the three engines used were: *Leonides* – 505 hours 40 minutes; *Maboré* – 24 hours 10 minutes; *Viper* – 281 hours 5 minutes.

So 1963 turned into a year where interest and development of the hovercraft concept was gathering pace. On a near freezing day in January, still with its 1-foot 5inch skirt on, the SR.N2 made a record-breaking run around the Isle of Wight in 51 minutes and 51 seconds at an average speed of 58 knots.

SR.N2 goes to Canada

In April she was transported to Canada by ship and swung over the side in Montreal docks following an agreement made with Autair Helicopters Ltd. to promote hovercraft in Canada. Bob Strath recalls:

> *For the first three years it was just Sheepy and me doing everything. Harry Phillips had then come on the scene towards the end of the N1 Mk.5 days. One day I was at home sick with a flu bug. The phone rang and my wife, Freda, answered it. It was Harry and he said, "I want to talk to Bob." "Oh he's in a terrible state – he's got the most dreadful flu." Harry said, "Get the bugger down to the telephone. I've got some news for him that's likely to make him forget his flu." So I staggered downstairs to the phone at death's door and he said, "Get yourself fit in a hurry – you're coming to Canada with me." Sheepy, as chief pilot, was supposed to be going on the Canadian tour, but, Harry said, "Sheepy's put up a 'black' and he's been taken off the Canadian project, so I'm going, but I'm not going without you." The 'black' was a story involving helicopters.*

Like many pilots and other professionals who are used to relying on their own individual skills and making their own decisions, Sheepy was not so good at taking orders. Coupled with a rather volatile temper, it made him not the easiest of people to get on with.

On this particular occasion, Sheepy was about to take off from Cowes in one of the company helicopters, when one of the Westland directors asked him to wait in order to take him across to the mainland. Sheepy however took umbrage at being treated as a flying taxi-driver and flew off without him. In turn the director involved took a dim view of Sheepy's attitude, took the helicopter away from Cowes and let it be known that Sheepy was not to be considered the right material to represent the company overseas.

Sheepy didn't lose his job as chief pilot at that time, but, sadly, ten years later this very talented and skilful test pilot, to whom the hovercraft industry owed so much, was to cross swords with the senior management once again with dire consequences.

SR.N2 being shipped to Montreal.

SR.N2 operating an experimental service on the St. Lawrence. Here it is seen shooting the Lachine Rapids in early summer 1963.

Meanwhile Harry and Bob carried out two weeks of demonstrations in the Montreal area, which included shooting the St. Lawrence River Lachine rapids at some 45 knots and carrying more than 1,250 passengers on trips from the Royal St. Lawrence Yacht Club. Canadian federal and provincial government representatives were joined by military and industrial delegates from the whole of the American continent and Japan.

On her return to the UK she was used for a six-week service in conjunction with the Welsh ship owners, P & A Campbell[6], from July to August across the Bristol Channel on an 11-mile route from Weston-super-Mare to Penarth, which was made in just 12 minutes. The end of 1963 saw SR.N2 fitted with a 4-foot deep flexible skirt[7].

P & A Campbell operating a trial route across the Bristol Channel between Weston-super-Mare and Penarth with the SR.N2 using the primitive 1-foot skirt. July and August 1963.

---

[6] P & A Campbell, whose origins date back to 1890 as a purely excursion steamer business trading in the Bristol Channel, became part of Townsend Ferries in 1959, which later operated an SR.N6 from Dover to Calais and had expressed initial interest in the SR.N4.

[7] Appendages called 'fingers' were not fitted to the skirts until 1966.

SR.N2 being fitted with the 4-foot flexible skirt.

The new Military SR.N3

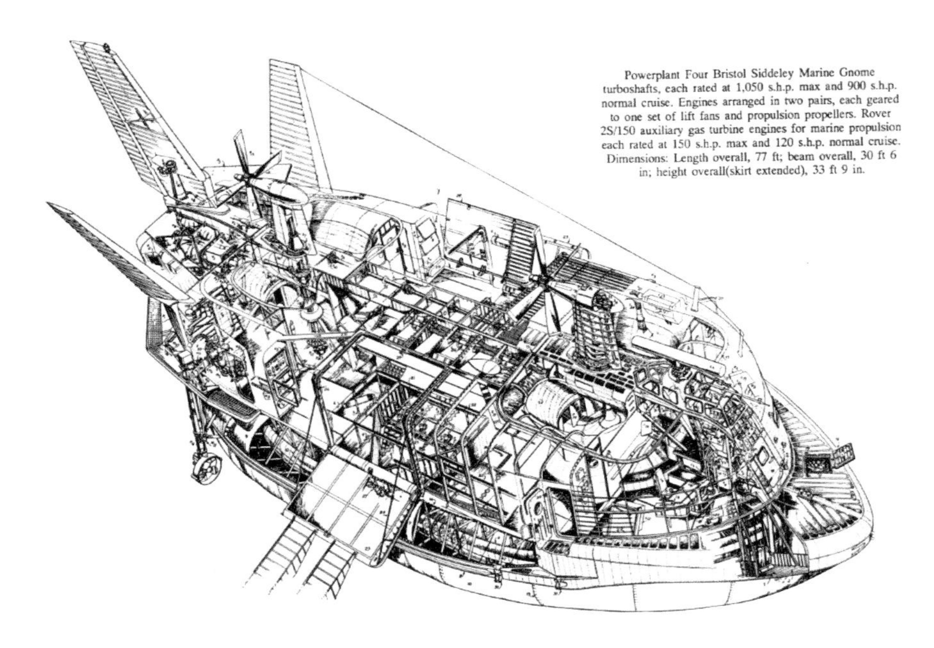

Powerplant Four Bristol Siddeley Marine Gnome turboshafts, each rated at 1,050 s.h.p. max and 900 s.h.p. normal cruise. Engines arranged in two pairs, each geared to one set of lift fans and propulsion propellers. Rover 2S/150 auxiliary gas turbine engines for marine propulsion each rated at 150 s.h.p. max and 120 s.h.p. normal cruise. Dimensions: Length overall, 77 ft; beam overall, 30 ft 6 in; height overall(skirt extended), 33 ft 9 in.

SR.N3 fitted with a similar 4-foot skirt to that of the SR.N2.

SR.N3 undergoing trials in 1964.

Construction of the 37-ton SR.N3 commenced in the spring of 1963 and the end of an eventful year saw its first sea trials. The power plant consisted of four Bristol Siddeley Marine Gnome turbo shafts, each rated at a maximum of 1,050shp. The engines were arranged in two pairs and, in a similar layout to that of the SR.N2, each geared to one set of lift fans and propellers.

Trials of the military SR.N3 were carried out at a hectic pace from its launch in December 1963 to when it was accepted by the IHTU on 2nd June 1964. At the end of July it made its first successful crossing of the English Channel.

Overseas Licence Agreements

In February 1963 Westland concluded an important licence agreement with Mitsubishi and then the following month another with Vickers and Mitsui. In July, an even more important and comprehensive agreement was concluded with Bell Aerosystems Company, giving them the right to build and sell Westland hovercraft in the United States, Canada, and Central and South America.

Bell had in fact already built and tested a small hovercraft called the *Carabao* and had just completed a skirtless 22.5 ton hovercraft called the *Hydroskimmer* (SKMR1) under a United States government research programme into 'triphibious capability', which they described as 'above the waves, over the shore and into the bush'! They demonstrated the craft on Lake Erie and it achieved a speed of 70 knots at a hover height of 1-foot 6 inches.

Bell's interest was in the flexible skirt technology, which Westland had so rapidly developed. It was therefore no surprise that the first contract they received under the licence agreement was to design and build a 4-foot flexible skirt for the SKMR1.

The SR.N5 (Warden Class)

The new 4-foot skirt had paved the way for a potentially viable smaller craft to be built. Mid-1963 saw the completion of the design of the SR.N5, the first craft to be fitted from the outset with these deep skirts and cushion compartmentation as on the SR.N1 Mk.5, and by the autumn a production line had been laid down. The 6.6-ton SR.N5 was 36-feet long and 23-feet wide and was intended for general purpose duties, including the carriage of 18 passengers, or up to 2 tons of freight, high-speed search and rescue, weed and pest control, survey and exploration and for military logistic support. Her calm water design speed was 66 knots and, with 265 gallons of fuel capacity, had an endurance of 3.6 hours at maximum continuous. Access to the cabin was through a bow loading door. The single 900hp Bristol Siddeley Marine Gnome engine was mounted aft of the cabin, directly in line with the rear-mounted 4-bladed Dowty Rotol 9-foot diameter propeller. Immediately aft of the engine was a company-designed reduction gearbox, which coupled an aft shaft to the propeller and a vertically downward drive to the 7-foot diameter company-designed and built centrifugal lift fan, mounted immediately behind the passenger cabin on top of the buoyancy tank.

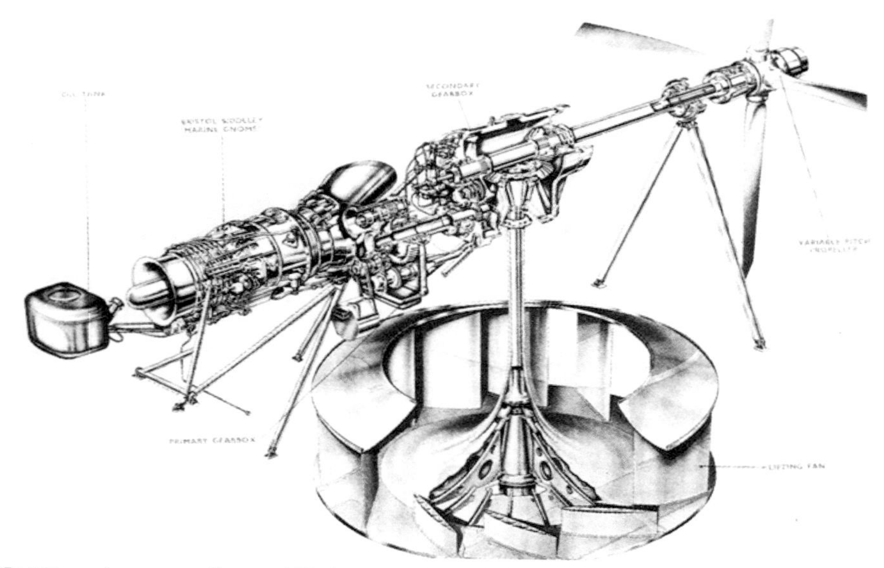

SR.N5 engine, propeller and lift fan transmission system.

The first SR.N5 on its initial sea trials in April 1964. Note the high tail fins.

The first SR.N5 with shortened fins and the addition of a forward fin to improve directional control. May 1964.

Directional control was provided by rudders mounted on twin fins and twin control by all moving tailplanes, which operated in the propeller slipstream. Additional low speed control was provided by a hydraulically-operated skirt lift system at the four corners of the craft.

During her early trials the SR.N5 proved highly directionally unstable and was quickly fitted with a fin above the forward cabin. The rear fins were cropped and a form of bucket was added to each rudder to increase their effectiveness. Apart from these modifications the craft proved an immediate success[8], although intensive skirt development had to take place in order to prolong skirt life. It must by now have been apparent that the hovercraft skirt was probably the major obstacle to the ultimate commercial success of the craft. The air jet at the bottom of the skirt was subject to constant wear and tear, requiring frequent repairs. In August 1964, the second SR.N5 was delivered to the IHTU and in September, Bell Aerosystems took delivery of one, which was the same month an SR.N5 was demonstrated at the Farnborough Air Show.

Thoughts turn to the SR.N4 (Mountbatten Class)

The design team could now turn its attention to the SR.N4, which it duly did in 1964.

Ray Wheeler recalls:

---

[8] During 1964 alone, seven SR.N5s were sold, two to the Ministry of Defence, and the other five went to Bell Aerosystems, Oakland Helicopter Airlines of San Francisco, Autair of Canada, Mitsubishi of Japan and Scandinavian Hovercraft Promotions Ltd.

I was involved in the project from the beginning because of my position in the structural side. I had intended to be the best aircraft structural engineer in the country – that was my aim. I did nearly four years as a post-graduate at Imperial College under a chap called Hadji Argyris who was the structural analyst of the day. It was Dick Stanton-Jones, Chief Designer, and Derek Hardy, Head of the Project Department, who came up with the first real design after the very initial N4 concept. At that time it was not the four-engine craft we ended up with. Christopher Cockerell did have an influence on that. It was a six-engine craft and we called it the 'Morphy-Richards' because it was like a flat iron with one system at the front and two at the back. Each system had two engines, but Christopher Cockerell said there were too many engines.

The six engine SR.N4 design nicknamed the 'Morphy Richards' because of its resemblance to a flat iron.

Westland's answer to the proposed Channel Tunnel

A 1963 Ministry of Transport white paper was published, entitled *Proposals for a Fixed Channel Link,* referring of course to the Channel Tunnel, which had been a serious on-going discussion since the 'Channel Tunnel Study Group 1959' produced a report dated 28th March 1960. In March 1964, in response to that ministerial white paper,

Westland produced what became known as *The Westland White Paper* on the future of transport across the English Channel entitled *A Proposal for a Hovercraft Channel Link*. The report concluded that a fixed link was not required and that:

> *A fairly modest fleet of 150-ton hovercraft could be expected to cope adequately with the predicted traffic growth up to 1975 and beyond. The fleet requirements to meet this traffic growth lie between 16 and 22 craft.*

A bold statement, *n'est-ce pas*, for an industry that only had the technology at that point to allow hovercraft to operate effectively in 3 to 4-feet waves, and a company whose largest working machine was just under a sixth of the proposed size of this cross-Channel craft, which had yet to be designed. It did, however, have the effect of postponing the 'Channel Fixed Link', which was, after all, the object of the exercise.

A Trial Service across the Solent

Initially the SR.N4 was to be used on a cross-Solent route so that development problems could be cured on the doorstep of the company. To this end a company was formed called Hovertransport Limited in the same month that the *Westland White Paper* was published. Christopher Bland, who was to play a leading role in the development of hovercraft services in the Solent and became chairman of the highly successful Hovertravel, tells the story:

> *My career was National Service, Rolls Royce as an apprentice, and then Rolls Royce took me on. There were four companies developing hovercraft at that time: Westland (Saunders-Roe), Vickers, Folland*[9] *and Britten-Norman. We had an aluminium V8 engine then that we wanted to sell commercially. I wasn't on the car side, but on the industrial engine side and I went down to Britten-Norman*[10] *and said, "You don't want a Coventry Climax engine for your cushion craft – this was in 1961 – you ought to have a Rolls Royce aluminium V8." So they bought a couple of aluminium V8s and subsequently offered me a job for all of £29 per week in 1962 – £1,500 a year. I was there like a*

[9] In 1959 Folland was acquired by Hawker Siddeley, who dropped the Folland name in 1963.

[10] John Britten and Desmond Norman started crop spraying in the 1950s and then went into aircraft design. Britten-Norman is best known for the twin-engine *Islander*, which first flew in 1965 and, following its success, the three-engine *Trilander*. Around 1,200 have been sold worldwide. In the 1960s, Britten-Norman became involved in the design and manufacture of small hovercraft.

hairy dog. I worked for Britten-Norman on their cushion craft, which weren't yet ready to be commercialised. Britten-Norman then decided in 1964 to try an experimental hovercraft service using the SR.N2 on the route between Appley on the Island and Marine Barracks up at the end of Portsmouth. It was a joint venture between P & A Campbell, whose governor was a man called S.C. – Mr. Cox – Britten-Norman and an engineer called Edwin Gifford. There was also Ministry of Technology involvement. I was given the job of running the service in 1964. We ran it experimentally for 11 weeks. The new company used the SR.N2 to gain experience in a cross-Solent service using the same two terminal sites at Eastney and Ryde, as in the 1962 operation.

The SR.N2 with its four Nimbus *and with 50 passengers was hugely expensive to operate. It was a bit unreliable and had a permanent engineer on board called Norman, who did manage to get all four engines alight, as in 'on fire', one day – I do remember that.*

Eleven trips per day of 11 minutes duration were scheduled between 17th June and 31st August 1964, although the average crossing time turned out to be eight minutes. The SR.N2 in fact proved to be very unreliable, mainly as a result of the *Nimbus* engines. Fortunately, the first 18-seater SR.N5 had been launched in April and was able to augment the service. In all, some 30,000 passengers were carried (20,000 by the SR.N2) at a fare of ten shillings (50p). The SR.N2 was never used again and, in 1965, became the first hovercraft to be broken up. Bob Strath recalls:

The SR.N2 was the first real development hovercraft, but it was the worst hovercraft we ever produced. Aesthetically it was lovely looking, but the performance was dreadful. We had those bloody awful Blackburn Nimbus *engines, which gave endless trouble, and structurally we'd thrown away our structural standards, such as rivet pitches from our flying boat days and we produced something that leaked – it was structurally weak.*

The systems engineering and control concept were poor – we had different control modes. Initially a designer called Chris Geer produced an arm rest with a throttle control. You could move your arm that way and the pylon would move with it. Then if you moved it forward it would increase pitch. It was pie in the sky. It was fine if you were sitting in a chair on a hard bit of concrete, but when you were pitching up and down at sea you were all over the place. The N2 then just had a yoke you moved backwards and forwards for pitch, as well as pitch levers, so that once you had set the pitch on the levers, you could adjust it with the stick. Then you had the yoke to move the pylons, depending on what mode you had the switch set.

It had no rudder bar. That was one of the things The Duke of Edinburgh picked up on. When he came in December of 1959 and flew the N1 it had rudder pedals. When we picked him up at Lee-on-Solent two years later in the N2 the first thing he said was, "Where are the rudder pedals?" Sheepy said, "We didn't need them, Sir – they were superfluous." The Duke said, "Bloody rubbish. What are all those scuff marks on the deck then?" Sheepy used to instinctively be pushing his feet back and forth even though there were no rudder pedals. We had quite a discussion about it. The Duke said, "I don't understand why you have dispensed with them. When I was a boy I built my own go-cart and steered it with my feet." From that time on rudder bars came back into the picture.

I know that whatever time I had in the pilot's seat I didn't enjoy it much. I did most of the time as a navigator. Sheepy, and then Harry Phillips, did most of the flying.

SR.N2 operating as Hovertransport between Eastney and Ryde, Isle of Wight in 1964. Note the 4-foot skirt.

A Look into the Future by the Military

SR.N5 of the British Inter-Service Hovercraft Trials Unit operating over marshland in Southern England during trials in the rescue role.

In September 1964, the Ministry of Defence got carried away with the euphoria of the advances in hovercraft design. Following a statement at the end of 1963 by the IHTU that there should be an investigation into the viability of an anti-submarine warfare 90-knot hovercraft frigate, the ministry made it known that they were indeed interested in the development of the hovership. Prior to that a comprehensive study, code named 'Black Lace', had been carried out by the Saunders-Roe team for the United States as to the best means of protecting ships from submarine attack, less than a year after the launch of the SR.N1. The US Navy, UK Admiralty and the Ministry of Aviation were the main participants in the study. It reported its findings in 1961, having investigated frigates, aircraft carriers, helicopters, aircraft, flying boats, hydrofoils and hovercraft and a combination of both where appropriate. The four-volume report concluded that the greatest potential for convoy and task force protection and barrier patrol missions would be provided by an 80 to 100-knot craft[11], capable of being able to operate in 10 to 12-foot seas and economically for long periods at 25 knots. In addition, it had to have the capability of carrying a small killer helicopter, long-range weapons and several types of long-range detection equipment.

---

[11] The US Navy first operated LCAC (Landing Craft Air Cushion) hovercraft in 1987. With a length of 88 feet, payload of 60 tons and speed of 40 knots, 91 were built and operated from the 'amphibious-well deck ships'. See Chapter 26: *The End*.

SR.N5 showing her ability to overload with 20 marines inside and 20 more on the external side decks. 1964.

HMS Intrepid was one of two *Fearless class* amphibious warfare ships of the Royal Navy. A Landing Platform Dock (LPD) at the stern allowed for the launching and recovery of craft. Here a Royal Navy SR.N6 has landed in the dock. 1967.

Royal Navy SR.N6 approaching the LPD of *HMS Intrepid*. 1967.

The initial conclusion was that a hovercraft in the order of 150 to 250 tons could meet these requirements. The study, however, was on-going and the last of four comprehensive reports was issued in 1966, which had investigated five specific hoverships from 160 to 1,200 tons. The final conclusion was that multi-thousand ton hoverships were entirely feasible. One drawing even showed a developed SR.N4 with 13 propellers and a displacement of 1,000 tons!

This was right up the street of the Saunders-Roe division of Westland given its past history of becoming involved in futuristic concepts. In fact, the Westland directors were so convinced that the Royal Navy interest was real that this became an important factor in deciding to buy the Falcon Yard facility, upriver from the Columbine Yard, when the John Samuel White shipyard ceased trading in 1966. It wasn't long before a sense of reality returned.

Projected 2,000 ton freighter, 1965.

The SR.N6 (Winchester Class)

Back to the end of 1964. The year concluded with the commencement of the design of the SR.N4 in earnest and the stretching of the SR.N5 by nearly 10 feet as a result of customers asking for a significantly higher payload, which gave an increase in passenger capacity from 18 to 38. Essentially, the SR.N5 was uneconomic for its payload, so the stretched SR.N5 became the new SR.N6. This huge 110% increase in payload was found to only decrease the performance by 10%, principally because the increase in payload was compensated by the increase in cushion area, leaving the cushion pressure substantially unchanged. The ninth SR.N5 was the first craft to be lengthened in just three months.

It was launched on 9th March 1965 and delivered to Scanhover in Norway in June. Later that month, a second SR.N6 was delivered to Scanhover. Clyde Hover Ferries[12] in Greenock, Scotland, took possession of their first leased craft, followed by a second in July.

---

[12] Clyde Hover Ferries was later to play a significant role in the early days of Hoverlloyd and is dealt with in Chapter 7: *The Early Hovercraft Operators*.

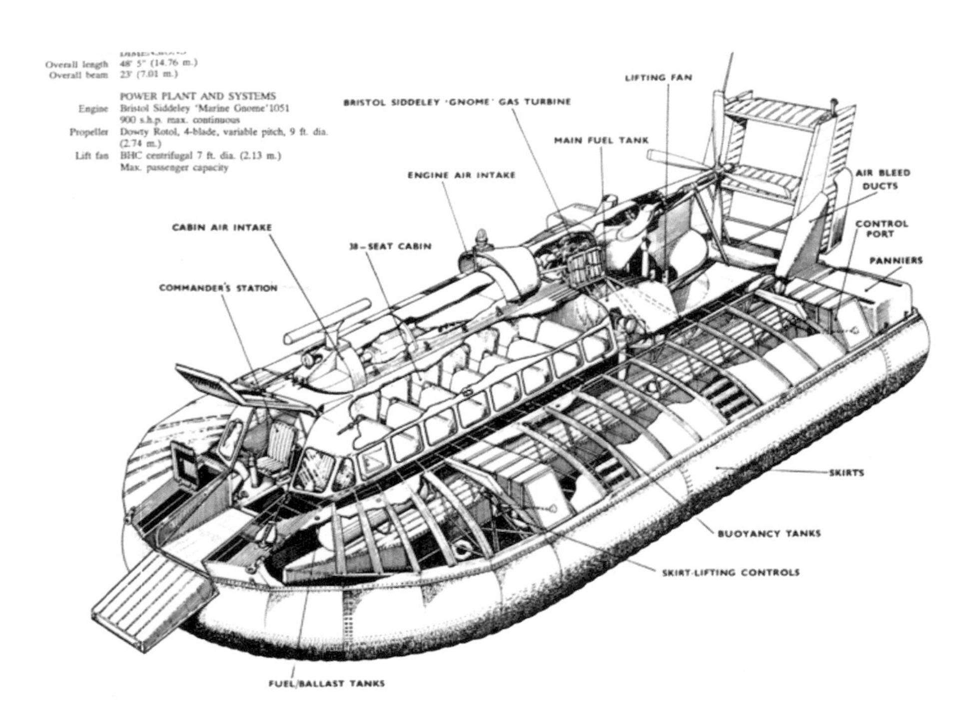

SR.N6 production line in the Medina workshop in 1965.

The first SR.N6 being wheeled out on to the Columbine apron in 1965.

The first SR.N6 at speed in the Solent in 1965.

Despite the industry still being in its infancy, with little concrete evidence as to the full potential of the hovercraft, it still attracted interest worldwide, although much of it through rose-tinted spectacles.

On 22nd April 1965 the *Financial Times* ran an article *Hovercraft Prepare to Leave the Ground* in which it stated:

> *In many ways it is a pity that the hovercraft, more than any other British development, has suffered from a surfeit of*

premature publicity. In the late fifties the headlines rang with promises of low-cost trans-Atlantic trips, 300mph air-cushioned journeys up the M1, and airborne tanks which would whisk armies across deserts. Like the circus barker and the bearded lady, the facts, impressive as they are, tend to be something of an anti-climax.

Nevertheless, it did herald 'The Dawn of an Industry'.

SR.N6 pilot controls and instrumentation.

Chapter 6

Vickers-Armstrongs Enters the Market

Towards the end of 1959, with the SR.N1 having proved that the hovercraft principle could be practically applied, Vickers-Amstrongs decided that their South Marston Works (near Swindon), at that time concerned with the last of the long line of Supermarine aircraft, should engage in the design and manufacture of hovercraft.

Vickers had started in aviation in 1911, but in 1927 had merged with Armstrongs so that jointly they had interests in aviation (civil and military), shipbuilding, tanks and guns. In their stable of well-known civil aircraft were the turbo-prop Viscount and Vanguard and the jet engined VC-10. In 1960, the aircraft interests were merged with those of Bristol, English Electric and Hunting Aircraft to form part of the British Aircraft Corporation (BAC)[1] and Vickers-Armstrongs continued with its other activities.

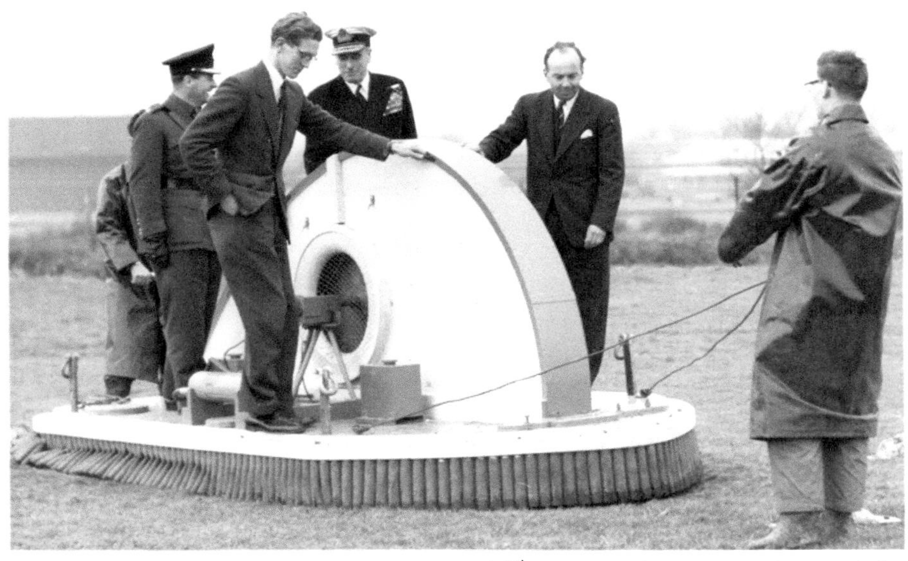

Lord Louis Mountbatten with Sam Hughes (2nd from right), Chief Designer of the Aircraft Division in 1959 and then Manager of the Hovercraft Division on its formation in 1960, during a demonstration of an early test model.

[1] BAC became British Aerospace in 1977.

So in November 1959, shortly before the aviation interests were hived off to BAC, and with the encouragement of Hovercraft Development Limited (HDL), they acquired the technical data vested in the hovercraft patents and obtained a licence to build a prototype. For the better part of a year the South Marston firm experimented with Vickers' first hovercraft, the VA-1, a simple research vehicle designed to explore the technical aspects of hovercraft development.

Les Colquhoun (right) at the controls of the open cockpit VA-1.

By no means as refined in concept as the SR.N1, the VA-1 was a simple laboratory tool designed to be rapidly modified for empirical experiment. As such, it was intended to offer more scope for the development of lift curtain systems, stabilizing devices and cushion controls than was possible with the SR.N1.

In 1961, a 44-year-old ex-fighter pilot, Leslie Robert Colquhoun, GM, DFC, DFM, became operations manager responsible for hovercraft development and trials. He had joined Vickers as a Spitfire test pilot in 1945. In 1950 he won his George Medal when, during the testing of an Attacker aircraft, a wing folded during the flight. In spite of this he remained in the plane and landed it safely.

The early day VA-1

Although Les carried out much of the testing on the VA-1, VA-2 and VA-3, in his capacity as operations manager, a Ray Old was in fact the senior hovercraft captain and accumulated over 1,000 hours on all three craft. It is interesting to note that Ray started his career in the Merchant Navy as a cadet and finished as a 2nd officer. He went on to become a flight observer with Saunders-Roe on helicopters before joining the Hovercraft Division of Vickers as a hovercraft pilot. So even at this early stage in hovercraft development, Vickers chose to involve both aviators and mariners in the piloting and testing of their hovercraft.

For its initial trials overland, the 1.47-ton VA-1 was fitted with bare essentials only, but, for protection on over-water trials, a cabin and various fairings were added, increasing the weight to 1.56 tons. An extensive programme of tests was put in hand, investigating static stiffness characteristics, internal flows and control effectiveness. During its life, changes were made affecting its appearance, its weight and propulsion engine.

The next step was the design and manufacture of a second generation craft. It was recognised at this stage of development that the next craft should be large enough to provide information relevant to the design of bigger craft for practical operations later on, but that at the same time it would be essential to have a craft of suitable size for transporting to distant trial areas in order to acquire experience in a variety of environments. To realise this double aim, it was decided to build two craft, the small VA-2 and the larger VA-3.

The 12-ton VA-3 was completed first and began its trials in April 1962, starting on the airfield at South Marston, Wiltshire, and a month later, transferring to the company's Itchen Works for trials on Southampton Water. But in less than three months it was to leave its base on the South Coast in order to be taken by ship for operational trials in the north-west of the country.

The VA-1 with cabin fitted for over water trials. Length 25 feet, beam 13 feet, lift 139hp – de Havilland Gipsy Major. Propulsion 133hp Continental C90.

VA-3 Protoype 26-seat. First flight 25th March 1962 at South Marston. Note the front fins and no skirt. Unlike the Saunders-Roe craft it had separate engines for the lift and propulsion. Length 54 feet 9 inches, beam 27 feet, 60 knots, 24 passengers. 4 x 425 shp Bristol Siddley Turmo Free Turbines – 2 for lift, 2 for propulsion.

Rhyl to Wallasey Trial Passenger Service

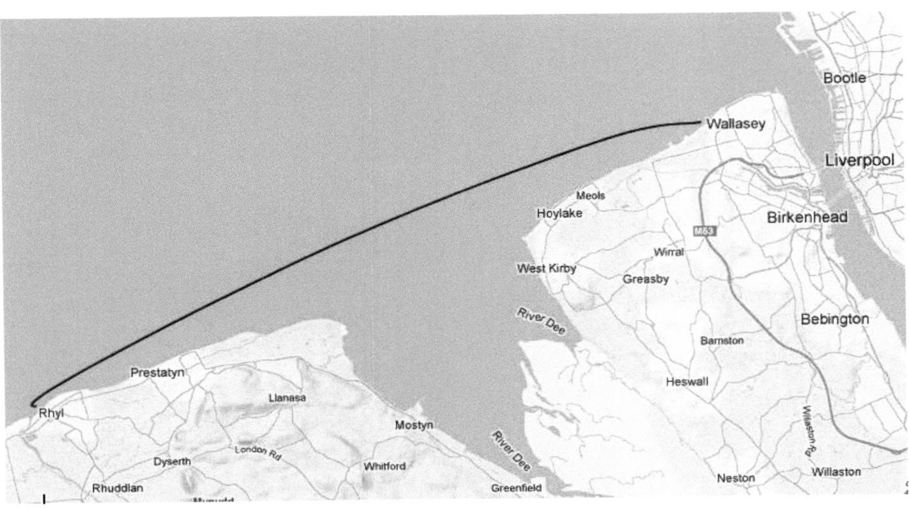

VA-3 Rhyl to Wallasey planned 8-week experimental scheduled passenger service. July-August 1962.

The trials were in the form of a service to carry 24 passengers across the estuary of the River Dee between Rhyl, Wales, and Wallasey, England. The service, operated by British United Airways with the active support of British Petroleum, began on July 20th 1962.

Although it could be considered the world's first scheduled hovercraft service for fare-paying passengers, it was in fact merely a passenger-carrying exercise as part of a prototype programme, so it could not claim to be a *bona fide* commercial passenger operation. That accolade had to wait for Clyde Hover Ferries some three years later.

During the eight weeks which followed, the VA-3 travelled 3,500 miles but only carried 3,765 out of an expected 10,000 passengers. Of those 59 days, the craft could only operate for 36 of them. Many of the passengers were holidaymakers from the neighbouring seaside resorts, but not unnaturally this first service drew visitors from all over the world. It also carried mail, thereby claiming to be the first 'Hovercraft Postal Service' with a special postmark to commemorate the event.

The primary objectives of the trial were to carry out an operational research programme to obtain practical experience of handling and maintenance in realistic conditions; to obtain experience of passenger handling and to assess their reactions; and to demonstrate to the public the 'safety with speed' characteristics of hovercraft.

VA-3 on the Rhyl to Wallasey service 20th July 1962.

VA-3 about to load passengers on the Rhyl to Wallesey service.

It was a brave move to start a trial service some 200 miles from the works at Southampton so early on, but the temptation to take the craft out of service back to the comprehensive maintenance organisation might have proved too strong. The terminals consisted of a caravan at each beach.

The operational maintenance base, a workshop-caravan and a roped-off area of sand dunes, was at Rhyl, some two miles west of the terminal. Refuelling also took place at Rhyl from road tankers which were parked on the promenade. In these circumstances, divorced from the immediate support usually associated with the trials programme, the VA-3 was subjected to a severe test, only three months after its first 'hover'.

Les Colqhoun, wife Katie and four daughters, Helen, Jane, Peta and Sally on the right at Wallasey July 1962.

The VA-3 service carried mail, thereby claiming to be the first 'Hovercraft Postal Service'.

A certain engineer, Emrys Jones, an ex-RAF wartime *Spitfire* and latterly *Swift* engineer, who joined Vickers-Armstrongs in 1951, had been working on the VA-3 hovercraft, and was sent with three pilots, Les Colquhoun, Ray Old and another Merchant Navy officer, Captain Banbury, who had been seconded for the trials to gain experience, to attend to the engineering needs of the VA-3 on the Rhyl to Wallasey trial service.

The weather in Britain is often a source of disappointment and the summer of 1962 proved no exception. The incidence of high winds and an accompanying build-up in wave height, increasing to 4 or 5 feet, had a limiting effect on the operations, especially without a skirt, and a hover height of just 8 inches. Naturally, passenger safety was of paramount importance, and in fact the Certificate of Construction and Performance issued by the Ministry of Aviation, through the Air Registration Board (ARB) – 'Permit to Fly' – in conjunction with the Ministry of Transport, stipulated the conditions under which the hovercraft could operate[2].

The passengers were seated in four groups of three either side in rear-facing seats for maximum safety and there was a steward on board. The first trip to Wallasey took 33 minutes on a calm day at a varying speed of between 40 to 55 knots.

Handling the craft in varying conditions of wind and water, which would not ordinarily be experienced in the Solent, was invaluable, but the craft was severely limited without a skirt.

There were constant changes in the coastal waters and, at low tides, in the exposed sand. In particular, the Dee has at least two areas of extremely deep water, so that although conditions may have been good at each terminal they were sometimes quite unsuitable at the mouth of the river, a fact not always apparent to the fare-paying public. Winds from the west or north of west were often an embarrassment. In fact almost all the cancellations imposed by the weather were because of westerly winds, blowing in from the open sea.

In addition to the limitations imposed by the weather, the operating schedule of six return trips per day was also affected by mechanical problems, almost all of which were connected with teething troubles with the engines. In fact, only on six days had the craft been able to operate the full schedule. There is no doubt that the design philosophy and use of lightweight materials employed in the VA-3 produced an extremely strong, seaworthy craft capable of withstanding conditions much more severe than those assumed when the project was originally conceived.

Unfortunately, some passengers were inconvenienced by the cancellations, most of which were attributed to either the weather or to mechanical faults in the engines, but despite this, passenger reaction was in the main enthusiastic. Even so, the VA-3 never operated another passenger service again.

Immediately following the ending of the service, a combination of weather and engine troubles resulted in an incident involving the VA-3 in

---

[2] The licensing of hovercraft and crew qualifications are dealt with in Appendix 2.

accidental damage. By Friday 14th September, and after 13 engine changes, the situation had been reached when there were no more spares left. Both lift engines were out of action and it had been necessary to cancel the last two days' operations. The VA-3 had limped into Rhyl and been positioned on the open beach with its anchor dug into the sand.

Unfortunately, on the night of 16th September 1962, extremely high spring tides coincided with gale force north-westerly winds, as a result of which VA-3 sustained a heavy pounding from the waves. All three captains were on board to keep an eye on things, but there was little they could do, as the whole anchor fitting to the bow was torn out. Les started the engines and tried to keep the craft out of danger, but in the end the Rhyl lifeboat was called out and Les and the other two had to abandon the VA-3 for the safety of the lifeboat. As the tide ebbed, the engineers were able to secure the craft and fortuitously there was a crane available to lift it on to the safety on the dock. The damage necessitated the return of VA-3 to South Marston for repair and for his superb efforts the coxswain of the lifeboat received a Silver Medal.

The modifications to the VA-3 included new propulsion engines, the fitting of a 3-foot skirt and the removal of the forward fins.

In addition to the repairs, modifications were put in hand. These included the fitting of a 3-foot deep flexible skirt, the replacement of the two Turmo 603 free-turbine propulsion engines with two Artouste 2Cs, and the removal of the forward fins. The contribution which the forward fins made in providing side force was limited to a speed range in the higher end of the scale, and their effect was proportionately reduced when the addition of the flexible skirt increased the side area of the craft.

The VA-3 on Trial with the US Navy

In February 1964, after a test programme at its Southampton base, the VA-3 was shipped to New York to be operated by Republic Aerospace Corporation on extensive trials. The major part of the trials, which lasted until early the following year, was on behalf of the US Office of Naval Research. Initially based at Montauk, Long Island, NY, the VA-3 started with sea-keeping tests combined with a training programme for new pilots.

VA-3 undergoing trials at Long Island for the US Navy in July 1963.

For the second part of the trials, the VA-3 was moved south to a new base at Norfolk, Virginia. Here the emphasis was on amphibious exercises and operations with a US Navy dock ship. Normally, the ship's dock would be flooded for such operations, but the hovercraft was able to enter the LSD *Fort Mandan* with its dock dry, while the ship proceeded at various speeds up to 15 knots.

At higher speeds, the increased turbulence aft of the ship made it unwise to attempt the entry into what appeared to the driver to be a long dark box. Once through the area of turbulence, the sudden decrease in drag would produce an alarming increase in speed. Having entered the dock, the propulsion engine layout of the VA-3 made it relatively simple to back out again. At the conclusion of these trials, the VA-3 was returned to its base near Southampton in May 1965.

From their earliest days in hovercraft, Vickers had been giving considerable attention to the military applications of Air Cushion Vehicles (ACVs), including the possibility of modifying military vehicles to give increased mobility through the use of an air cushion system. After some detailed study it was decided to modify a Land Rover quarter-ton truck. This would provide a practical research vehicle to investigate the effect of varying the weight bearing on the wheels, and at the same time be a convenient rig for testing the design of flexible skirts. It was perhaps appropriate that these two aspects of overland ACVs and flexible skirt design should be coupled in one vehicle, because these were particular areas where Vickers had made substantial and specialised progress.

The VA-2

The manufacturing programme of the 3.7-ton, 4-5 seat VA-2 had been re-phased so that the VA-3 could be made available for its first major trials in the summer of 1962. So the VA-2, intended to be the first of the Vickers second-generation hovercraft, had its initial testing in October 1962, shortly after the completion of the VA-3 Rhyl-Wallasey trials programme. The testing went on through the winter, and included testing over frozen snow in January 1963. A special feature of the craft was its retractable undercarriage, a pair of wheels amidships which could be selected for use at modest speeds over reasonably firm regular surfaces. They took about 10% of the craft weight and, by differential braking, they were used to give much finer directional control than would otherwise be possible. Their effectiveness was very quickly proven when the craft was driven between rows of parked cars giving only 6-inches clearance on either side.

But if the planned date for the completion of the VA-2 had been delayed, time was not lost in embarking on the overseas tours for which it was intended. With the initial testing at South Marston and Southampton completed by March 1963, the VA-2 began what was to become an extremely busy 12 months of environmental testing. In mid-March, it was loaded onto a truck, shipped across the English Channel and transported south to Ingolstadt, between Nuremberg and Munich. The craft designers had anticipated such an eventuality by arranging that the structure outboard of the cabin-line could be easily removed and reassembled by a small team of men. Thus the width could be reduced for transportation and, when necessary, the height by removing the fins and rudders. The purpose of the visit was to take part in a military demonstration to a large audience of NATO officers. The VA-2's contribution, which was under the auspices of the Inter-Service Hovercraft Trials Unit, was to demonstrate the high speed-over-water and amphibious capabilities of hovercraft, which Les Colquhoun reported as a great success.

VA-2 on trials in October 1962. Length 30 feet 4 inches: Beam 15 feet.

Lift 2 x 133hp Rolls Royce/Continental O-300-B aero engines: Propulsion 310hp Continental O-470-I, replaced in 1963 with a Continental 470A: Top speed 69 knots.

After two weeks in Germany, the craft was transported by road to Amsterdam where, at the invitation of British Petroleum, it was to operate in conjunction with the opening of a new petroleum installation. During the fortnight of its stay, demonstrations were given to a large number of visitors, mainly with commercial interests in the Netherlands and neighbouring countries, and it was in Amsterdam that the VA-2's first 'Royal Driver', Prince Bernhard of the Netherlands, took the wheel.

On return to England, a 1 foot 10-inch flexible skirt of convoluted design was fitted in place of the original 8-inch flexible nozzles, in readiness for more ambitious trials overseas.

But before the overland trials, another overseas visit was put into the programme, this time in Scandinavia. For two weeks at the end of September, the VA-2 was based at Malmo, Sweden, for demonstrations and trials on the Øresund. Visitors representing commercial and government interests from all the Scandinavian countries and Germany came in considerable numbers, and special days were reserved for military visitors.

VA-2 in Holland 17th April 1963.

For some, the most striking demonstrations were the high-speed runs across to Amager Strandpark, near Copenhagen, during which the ability to operate in the shallows around Saltholm Island, and thereby take the shortest route, plus the transition onto the beach adjacent to the Copenhagen-Kastrup Airport road meant that the Øresund was crossed in about 15 to 20 minutes compared with 35 minutes by hydrofoil or 90 minutes by normal ferry.

The two final trials of this hectic 12 months were devoted to something never before attempted, namely operations over desert and over ice, which duly happened with some degree of success in North Africa and Sweden.

The next visit overseas was a remarkable one, but for the wrong reason. Two months after its successful winter trials, the VA-2 was taken to Dusseldorf for British Week, but, alas, stubbornly refused to perform; a component of the propulsion engine was at fault. Honour was restored on the craft's next visit to Germany, but in the meanwhile a long programme of testing at home was due, which lasted through the remainder of 1964 and most of 1965.

A Change of Direction

Towards the end of 1965 Les Colquhoun was head-hunted as the operations manager for a new commercial hovercraft operator, financed by Swedish interests, called Hoverlloyd, with the intention of setting up a cross-Channel hovercraft service between England and France.

Les would need an experienced chief engineer and asked Emrys Jones to come with him, to which he readily agreed. In turn, Emrys recruited two other first-class engineers from the Vickers stable. One was Monty Banks, who became Emrys' deputy, and Bob Taylor an electrical engineering specialist. On the flight crew side, another member of Vickers, David Wise, who worked in flight testing and knew Les, applied for a pilot's job and was accepted.

The Land Rover 'hovertruck'.

There was a concept plan for a 150-ton VA-4 to be built, similar to the original 'Morphy Richards' SR.N4, capable of carrying 24 cars and 200 passengers, but the reality of the difficulties in developing the hovercraft to a point where it would be economically viable must have been beginning to dawn on both Vickers and Saunders-Roe.

Furthermore, there just wasn't the market for two manufacturers of large hovercraft. So, early in 1966 there was to be a major reorganisation of the hovercraft industry with the merging of interests between Westland and Vickers-Armstrongs to become the British Hovercraft Corporation (BHC), based at Cowes on the Isle of Wight, of which Vickers would be a 25% shareholder.

Only three Vickers craft had ever been built, in addition to the Land Rover 'hovertruck'. None of the Vickers personnel went to BHC at Cowes and thus ended Vickers-Armstrongs' five-year affair with the hovercraft, although the 'hovertruck' was taken to Cowes, and the remaining VA-3 did appear at the Browndown Hovershow in 1966.

The VA-3 was subsequently used by the Royal Navy for marine mine tests on the basis that it would fare better than a conventional vessel if struck by a mine. Things didn't work out as planned and it was blown to smithereens and scattered across the Solent. Fortunately there wasn't the confidence to put a pilot on board.

Chapter 7

The Early Hovercraft Operators

By the spring of 1965, hovercraft development at BHC was forging ahead; nevertheless there was still much to learn concerning this novel form of transport. There occurred in quick succession a series of accidents involving the SR.N5, all three of which could be put down to pilot error with two of them involved capsizing in plough-in situations.

The first was to Scanhover's 001, which overturned at Ålesund Harbour in Norway, while the pilot was demonstrating an emergency stop.

The second was six days later when, operating a trial service across the Elbe Estuary (Germany) in a strong tail wind[1], the craft overshot the base and hit two cars.

The third incident was in May when an SR.N5 was being piloted by an experienced airline pilot, but with little hovercraft experience, overturned in San Francisco Bay.

Fortunately, none of these accidents involved casualties, but the first and third incidents required further and urgent investigation by the design and development teams at Saunders-Roe.

The problem was revealed to be the hydrodynamic suction at the lower edge of the peripheral jet on the side of the bow. In other words it was the classic case of a semi-sideways plough-in, whereby a section of the skirt would catch the water and the compounding effect would eventually drag the bow down on one side, producing a turning moment, which would flip the craft.

'Fingers', or appendages to the skirt, had not been fitted at this stage but the problem was solved by the addition of vertical rubber breaker strips in the bow area of the lower part of the skirt with holes between them to introduce air lubrication (see page 177).

By July, all SR.N5 and N6 hovercraft had the approved modifications fitted, but it is fair to say that after solving the problems associated with the annular jet principle, the hovercraft skirt was for some years thereafter to prove the most difficult aspect to resolve in order to turn the hovercraft into a viable form of commercial transport. It was also to become one of the major cost items in the operation of the hovercraft.

---

1 Controlling a hovercraft downwind was one of the most difficult aspects of piloting to master.

Clyde Hover Ferries

Any history of hovercraft cannot be told without mentioning Clyde Hover Ferries. An operational failure it may have been, but this small Scottish enterprise was the very first commercial hovercraft passenger service and had a significant influence on Hoverlloyd as a company and its initial development.

On 16th June 1965 Clyde Hover Ferries commenced a daily service, using the very first production SR.N6 craft, leaving the base at Tarbert, Loch Fyne at 6am, stopping at Tighnabruaich, Rothesay and Dunoon. This was what might be termed a 'positioning run'.

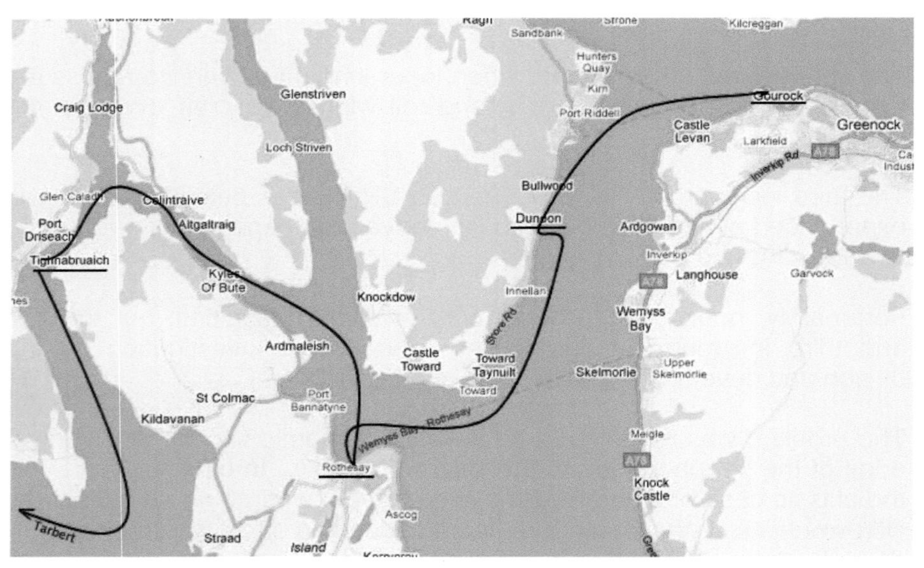

Clyde Hover Ferries major routes. 1965.

The major potential money earner was between Dunoon and Gourock, there being a large daily commuter demand via a railway link to Glasgow.

Right from the start, apart from the expense of unnecessary extra mileage from Tarbett, the maintenance base, to the first real potential for commuter traffic at Rothesay, the viability of the enterprise was questionable. The delivered craft, yard number 010, was the first production craft to be built, coming a mere two and half months after the prototype number 009[2]. It could be said, therefore, to be largely a prototype model itself, and maintenance costs for such a minimally-

[2] The second SR.N6 for the service, yard number 012, started work with CHF on 24th July 1965.

tested machine were well over the projected budget. The unforeseen operating and maintenance costs of such a new form of transport, coupled with a poorly-constructed business plan, leading to an over-estimation of the potential market, all contributed to its eventual failure.

Clyde Hover Ferries SR.N6 approaching Tarbert.

Following the demise of the company after only one season in November 1965, four of the seven Scottish N6 pilots migrated south to form a training cadre for the fledgling Hoverlloyd, planning to start an N6 operation of its own from Ramsgate to Calais. After the brief summer operation on the Clyde, Bill Williamson, Tom Wilson, George Kennedy and Roy Mortlock, with a collective 600 hours of operating the N6, could count themselves amongst the most experienced hovercraft pilots in the world. It was fortunate for Hoverlloyd that such expertise became available at just the time it was needed, not only the driving skills, but invaluable lessons learnt from an overall operational viewpoint.

Clyde Hover was the brainchild of local flamboyant entrepreneur, Peter Kaye, whose somewhat, at times, eccentric view of the operation – stopping the service to run day trips for his friends, commandeering the craft to run sheep between his property on Little Cumbrae and the mainland – could not have enhanced the public's view of the professionalism of the operation, or the budget for that matter.

The base for the company's operation was located at Tarbert, which at first sight appears an odd choice, but unfortunately not an unusual one where planning permissions are involved. Like so many such decisions, the authorities tend to look first and foremost at the political implications and rarely at the economics. Tarbert was a typical example. Bill Williamson, who was CHF's operations manager explains:

> *It was placed at Loch Fyne because it was a development area. It should never have been there but Peter Kaye was getting all sorts of government money, all to do with employment in the Western Isles. Of course it was miles from where we were supposed to be doing all the work.*

'From where we were supposed to be' vies for the prize of understatement of the century. Tarbert was roughly 100 miles by road from Glasgow, the nearest big city, and even more problematical, it was 24 and 32 nautical miles respectively from the only real commercial prospects, the small towns of Rothesay on the Isle of Bute and Dunoon on the Argyll mainland. Both the populations of Tarbert itself and Tighnabruaich, the next port of call, were miniscule and not worth the fuel to visit. Such a totally misguided decision to site the operational base so far away most probably doomed the whole project to failure before the hovercraft even took to the water.

George Kennedy's whimsical tale of setting up the base in Tarbert epitomises the whole *ad hoc*, 'seat of the pants', nature of the business from its inception:

> *Peter Kaye was looking for somewhere and he bought a very famous boatyard in Tarbert, called 'Dickies'. There was a big boathouse that could probably take a fishing boat, with rails which came straight up the slope. Kaye decided we would have to level it inside and concrete it for a lifting gantry for the craft.*

So together with a couple of engineers we went up to Tarbert where Kaye had got hold of a tracked mini-bulldozer. He said, "Right lads. It's three weeks before the craft comes. Go ahead and level it."

So seven of us, including the engineers, started levelling it. Eventually we got it fairly level except at the end there were a lot of large rocks. Kaye came down, looked at the rocks and said, "We'll get some explosives." He hopped into his Bentley and went up to the local quarry. He came back later saying they wouldn't let him have any 'jelly' but he'd got some black powder and cordite. He also hired a rock drill. So we drilled the rocks, put the black powder in and rammed it down with the cordite and clay.

He showed us how to light the cordite – it needs a very hot flame. You need to hold it close to the match to get the initial very hot flame. We lit it and all ran for our lives. There was a big bang and the boathouse jumped a foot in the air. We did this about four times until eventually the council came along and said, "You can't do this. You haven't got a licence." So we had to build up a bit on the outside to get the thing level.

Clyde Hover Ferries' brief one-season existence characterises the pioneering nature of these early attempts to make a brand new concept in transport pay. Both operators and the manufacturers were equally ignorant of the problems involved. A prime example is the suggestion by Westland that the craft could be easily maintained by one engineer and a 'boy'. Also, at Westland's suggestion, apart from a maintenance base at Tarbert, which used the existing boatyard, all the other ports of call on the itinerary were simply cleared areas of beach. As George points out, the problem there was that 'there are beaches and there are beaches':

Westland claimed that the craft could be used on unprepared beaches, by which I suppose they meant 'sandy beaches', but most of our beaches up there were cobbles. Some of the stones could be 8 or 9 inches in diameter and when kicked up by the skirt could puncture the buoyancy chamber. When the craft was lifted at night, the engineers were forever finding large holes.

There were certainly learning curve problems, in addition to the maintenance costs of this new technology, such as inexperience of the minimally trained crews, and going solo with less than 5 hours of tuition[3], which caused a number of expensive accidents.

---

[3] Most of the training was undertaken by Harry Phillips and Bob Strath. Sheepy Lamb had little involvement as he was required on other matters back at Cowes.

George Kennedy, who was one of the four pilots to move to Hoverlloyd from Clyde Hover Ferries, at the controls of an early SR.N6.

Clyde Hover Ferries customers queuing on the beach.

By the end of the summer it was obvious to all involved that the project was a failure. The 'soon to be Hoverlloyd men' were already making plans[4] to move on and later in the year, one of the craft (012) was returned to Westland's factory in Cowes. The whole thing then degenerated into a lengthy dispute between Kaye and BHC concerning the lease payments of the two craft, of which Kaye had made only one lease payment for one craft, refusing to pay anymore until the machines came up to specification.

The impasse dragged on through the following year until in October 1966, Kaye put an end to the saga by returning the other craft to BHC, dumping it on the pad and walking off. If the story had elements of tragic farce in its few months of existence, it most certainly ended very much in pantomime. The story of the final chapter goes, according to George Kennedy, something like this:

> *Once the decision had been made to end the dispute by returning the last hovercraft to BHC, Kaye decided they should have a holiday with it in Ireland first. In due course Willy Clements, the last commander to be employed, took Kaye and his crowd across to Ireland and then they worked their way down the Irish coast, across to England and drove up the big ramp at Cowes on the Isle of Wight. Kaye and everyone stepped out and asked BHC, "Can you drop us off on the mainland?" They said no, they would do no such thing. So Kaye said, "Right lads – back on board." Everyone piled back on. Willy started it up and took it across to the mainland, parked it, and they all went home.*

Thus ended, all too soon, the first serious attempt at using the hovercraft as a passenger ferry. Nevertheless, valuable operational experience was gained, experience that was to benefit Hoverlloyd enormously.

Hovertravel

The second and most impressive of the early operators was Hovertravel Limited, the sole surviving British-based scheduled hovercraft service to this day, formed on 14th June 1965 by some of the original partners of Hovertransport Limited (which had been disbanded) to operate from Ryde on the Isle of Wight to Stokes bay, near Gosport, a route length of just 4 miles.

It was Don Robertson, a City stockbroker, who put Hovertravel together and became its first chairman. In addition to Don, the main founders,

---

4 Head-hunted and encouraged by Les Colquhoun in his original capacity as Hoverlloyd Operations Manager.

Christopher Bland, Chairman of Hovertravel from 1970 to 2008.

among others, of the new company were Desmond Norman and John Britten, founders of the Britten-Norman Aircraft Company of Bembridge, Isle of Wight; Edwin Gifford, a civil engineer from Hampshire; David Webb, an accountant from Webb Acres and Hayes; Frank Mann, Desmond Norman's partner in a crop spraying company; and Christopher Bland, who recalls:

I was one of the original shareholders of Hovertravel, which was started with an initial paid-up share capital of £62,000, but we realised that wasn't enough and so we increased it to £134,500. It was still £134,500 paid up when we sold the company in 2008 to the Bland Group – no connection with me whatsoever – so we never raised any more capital. They were £1 shares and we sold them for £60 each in 2008, but there was inflation in between.

Don Robertson was a great hovercraft enthusiast – so much so that he had no fingers on one hand where he had accidentally chopped them off. He had a little hovercraft which he ran on his lawn at Redhill and the engine was in the air intake. He was adjusting the mixture when his fingers went through the fan. He went round the garden, collected up the fingers and took them to the hospital, but they couldn't get them on again. He was a wonderful chap. He had flown in Canada before World War Two and was the Chief Naval Test Pilot at Boscombe Down during World War Two. He had also flown pre-war mail planes in the Arctic. Westland, then BHC, carried out a lot of the development of the SR.N6 in conjunction with Hovertravel because they were only four miles down the road at East Cowes.

Passengers arriving at Ryde Town.

Hovertravel probably had, and still has, the best hovercraft route in the world. The low tides in the area around Ryde take the sea far out beyond the town's esplanade and it has always been impossible for any ship to berth anywhere near roads or bus routes. Until 1965 the only passenger ferries from Portsmouth stopped at the end of Ryde's ½ mile long pier. Chris Bland noted:

> *It was very shrewd of Don Robertson and Desmond Norman to spot it – it was on our door step. Anywhere else in the civilised world they would have slung a bridge across – just four miles.*

Not only that, but the route ticked all the right boxes for the perfect operation – namely, a short sea route inaccessible by ship or boat, high density traffic and high-yield revenue. Even to this day it competes very effectively on speed and price with the Wightlink catamaran service, which takes 22 minutes to Portsmouth against Hovertravel's 8 minutes[5] to Southsea. Wightlink has recently been obliged to spend nearly £4 million on repairing the pier, which all has to be added into the equation. The hovercraft just runs up on to a relatively inexpensive piece of concrete.

Hovertravel[6] took delivery of their first SR.N6, No. 130[7], on 24th July

---

5 In 1970, a record crossing time of 4 minutes 53 seconds was set in an SR.N5.

6 Hovertravel has become the oldest and only passenger hovercraft company in the western world still in operation.

7 The craft was the thirteenth on the production line, but to placate the superstitious, all records show the craft as number 130.

1965 and commenced their Ryde to Stokes Bay (Gosport) service, operating from a caravan at both ends, just one month after Clyde Hover Ferries started theirs in Scotland, with fares ranging from 7s 6d and 10s (37½p and 50p), carrying 805 passengers on the first day. The most ever carried in one day was 6,200.

The start of the Clyde and Solent hovercraft services, together with the earlier announcement by Swedish Lloyd of their order for two SR.N4s, created huge public and press interest in the hovercraft, to the extent that anything remotely interesting was being published without much thought given as to what the reality might be. *The Financial Times* reported on 27th July, three days after Hovertravel's inaugural service:

> *The first British operator of the giant £1m Westland SR.N4 hovercraft is likely to be Hovertravel it was learnt today.*
>
> *In three year's time it plans to introduce a car ferry service across the Solent from Stokes Bay to Ryde, where on Saturday it began a passenger service with the smaller SR.N6 craft. Negotiations for Government financial backing for the project are already at an advanced stage, declared Mr. E. Gifford, a director of the company. But he emphasised that the arrangements would involve only a 'partial loan' and no form of subsidy would be involved.*
>
> *Mr. Gifford said his company would also be interested in a Channel service, but, "It is our view that this being a new form of transport, one ought to operate first across a sheltered route." The Solent, he thought, provided an excellent area for gaining experience [with the SR.N4], although once a service was introduced there, the company hoped to keep it going. The total cost including terminals would be about £1.2m. If the decision was taken to expand further new craft would be ordered.*

It soon became apparent that this was never going to work and the idea was quietly dropped.

Hovertravel's second SR.N6, No. 026, arrived shortly after the first, both craft having been leased from Westland Aircraft Leasing and with it a new route, Ryde to Southsea, was started. The Stokes Bay service was closed in 1967, but the Southsea service became increasingly popular and flourishes to this day. The timetable announced a departure every half hour from 8.30am to 6.00pm. At the end of each day the craft went back to Cowes to have their skirts looked at and repaired. For the first year of operation, when the 4-foot flexible skirts were without fingers, the level of skirt maintenance was extremely high. Another difficulty was that the first SR.N6s did not have forward puff ports for better control of the front end. Both the fingers and puff ports were introduced in the second year. There were the inevitable teething problems, but a backup 18-seat SR.N5 was available for charter from Westland as and when required.

The SR.N6 with the 4-foot flexible skirt before fingers were fitted. Note the vertical rubber breaker strips in the bow area of the lower part of the skirt with holes between them to introduce air lubrication to help prevent capsizing, as in the case of the SR.N5s in Norway and San Francisco Bay.

A close up of the strips and air-holes inserted for better lubrication to prevent plough-ins.

The payload of the SR.N6s was 38 seated passengers, two more than on an SR.N6 equipped with radar, because initially Hovertravel's radar was on the roof of a hotel with a shore-based operator guiding the pilots across the Solent, negating the need for installing expensive radar on each craft and taking up valuable payload space. Even so, in order to increase the payload still further, Hovertravel decided to install strap hangers for the commuters, but the ARB (Air Registration Board) wisely put paid to that.

On the issue of the shore-based radar, Barrie Jehan, Operations Manager, who joined at the age of 18 and eventually became a pilot, explains:

> *We had a big radar on the roof of the Royal Esplanade Hotel, situated across the road from the Ryde terminal, and the operators cabin was just below the scanner. The room measured about 5 feet by 8 feet. It was cosy! We used to control the craft from there rather like an ATC (Air Traffic Control) radar control. After the appropriate training, the other pilots and I used to play ATC controllers at night and in fog. Things got interesting when you had a craft on the Ryde to Southsea route and another one on the Ryde to Gosport route. Things got a little tense at times. You could not do that now as there was a fraction of the traffic then compared to what we have these days. I think we fitted radar to the Solent craft around 1971.*

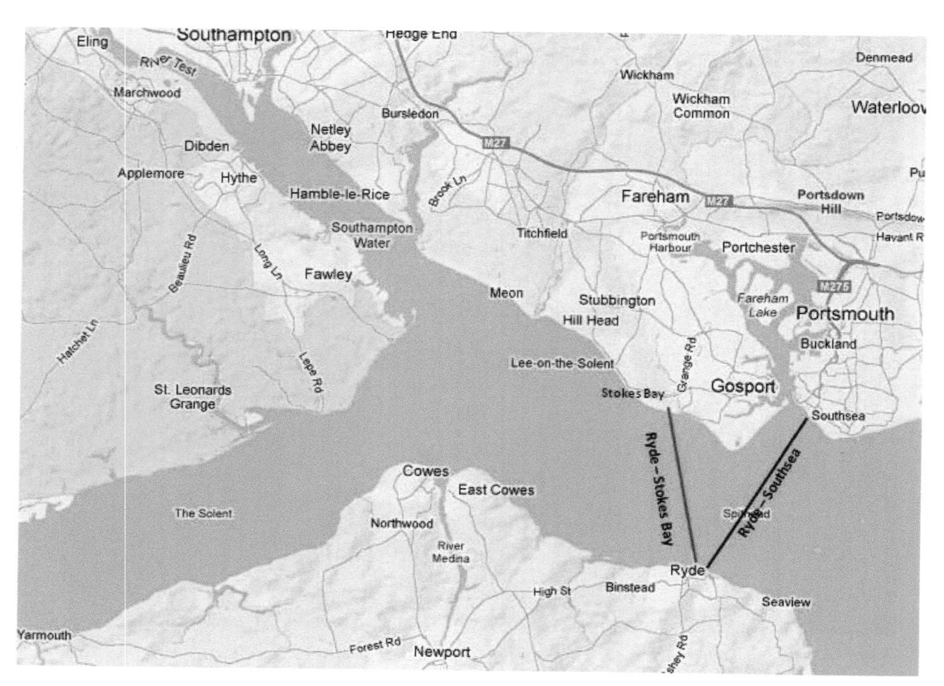

Hovertravel's Southsea Terminal.

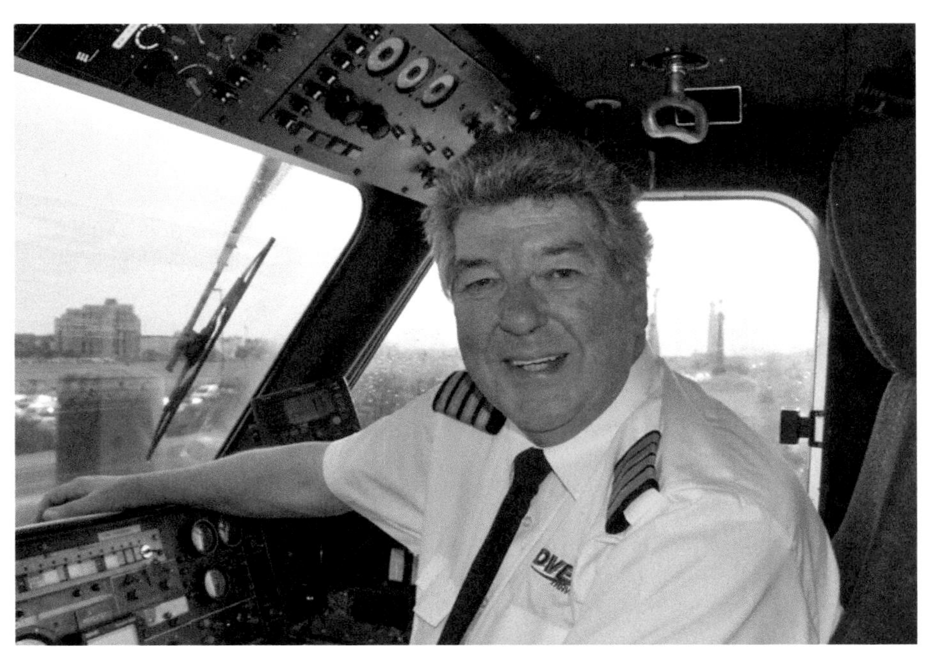

Barrie Jehan, Hovertravel's Operations Manager in an AP1-88, who, after 43 years with Hovertravel, retired in June 2012.

This seems to illustrate the early dilemma of how to deal with high-speed two-dimensional navigation and anti-collision in reduced visibility. Most mariners would be very wary of such an arrangement, but as Hovertravel's pilots were aviators, this probably seemed the most logical solution, as, unlike mariners, they had not been trained in the use of radar at sea. But they were also dealing with a very unfamiliar environment; they simply weren't used to uncontrolled space. In the air, everything that flies is controlled in some manner or other. The idea that you have boats swarming around with very little constriction as to the area, or control as to direction, is totally alien to air traffic.

The ½ mile long Ryde Pier next to Hovertravel's hoverport that underwent a £4 million refurbishment to enable conventional ferries and catamarans to continue to dock there.

On the subject of the economics, Chris Bland comments:

> *We had Gnome engines, which were terribly expensive, but we had our own overhaul shop for the Gnome engine at Ryde because we couldn't afford the Rolls Royce prices. Our first engineer was R.B. Stratton, who had been the flight engineer on the* Princess *flying boat that flew. Gnomes were gold dust in those days, but now you can buy them cheaply although they still use them on the Sea King helicopters. We had some terrible times. We couldn't afford to pay Rolls Royce for engines – at one time we were down to one engine and one hovercraft and that was it. If the engine had gone we would have been completely out of business. We had a second*

engine in overhaul in our own workshops, but we didn't have a compressor for it. I always remember when someone from the Royal Navy met me in a mini-van at a pub in Lee-on-Solent with a brand new compressor and we went on from there. We had very dark days.

We always made an operational profit because we never had a negative cash flow. We washed our faces from day one, but I think we had to be helped by Westland at one time because they sent us and charged us for an excessive amount of unnecessary spares for the hovercraft. We just didn't need them so they took them back. They also helped us by repairing the skirt every night at the factory and charged us so much an hour. We were getting about 500 hours out of the skirts towards the end of the SR.N6, but it still represented about 20% of operating costs. Even so the economics of the N6 were such that it would never have made a proper profit to include depreciation. I think the SR.N6 skirt cost the same as the fuel in the early days. In fact we didn't make a decent profit for about 20 years.

The Pilots

The pilots from left to right: Barry Goldsmith, Tony Smith, Peter Ayles, Phil Phillips, Peter Atkinson. 1970.

The early pilots were mainly aviators from the RAF and Fleet Air Arm. Peter Ayles was the senior captain, who was to set a record of 77 crossings in one day before the schedules started, followed by Tony Smith and Peter Atkinson. Others joined and all but one of them were aviators. Barrie Jehan:

What we have found is that when we have trained aviators their learning curve has been a lot quicker and smarter than mariners, but when they plateau they plateau lower than the mariners, who take longer to get started, but once they do get started they're there. Currently the trainee has 25 hours one-to-one instruction on the craft and during that time he will be checked by a Type Rating Examiner. Once he's cleared to take passengers, he will then do a further 100 hours with a qualified pilot. So the pilot will have at least 125 hours under his belt before what we call him 'self certifying' and he can take the craft under his own command.

Hovertravel purchases the Townsend Hovercraft

Townsend Ferries, in conjunction with P & A Campbell, was the only operator to purchase a brand new SR.N6 – No. 024. They started cross-Channel operations from Dover Harbour in 1966 to a new hoverport in Calais, which they shared with Hoverlloyd's SR.N6 proving services, built by the Chamber of Commerce just inside the western breakwater of the harbour. After two months, as a result of weather and breakdowns, Townsend had only carried around 400 passengers and their ill-fated cross-Channel venture was curtailed. As P & A Campbell was an excursion operator it was decided to run trips for the holidaymakers off the Kent and Sussex beaches. Brian Laverick-Smith, who joined Townsend as their hovercraft navigator on the cross-Channel service recalls what followed:

We went from Dover to Margate and then Margate to occasionally Sheerness, Isle of Sheppey, and then back round again to Dover. On alternate days we'd go to Hastings and we'd stop halfway at Dungeness Power station at Greystone on the beach. You don't navigate doing this and for the first year I just sat there doing nothing. I would just walk up and down the aisle selling tickets, but in 1967 as a pilot it was totally different.

When Townsend's two year spasmodic affair with the SR.N6 came to the end in 1967, and with it any thought of operating an SR.N4, 024 was put on board the Cunard liner *Sylvania* as a cruise experiment to give government officials of various countries bordering the Mediterranean a practical demonstration of the hovercraft's capabilities with BHC's Harry Phillips[8] as the pilot. It was not there to provide sight-seeing trips for cruise passengers, although occasional short trips for passengers were made 'around the bay' between the official runs. At the end of the voyage, 024 was sold. Chris Bland:

---

[8] It was on the *Sylvania* that Harry Phillips met the second 'Mrs. Phillips'.

The first hovercraft we bought was Townsend's in 1968 for £37,500 (About £62,500 less than Townsend had paid for it in 1966). We paid off whatever we owed on it. It was very difficult to make a profit on the N6, bearing in mind we were under capitalised. We had a £1 million bank borrowing at one time, unsecured, on an issued capital of £134,500. What hasn't been mentioned was that 10% of the company was owned by Glyn Mills Bank and 10% was owned by C.T. Barrings from inception. Those were the days when banks were banks. Glyn Mills said they would put some money in and take 10% of the equity, which they did. Glyn Mills was then bought out by William Deacons and it became Williams and Glyn after which the Royal Bank of Scotland bought them. RBS never cashed the dividend cheque because they didn't know where to put it because we were the smallest company they had any interest in. Over 40 years they were hopeless to deal with, but Glyn Mills had been great.

The Townsend SR.N6, 024, on Margate Beach. The craft was subsequently purchased by Hovertravel.

The VT-1 Fiasco

Vosper was established in 1871 by Herbert Edward Vosper, concentrating on ship repair and refitting work. Vosper would become famous as the builder of small 60 to 70 foot un-stepped planing hull-form naval Motor Torpedo Boats (MTB) and Motor Gun Boats (MGB) for the Royal Navy in World War Two.

The Vosper VT-1 'semi-amphibious' hovercraft. It had 2 flight crew and could carry 146 passengers plus 10 cars at 40 knots. It was 95 feet 6 inches long with a beam of 44 feet 6 inches. The power plant was 2 x 2000 hp Avco Lycoming Gas Turbine. The hovering draft was 3 feet 9 inches.

In the 1960s the company began to move into producing larger vessels, especially for the many emerging navies of post-colonial countries, including their *Alvand-class light patrol frigate*, with Vickers Shipbuilding and Engineering Company, for the Iranian Navy. Vosper alone, however, was unable to produce craft of this size, and in 1966 a merger with John I. Thorneycroft & Company provided the shipbuilding capacity and experience to produce the larger vessels being designed by Vosper.

In 1965, Vosper decided to enter the hovercraft business and applied for a licence from NRDC to build the VT-1, a semi-amphibious craft with underwater propellers, which they eventually acquired, and launched the prototype in 1969.

Although two production VT-1s followed, it was never really accepted and in 1972 Vosper decided to convert it to a fully amphibious hovercraft, which they called the VT-2. The two production VT-1s were scrapped in 1973. The VT-2 was sold to the IHTU and in 1981 when the IHTU was disbanded, the VT-2 was sold with her two Proteus engines to Hoverspeed (the merged Hoverlloyd and Seaspeed), parked at Pegwell Bay at Ramsgate, stripped of its spares, of which the Proteus engines were the most useful, and eventually scrapped. So ended Vosper's brief and unsuccessful involvement in hovercraft.

But in 1969 Hovertravel decided to take delivery of the first VT-1. Chris Bland:

> *We bought the VT-1 but got our money back on it. It was my very early days, 1967 or '68 and for some reason Gifford said, "Yeah, it won't be amphibious, it will have a draft." Everybody at Hovertravel at that time must have had complete amnesia because Gifford said, "Oh, but we can engineer it – we can do something about that." I was by far the youngest – they were all about 20 years older than me – and I said, "How are you going to get this thing into Ryde?" Vosper was physically building it and eventually Hovertravel came out of their dream and realised it wasn't any good at all. We took it to the Channel Islands and it was a complete waste of time. We paid £350,000 for it and Vosper gave us £350,000 back, or however much we paid – I can't remember exactly. Everyone had had a complete mental blank. It was eventually bought by the IHTU and they made it amphibious.*

The VT-1 off-loading 10 cars at Southsea. The description 'semi-amphibious' referred to the fact the craft could nose up onto a beach of suitable slope to allow passengers and vehicles to embark. The VT-1 never carried any cars or passengers for Hovertravel.

Barrie Jehan:

When we were testing the VT-1 in the Channel Islands we had two wave rider buoys. One was deployed in the test area and the second was kept as a spare in a cargo shed in St Helier Harbour just in case the first one went down. We spent four hours one day passing the buoy in a square pattern situated off Jersey with all the testing gear on board. There was some wind and waves and as we changed direction the craft was recorded as doing different speeds, but the wave rider buoy was subsequently found to be showing the conditions were calm. Later that evening a very sheepish trials engineer came to us and admitted the wave rider buoy he had used was the spare one in the shed back at harbour. Unfortunately the buoy in the shed had been switched to 'operate' while the buoy on the test site was switched to 'off', so we had spent four hours going around a wave rider buoy that was not functioning – a complete waste of time.

The advantage of the VT-1 was supposed to be that water propellers were quieter and more efficient than air propellers and also that the drift typical of hovercraft was reduced due to the skegs, which acted as keels in the water. The major disadvantage, however, was that the craft was not fully amphibious.

For a brief period in the summer of 1972, the only two VT-1s built were put into service between Malmo, Sweden, and Copenhagen, Denmark, in a 270-seater configuration.

The Baltic is one area where the non-amphibious craft could have had some chance of success; the tides in the area are minimal compared with those in the Channel. Malmo for example has less than a metre tidal range[9] whereas Ryde has probably more than three metres.

Unfortunately the enterprise failed for very different reasons, their competitor, a state-operated ferry system, simply dropped its fares below cost.

Tragedies

There were only two incidents involving loss of life in the UK, one of which was on the Ryde to Southsea route. Barrie Jehan:

There was one horrible accident on 4th March 1972 on the approach to Southsea in an N6. It was a Friday afternoon, high

[9] Tidal range, the difference between high and low water.

tide had already passed and the tide was ebbing. We had a very strong south-easterly wind, which, as a local phenomenon, tends to hold the water up in Portsmouth Harbour. The tide levels in the harbour will remain a lot higher for longer, but the time comes when the pure weight of water in the harbour cannot be held by the wind any longer and it suddenly comes gushing out, resulting in a tide race. You've then got a very fast moving water stream against a very strong wind stream and you get the wind-over-tide effect, which culminates in very short steep seas. It was at this stage that the hovercraft was crossing from Ryde to Southsea in these poor conditions. There were 25 people on board and the craft was less than half a mile off shore. Part of the problem was that a wave smashed open the yaw door (puff port) thereby flooding the inside of the plenum chamber, adding to the turning moment and the craft just very slowly turned over.

Chris Bland takes up the story:

I remember it was 4 o'clock in the afternoon and I went straight over in the other hovercraft. By then the weather wasn't that terrible. Not everybody got out because, although the craft was floating stern down, the fire brigade, unfamiliar with the structure of the craft, hacked the wrong end and the craft filled with water. Had they hacked into the front end they might have all got out. No one had any injuries – the five drowned.

It happened on a Friday. Beaumonts was the legal firm and Barrings was the insurance company and they came down with a cheque on the Sunday – those were the days. They came down to see us and there was the cheque for the hull 48 hours later.

There was an inquiry and, after extensive tests by BHC, modifications were made to ensure there was no reoccurrence. Barrie Jehan:

From then on the pilot had to wear a seat belt. Weather limitations were also imposed on the craft, which are there to this day and similar to those on the N6. On the Solent we can now operate in winds up to 40 knots, gusting 45 knots – outside the Solent it is lower, 35 knots gusting 40 knots. The sea state has not been changed. It is still 1.5m significant sea state, which equates to about a 2.4m isolated wave, which is quite a big lump of water. We have the same limitations for day and night.

The incident did not have a lasting effect on the fortunes of Hovertravel with only a slight dip in passenger numbers for that year.

There was, however, just one more potentially serious incident in Hovertravel's history, described by Chris Bland:

> *We hit the Isle of Wight ferry head on in fog, but fortunately there were no injuries. We had Sir Matthew Slattery on board, who was Chairman of the ARB (Air Registration Board) making a special trip and I was on it with him. It was very scary, I tell you. The noise hitting the ferry was unbelievable. The collision was totally inexcusable as by that time we had the radar on the craft with an operator. I'm telling you the ferry looked absolutely enormous, but the noise of the propeller hitting the side of the ferry was just like a machine gun being fired.*

A New Route

The success of the Solent services – by the end of 1972, 2.5 million passenger journeys had been made – encouraged Hovertravel to open up elsewhere. Chris Bland:

> *We started Blackpool to Southport in 1973 across the Ribble, the idea being to enable people to get to and from Squires Gate, Blackpool's airport, to Southport. It was wonderful hovercraft country because the whole thing dries out at low tide and by road you've got to get to Preston to go across the bridge. We had an SR.N5 and N6 up there, but it didn't work because no one wanted to go from Southport to Blackpool.*
>
> *The Southsea-Ryde route is the only one in the UK that works. If we haven't found other routes yet, we're not going to find them now.*

The Stretched SR.N6

The clever design of the SR.N6, unlike the later AP1-88s, enabled them to be stretched from 36 to 58 seats without a major redesign and build cost. In fact, the SR.N6 was a stretched SR.N5, so basically the initial design went from 18 to 58 seats without a change in power unit. Even so, for Hovertravel, it was beyond their financial capability, but with some improvisation they got one. Chris Bland:

> *What we did was to get the wreck of our SR.N6 that had turned over at Southsea and two SR.N5s. We put them into Woodnut's*[10] *yard, the old Cushion Craft factory at Bembridge,*

---

[10] In 2011 Woodnut's was where an AP1-88–400 costing £8 million was being built for the Canadian Coastguard. The equipment on it represented a large part of the cost.

and made our first stretched SR.N6 out of those three hovercraft – BHC helped us. It was the only way we could afford to get a stretched SR.N6. Seaspeed was running two by then, which BHC stretched for them.

Ray Wheeler, Chief Designer of BHC, had been responsible for stretching the SR.N6 and ultimately converting a second stretched design into a twin engine craft. His team had come to the conclusion that most of the noise generated by the single propeller was tip speed, so by incorporating two propellers they managed to reduce the noise to a more acceptable level.

The twin prop 'Super 6' powered by a Gnome RR 1452.

As Ray recalls they were in the throes of designing it when the Iraqis ordered six, as a result of which it was designed with squared-off sides in order to enable more troops to be carried on deck.

Barrie Jehan was involved with the twin prop N6 from the early days:

I spent nine months in Iraq training pilots there. The idea was that the craft were going to be used to patrol the borders between Iraq and Iran, but when they were delivered the Iran-Iraq war had started, so our deployment there was interesting to say the least. They had a Gnome RR 1452 main engine and a Lucas gas turbine as a generator to run the air conditioning and provide power to start the main engine, except we started the main engine to provide power to start the generator!

On the converted commercial craft, 025 (at one time Hoverlloyd's SR.N6 *Sure*), Barrie commented:

The commercial twin prop we had, nicknamed the 'Super 6', was operated by Solent Seaspeed Ltd., a Hovertravel Company. When British Rail Hovercraft (Seaspeed) pulled out of the Cowes to Southampton operation, BHC asked Hovertravel to take it over as it was BHC's shop window. We ran the service from 1st May 1976 until Christmas Eve 1980 when the route closed. I remember driving the last service. The craft also operated occasionally on the Ryde-Southsea route.

A New Generation of Hovercraft – AP1-88

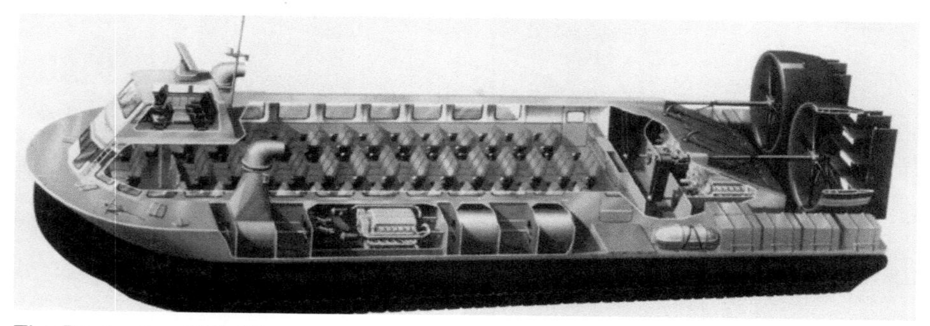

The Production AP1-88.

The jet-engined SR.N5 and N6 had served Hovertravel well in the early days, but by the end of the 1970s, as diesel engine technology advanced and the service became ever more popular, there were demands for more passenger capacity but less noise. The introduction of the first AP1-88-80, named *Tenacity* by Sir Christopher Cockerell in March 1983, proved the solution to these requirements. It was joined three months later by her sister craft, *Resolution*. When that was sold to the US Navy, a third AP1-88, *Perseverance*, made its debut on the company's 20th anniversary in 1985. By that time, over 8.5 million passenger journeys had been made across the Solent.

The AP1-88 came into existence because BHC and its customers wanted a replacement for the SR.N6 at half the build cost and with half the operational costs. Ray Wheeler, by now chief designer at BHC, was instructed to get on with it:

The Advance Project Office with five blokes in it was next door to me. Together we came up with the design of what was then AP118. The 118 was just the next number on the Advanced Project register, but Dick Stanton-Jones, the managing director, couldn't say that number for some reason or other and kept on calling it '188'. He did this in meetings whenever we were discussing it with various people, so in the end I capitulated and said, "Alright, Dick, we'll call it the 188."

In order to achieve the cost savings everything had to be simple. For example, the fans which were on the N6 and N4 were very complicated because of the centrifugal force. You have to make them very light otherwise they just accelerate, if you see what I mean. So we looked at the maximum size of simple welded fans. You know the volume you want and from there you know how many fans you need. That turned out to be four, two each side.

The AP1-88 went into service in 1983.

We'd carried out a lot of research on the noise created by the propellers, in addition to which I'd given a presentation to the Portsmouth Council. We came to the conclusion that it was just tip speed, which was by far and away the most important thing. We'd already made the twin propeller N6 to check that the noise really would be reduced and had demonstrated that noise level to the council involved in the route. We'd decided on that tip speed, but the craft was getting quite heavy by this time as a result of choosing diesel engines, because their initial cost and fuel consumption were nearly half those of the Gnome engine. We now had five tons of diesel engines compared with 500 pounds of gas turbine. So the speed, instead of 70 odd knots maximum, was down to 50 odd knots.

We also considered ducted propellers. If you compare the thrust for the same outside diameter of a ducted propeller and a free propeller, the thrust of the ducted propeller is something approaching twice that of the free propeller at zero speed. At about 40 knots they cross over and the propeller becomes more efficient – more thrust. So as we were only just a little bit above that, and we needed to go up slopes, we decided to go for ducted propellers. So that's the engines, the structure and fans. So now you've got a package to decide how big it should be. You then work backwards – it's an iterative process of guessing the weight, going through the engine size and fan size. We went round and round, and ended up somewhere with about a 100 passenger craft, which, according to our estimates did have half the initial costs and half the operating costs of the SR.N6.

There was a tripartite contract between NRDC, BHC and Hovertravel whereby NRDC had paid for the BHC design and Hovertravel was given the manufacturing contract, of which they sub-contracted the basic hull to Bill Allday of Fairey Allday at their Gosport boatyard, for considerably less than BHC could have built it. The hull was built in Gosport in 1981 then taken over to Hovertravel's associate company Hoverwork workshops at St. Helens on the Isle of Wight to be fitted out. The first two AP1-88s cost £400,000 each.

The production AP1-88 was 78-feet 4 inches long with a beam of 32 feet 8 inches, 8 feet longer than the prototype. The hover height was 3 feet. It was powered by four 428hp Deutz diesel engines, two for propulsion and two driving the lift fans. Barrie Jehan:

It was the first commercial hovercraft to have diesel engines and most incredibly it didn't have a gearbox on the props at all, but had a toothed belt and when you look at the design you think, "My gosh, that will never work." It is probably the most reliable thing on the craft over all these years.

Unfortunately, the early AP1-88[11] production craft were hopelessly underpowered, as Chris Bland explains:

---

[11] The second and third production AP1-88-100s in an 81-seater configuration were bought by SAS, the Scandinavian Airline, who operated a service on the 14-mile route from Malmo to Copenhagen Airport from June 1984 to August 1994, at which time the AP1-88s were replaced with catamarans.

The trouble with the original AP1-88 was that if BHC had told us that the thing would not go into more than a 20 knot wind over hump we would never had bought it. We weren't smart enough to work that one out until we got the thing on the route. If you got a headwind of more than 30 knots, it wouldn't go over hump and the Deutz[12] engines melted. The engines were all buggered.

The 70 ton 133 seat BHT-130 currently in service on the Ryde to Southsea route.

Hovertravel eventually went on to operate two AP1-88-100s, but after some 15 years decided to overcome the problem of them being underpowered by replacing the Deutz propulsion engines with two powerful MTU diesels. Chris Bland:

The reason for putting the extra 3 feet in the craft length in 1999 was that the Deutzs weighed 1 ton 7 cwt and the MTUs weighed nearer 1¾ tons. It was a question of altering the centre of gravity to get the cushion pressure in the right place with the skirt pressure remaining the same. The engines were also physically longer.

This was a Hovertravel modification. It was a stupid modification really because we put in an extra three feet, then

[12] Ray Wheeler's view was that the Deutz didn't produce the power used in the estimates and they were not as reliable as they were said to be either.

*put in engines 60% more powerful, but not necessarily more economical, and it didn't carry any more passengers. I always remember that when we put the MTUs in it the chaps called it the 'GLF' – 'Goes like ****'!*

In August 2005 the first aluminium welded hull, fabricated by Aluminium Ship Builders, Fishbourne, Isle of Wight, arrived at Hoverwork's St. Helens works for completion. It was the hull of the new 70-ton 133-seat BHT-130 *(British Hovercraft Technology)*, designed by Hoverwork with a hover height of 4½ feet. After sea trials and certification it entered the Ryde to Southsea service on 14th June 2007 and is still operating on that route. Chris Bland:

We still get problems with the skirt – it rips occasionally. It all gets weaker with age and keels and things come off. I remember once when the keel came right off in a 188 and nobody knew it had gone – it's lying somewhere at the bottom of the Solent.

In early 2008 the internationally diversified Bland Group acquired Hovertravel, having previously held a 25% shareholding, which dated back to the early days of Hovertravel. With two smart terminal buildings at each end at Ryde and Southsea and large modern passenger-carrying machines, it is not only a highly efficient and successful hovercraft operation, but the only one in existence in the United Kingdom.

Hoverwork

Hoverwork came into being in 1966 as a wholly owned subsidiary of Hovertravel. It was the idea of Desmond Norman, whose father started Airwork in the 1930s. He thought it would be a very good idea to fill the gap between operator and manufacturer to serve the charter market – people who wanted a hovercraft temporarily. Hovertravel's second craft was available for charter, but in the early days business was slow. In the end it did well and between the companies they had eight or nine SR.N6s employed in many places throughout the world with the staff inter-changeable between the two companies.

The first charter was to a film company that required a hovercraft for a super-charged scene in a film called *Murders Row*, starring Dean Martin. Just one craft ran the service while this was going on. Barrie Jehan:

As soon as we started Hoverwork, we started getting some work in. One interesting story was that a company called Seismograph Services Limited thought that the hovercraft would be ideal for looking for oil in what they call the 'twilight zone', which is five miles out to sea and five miles inland. There are many parts of the world where you can't use buggies, or shallow draft boats, but the hovercraft was ideal and they wanted to try out their equipment to see if it was feasible with a hovercraft. So they installed all their listening devices off the pad at Southsea and we hovered the craft over these listening devices. It was two or three hours and they couldn't get their equipment to work properly and it was only then that it dawned on them that their equipment was working correctly and the hovercraft underwater signature was so low. It was from that point on that we started getting seismograph contracts. That's when we started to build up our fleet. One of the early contracts was in the Waubonsee in Holland. I think we completed the job in a quarter of the time that it would have taken them otherwise. It is extensively chronicled and we did a lot more work in Holland around the Frisian Islands in the North Sea off the Dutch coast.

A Hoverwork SR.N6 on seismic survey work in Abu Dhabi.

In 1967 two SR.N6s were sent to Expo in Canada, the international trade fair hosted by Canada on an island in the St. Lawrence River. Hoverwork ran a service between Montreal and the exhibition site, operating for 12 hours daily and carrying 366,633 passengers during the six months that Expo '67 was open. Hoverwork was obliged to take out an Air Operator's Certificate and an Air Transport Board licence as well as satisfying Montreal Harbour Pilotage requirements by having land/sea/air radio licences issued to their commanders by the Department of Transport.

The company reportedly carried the world's first hovercraft stowaways. Two Canadian boys were found hiding under the back seat of one of the craft on the Montreal to La Ronde run. The boys had been having a free ride for two hours. Captain Peter Ayles, who was head of operations, said, "It must have been very uncomfortable under there. The kids have been going delirious over the hovercraft. They all say, 'Gee, we've never seen anything like this before.' That must have been true of many thousands attending Expo 67."

A seismic explosion behind the SR.N6. The effect as the result of a mistake is all too obvious.

In 1968 Hoverwork provided an SR.N6 and crew for the scientific Amazonas Expedition, organised by *National Geographic* magazine. The route covered 2,400 miles of treacherous jungle rivers in the heart of South America, including a 50-mile stretch of rapids never previously navigated by any powered craft. The SR.N6 left Manaus in Brazil on 11th April and, after four stimulating weeks of speeding its way through the wilds of South America, it emerged triumphant in Trinidad on 9th May.

This Hoverwork venture was not without its problems. Barrie Jehan:

In 1973 we got a Middle East contract that lasted for three years and we surveyed the whole of the Saudi Arabian Gulf Coast. At one stage we had a contract running in Saudi Arabia, which took four pilots; another one in the Torres Strait in Australia that took two pilots. We had another contract running on the east coast, which was a single pilot operation and then we had to run the Solent service as well. So we were desperately short of pilots and we were getting them from wherever we could. At that stage we had an SR.N5 in Australia, which was actually an Air Vehicles' hovercraft, which was a company owned by Charles Eden and Christopher Bland. We had an SR.N5 in The Wash, which was an ex-military SR.N5. We had two SR.N6s in Saudi Arabia and we had two SR.N6s here.

Unfortunately there was one tragic accident in the Persian Gulf. Chris Bland:

We were carrying out seismic surveys in the Persian Gulf and there was one accident in 1972 when someone wired the explosives, which were carried on the craft, incorrectly and pushed the wrong button, blowing up the hovercraft and killing 14 people. It killed five of our blokes and nine others. Fortunately, there was hardly a word about it in the English press.

It was just before the Solent accident. Barrie was out in the Gulf sorting out the dead bodies – he was picking people up and finding there was only half a person. It was horrendous. Explosives are funny things – they cut people in half – they'd look alright on the top – you'd pull them out and there would be no bottom end on them. So Barrie went through that and just blanked it out of his mind – he was there. He was there when it went bang. One of the pilots died as a result of getting burnt, but lived for about three days. Tony Smith and Barrie were the other two pilots. There were two hovercraft out there and Barrie would have been driving the second craft. We stopped the operation after that. You can still see the remains of the machine. It is a white marked SR.N6, which was one that the Army had – one that you could drive a vehicle onto – and apparently it is still in the clear water, so I am told, off Abu Dhabi. You can still go over the spot and see it sitting down there.

Hoverwork became the manufacturer of the AP1-88 and BHT-130 for Hovertravel. Barrie Jehan:

We had a contract to build a BHT-130 for the Americans for use in Alaska. We negotiated for a company in Seattle to build this craft under licence, but before we did we thought it best to separate Hovertravel from Hoverwork, on the basis that if something went wrong they would hammer the company through the courts with possible serious consequences for Hovertravel. So we completely separated the two, but with the same shareholders and very inter-changeable in all operational aspects. I am still operations manager for Hoverwork and Hovertravel and I just have a different looking business card. One day you would see pilots taking a craft across the Solent for Hovertravel and then two days later they could be operating in Australia for Hoverwork. When Jo Gaggero's Bland Group (no family connection to Chris Bland) bought Hovertravel they bought Hoverwork too. Jo was a founder member of Hovertravel. Shortly afterwards The Bland Group took over Griffon Hoverwork.

Griffon

From left to right: The Griffon 8000, 4000 and 2200.

In order to complete the story of where the British hovercraft industry is in 2012 it is worth briefly mentioning Griffon.

Griffon was started by Edwin and John Gifford, father and son, founders and directors of Hovertravel. Griffon was a hovercraft manufacturer, building commercial, military, paramilitary, rescue and bespoke hovercraft capable of carrying payloads from 840 lbs (380kg) to 22 tons (22.5 tonnes). Eventually it made sense for Hoverwork and Griffon to become one company, which they duly did, to become Griffon Hoverwork, based in Southampton in Woolston at the ex-Hovermarine site, with an off-shoot on the Isle of Wight.

One of the differences between the two companies was that they had different skirt systems. The skirt on an AP1-88 and BHT-130 is a scaled-down version of the last SR.N4 Mk.3 skirt. Although scaled down in thickness and size, the technology was very similar in terms of the shape of the BHC longitudinal and athwartship keel bags. So the larger the hovercraft the better the stability is retained in rough weather.

Griffon employs the open loop skirt, in that it isn't like a sausage, but is completely open inside. It is held together with lots of cables and wires and fingers. The fingers are attached to the skirt in a similar way to the fingers on the BHC skirt, but it is more complicated. Having said that, the Griffon skirt system uses a lot less power so it is cheaper to operate, whereas the BHC type skirt is quite power hungry, but nevertheless the better choice for the rough water environment. Skirts are not cheap, but the materials have improved and the technology has also improved over time.

Griffon currently has a contract to build 12 Griffon 8100s for the Indian Coastguard Frontier Force, which are a similar size to the AP1-88s, but of a slightly smaller and different construction. Hoverwork is building another AP1-88-400 for the Canadian Coastguard.

So without doubt, Griffon Hoverwork is a world player in the manufacture of hovercraft and the clear leader in Britain.

Chapter 8

1965 – The Birth of Seaspeed

British Railways' Reluctance

If the title of this chapter employs the usual metaphor to describe the arrival of a brand new enterprise, then it has to be added that British Railways (BR) was a somewhat unwilling mother and the birth itself was in the nature of a forceps delivery.

From the start, BR was a very reluctant player, and if Swedish Lloyd hadn't announced in the spring of 1965 their order for two SR.N4s and two smaller SR.N6s, the latter for proving trials across the Channel route, it is highly unlikely that the railway hierarchy would have entered the cross-Channel business at all.

Even then the first reaction from BR Southern Region General Manager, David McKenna, was hardly enthusiastic:

> *We have no plans to go into the hovercraft business by ourselves. We are, however, quite prepared to act as agents for the government if they so think. In the present state of hovercraft development it would be wrong for a nationalised undertaking required to pay its way to take a large risk of this kind.*

That the government of the day *was* 'thinking' was openly confirmed by Frank Cousins, the current Minister of Technology, in answer to a parliamentary question in July:

> *Discussions are at present in train between us on the possibility of a Solent ferry service being operated by British Railways. It is necessary to remember when talking of the hovercraft, as I made clear in the recent debate, they were not simply created for internal use. They were also designed to have export possibilities, and if people buy them they have to have somewhere to use them. There have to be some easy waterways in order for experiments to be carried out. We hope that the hovercraft will be a commercial success, which is why we are asking British Railways to continue its experiments in this direction.*

The Power of the Press

What was really motivating the government in this direction was undeniably the British press, most of which took delight in goading the government for having the turpitude to allow a foreigner to take charge of a precious British invention. Some were more self-righteous and jingoistic than others but the tenor of their criticism was much the same, and it is fair to say that the feeling of the majority of the general public was probably in agreement.

This short quote from the *Daily Mail* is typical:

> *Unless the government now generously supports the development of hovercraft – a British invention – Britain will proclaim herself too timid, careless or dilatory to take advantage of her own first-class brains.*

Presenting one of the few more sober assessments was the *Journal of Commerce:*

> *The fact that Swedish shipowners should be the first to give full-scale recognition to the commercial opportunities opened up by the hovercraft does not detract from the fact that a triumph has been scored by British engineers. It is not much more than a year since Sir Eric Mensforth, Chairman of the Westland Aircraft Group, made the declaration: "The fairy tale element has now gone. Now we can go abroad on real business."*
>
> *There is little to be gained in suggesting that British ship owners might have participated in this pioneering of the hovercraft in the English Channel. If they had, they might have been accused by shareholders of being influenced by sentiment. As it is we have the very efficient and very businesslike Swedish ship owners giving the hovercraft the breakthrough as a competitive form of sea transport – and that is the best boost a British product could get if it is to make an impression on the world market.*
>
> *Already there are about 20 Westland hovercraft of smaller sizes than the SR.N4 operating in areas as distant as the Clyde and Tokyo, but generally these have been regarded as off-beat services. The English Channel, with its fleet of modern car ferries, must be regarded as one of the world's most competitive short sea crossings. If the hovercraft can prove successful here, and the Swedish ship owners are confident it will, then there is no limit to the hovercraft's export potential.*

However, apart from this well-judged assessment, the rest of the public comment was decidedly anti, and no democratically-elected government could afford to ignore it for long.

By the end of 1965, in November, Mr. Cousins had more encouraging news to impart to the House, the crux of which was that, after all, not only would BR take part in the development of one of Britain's major inventions, but would in fact take delivery of the first SR.N4 off the production line:

> *During the last year a number of undertakings both in the UK and abroad have been operating small hovercraft carrying up to 38 passengers on regular services. In particular, over 100,000 passengers have been carried across the Solent in little more than three months, and the advantages of convenience and timesaving have resulted in the hovercraft becoming the preferred method of travel on this route for many businessmen.*
>
> *Licences have been acquired by American and Japanese companies to manufacture hovercraft to British designs and patents*[1]*, and other countries are starting developments of their own. But this country has by far the greatest experience of design, manufacture and operation of hovercraft, and I believe we still have a lead thanks largely to the ingenuity and determination of the inventor and the faith shown by the National Research Development Corporation and by the manufacturers and transport operators who have taken a stake in this field.*
>
> *A major step forward is about to be taken with the production of the 150-ton SR.N4 – four times the size of the largest hovercraft so far built. A Swedish consortium has ordered two SR.N4s for a cross-Channel service. I am now glad to announce that the Railways Board has decided to acquire experience of this important new means of transport, and they have accordingly obtained the approval of my right Hon. Friend the Minister of Transport to negotiate for the first SR.N4 to operate as a combined car and passenger ferry across the Solent starting in 1968.*

---

[1] By the end of 1965 Westland's justly deserved euphoria was embodied in the knowledge that it had in fact taken orders for a total of 24 SR.N5s and SR.N6s, including three SR.N5s acquired by the US Navy through Bell Aerosystems, and three SR.N4s. The IHTU had carried out numerous successful trials on the SR.N3, which included comprehensive anti-submarine warfare trials in Northern Ireland. A Far East unit had operated two armed SR.N5s for almost a year in a successful evaluation, which was to lead to the formation of a Royal Corps of Transport ACV (Air Cushion Vehicle), 200 Squadron in May 1966, in which Hoverlloyd was later to play a major role in their training. This was shortly followed by an order for three SR.N6 hovercraft.

So a British enterprise, albeit a government-owned one, would operate the first SR.N4 and national pride had been assuaged.

Ironically, as it turned out, despite all the fuss, it was always inevitable that some other entity would take delivery of the first craft to be built. The truth was that right from the start of negotiations the Swedes had declined to take the prototype. It was not generally known at the time but Torgeir Christoffersen, chairman of the Swedish consortium, had made it clear to Westland that their SR.N4 deal was conditional on the fact that they wouldn't take the first prototype craft, and so it was essential to find another operator in order that the SR.N4 production could proceed[2].

The other interesting point was that the Solent was still being considered as a viable route for the SR.N4. In retrospect it is an amazing revelation how very little those making the important decisions understood the simple basics of hovercraft economics.

Tony Brindle, appointed to set up British Rail's hovercraft enterprise recognised at once that the idea of the N4 on the Solent simply didn't add up:

> *The original plot from Westland and Don Robertson was a route from the mainland to the Isle of Wight for our N4. I looked at it and said, "This whole concept is ludicrous. You can move all the traffic for the year in a week. It won't prove anything by running an N4 about in the Solent."*

Looking at Hovertravel's annual traffic today, just short of 1 million passengers a year and extrapolating back to 1968, Tony was probably not far wrong with his estimate.

Seaspeed Takes Form

The political decision by the Ministry of Technology made in mid 1965 that an organisation would be formed, which would operate the first SR.N4 was not announced publicly. It was also considered that the finance for it would be provided by the four nationalised transport undertakings, which consisted of the two government-owned airlines, BOAC and BEA, British Rail and the Transport Holding Company. The

---

[2] This condition, for some unexplained reason, was never made known to the British management of Hoverlloyd, who at one point was hoping to operate the first SR.N4 until persuaded that this might not be a good move. In the event, the delays caused by the Pegwell Bay Inquiries would have precluded them from doing so even if they had wanted to.

latter was made up of British Road Services and the National Bus Company. Each of them was asked to put up two candidates to run this new company, with ownership equally shared between them.

Tony Brindle - Seaspeed

Anthony (Tony) Brindle entered the Royal Navy on leaving Droitwich School in 1942 and there became qualified in mechanical engineering. A desire to transfer from engineering and become a 'salt horse' was not possible within the Navy, so he joined the merchant service, initially as a 4th mate. Following service afloat and ashore, he joined a Swedish marine consultancy in 1958, which, at one juncture, investigated hovercraft with Vickers.

C. Anthony Brindle, first general manager of British Rail Hovercraft Limited – Seaspeed.

In 1962 he was invited to join British Rail under their new chairman, Dr. Richard Beeching, who decided to recruit some management expertise from outside the rail industry to assist with his reorganisation. Although assuming that British Rail's interest in him was because of his marine background, Brindle ventured that little would be achieved by the four newly-appointed London-based

'Beeching Boys' unless there was a willingness to learn something about the traffic management of an organisation with a century of running railways behind it. This resulted in him being sent to Birmingham to aid commercial development. During the following year Brindle introduced new passenger and freight services, including the London Heathrow Rail Service, and the marketing of Liner Freight trains for bulk movements.

With the introduction of the famous 'Beeching Plan' in January 1963 he was promoted to what he termed 'a deputy gaffer', with some thousands of staff under him. He achieved some notable advances in express services, in railway-owned port operations and in reducing costs. After nearly three years, Beeching departed to pastures new and the House of Lords. Sir Stanley Raymond, Beeching's deputy, took over, and instigated another reorganisation, at which point Brindle was appointed Regional Chief Public Relations Officer in London:

> *I said, "I do not want to be a PRO", but the Chairman said, "But you always seem to get us a good press". "Yes," I said, "because I can do something about it. Just being a PRO is not my scene." I was then told to do it for a bit as there was something else in the wind. I didn't know it at the time, but it turned out to be the hovercraft.*

In November 1965 Brindle was appointed to set up and run the hovercraft. Having inaugurated the Cowes to Southampton service, and later the Cowes to Portsmouth service, in addition to finalising the SR.N4 programme, Brindle was advised by the Seaspeed chairman, Dr. Sydney Jones, that a general manager (CEO) was to be appointed, and that he should remain responsible only for dealing with 'outside bodies', i.e. BHC, CAA, Board of Trade and the operation of the craft, including the introduction of the SR.N4 and sidewall HM2.

There were then two candidates formally interviewed for this new post: Tony Brindle and Don Bartlett, the latter being Dr. Jones' choice. The Seaspeed board ruled that Brindle should continue, as a result of which he was promoted to the new post with an increase in salary and status, doing exactly the same job as before. Perhaps, not surprisingly, with Dr. Jones' choice of candidate having been turned down by his own board, relations between Brindle and Jones were never going to be easy.

In November 1965, Tony Brindle was appointed as manager of an organisation yet to be inaugurated (it was established as British Rail Hovercraft Limited in March 1966), and attended an eventful first meeting:

The discussions were interesting because nobody there knew anything about hovercraft at all except me, and I didn't know very much, but I'd read voraciously everything I could put my hands on. And by this time, of course, I'd been to Westland and had a look at what they had.

At this first meeting the then General Manager of British Rail Shipping was one of the people representing British Rail. He was chief executive of the shipping service and a delightful man. He said something to the effect, "If these machines are successful it may have an adverse effect upon our traditional shipping services", which at that time was the only bit of BR that was making a profit. During the discussions he said he would expect the other three corporations to contribute to any loss which British Rail might incur as a result. There was no figure ever discussed – it was merely a principle.

At that point the then Chairman of BOAC was somewhat forthright and said, "I don't think this is for us", got up and left. A few minutes later BEA's Chief Executive, who was also there, looked around and said, "I don't think this is for us either", and he left. So by the first half of the first meeting I'd lost half my shareholders and there was still no company formed. It was now left to British Rail and the Transport Holding Company. Their chairman, a delightful chap who was also Chairman of London Transport, soldiered on for about two months and eventually he decided it wasn't for him either. So I finished up with one shareholder, which was British Rail. There was no decision ever to say this would be done by British Rail. The decision was that this will be done by the nationalised corporations, which ended up with only one of us left on the bandstand.

The directors of what was to be inaugurated as British Rail Hovercraft Limited, were appointed by the British Railways Board, with Dr Sydney Jones[3] as chairman, together with the General Manager and Chairman of the Southern Region, the General Manager of the Shipping Services, the chief accountant and two independent members.

Tony had already met Dr. Jones previously when he was interviewed for the job:

I had quite an enjoyable interview, but I disagreed strongly with one chap on the board. He turned out to be Dr. Sydney Jones,

---

3 Dr. Sydney Jones had a distinguished career, including being a ministry scientist at Malvern, specialising in radar, before joining British Rail as Director of Research, 1962-1965, and continuing as a Member of the Board of British Railways until 1976.

who subsequently become my chairman and, unbeknown to me at the time, was later to cause me a great deal of grief.

Starting from Scratch

By the time of the first board meeting, Tony's investigations in how to set up his operation had necessitated spending a fair amount of time at Cowes and at Westland's head office in Yeovil. He had even met with Les Colquhoun, who was going through much the same exercise with his preparations for setting up Hoverlloyd at Ramsgate.

Les at least had the advantage of knowing what he was supposed to be doing – i.e. setting up a cross-Channel service between Ramsgate and Calais. The aim was to use two SR.N6s to gain experience of the route prior to the arrival of the giant SR.N4s.

Tony had no idea at that stage where he would be operating from. All he knew was that he had to set up an SR.N4 service. Coming from outside British Rail enabled him to exercise some entrepreneurial flair and think outside the box, but he had to tactfully navigate his way through a Civil Service mentality to achieve his aims – not an easy task:

The raison d'être for forming a national hovercraft operation started with the N4 – nothing to do with N5 or N6. When the company was actually formed it was for the purpose of developing services for new hovercraft – not just the N4, but what might come afterwards.

The obvious route was the short sea route across the Channel, which was exactly where Swedish Lloyd wanted to go. Their motivation was to get into the cross-Channel business, which was growing fast and looked to be very remunerative (and always was), but they couldn't get room in Dover or Folkestone. Folkestone was certainly a 'no go area' because it was owned by British Rail. The Dover Harbour Board, who had one or two members of British Rail on their board, wouldn't let Swedish Lloyd get a hold in Dover either. Townsend had got in for sundry and various reasons, but Swedish Lloyd couldn't manage it, so the N4 concept allowed them to look at the cross-Channel route, from elsewhere.

Tony decided that the SR.N4 had to operate on a short sea cross-Channel route. Already a conflict of interests was beginning to emerge between the shipping division and the newly-formed hovercraft company, with British Rail Shipping proclaiming their disapproval of the idea because they saw it as an in-house competitor for their own services on the Channel:

I wasn't the most popular citizen in town for coming up with this idea, but I managed to persuade my board that I believed this was the right way to go. Quite where we would go from and to was a moot point and I looked at a number of alternatives. It wasn't only Pegwell Bay or Dover. We looked at Lydd, we looked at Le Touquet, north Kent and all over the place. I also looked at going from Brighton, but I didn't think the N4 would cope with the weather. Some interests within British Rail wanted to serve the Isle of Man.

As well as trying to decide on a suitable route, Tony had then seen the SR.N6s and was trying to formulate a plan as to how he would train the pilots for the SR.N4, which was now expected to arrive sometime in 1968:

It was perhaps January 1966 when British Rail Hovercraft Limited was planned. The Seaspeed name didn't exist at that time – I later thought up the name in my bath.

Pilot training became a major issue. I could train engineers by sending them to Westland, but where could I train the pilots and navigators? So I looked at a number of possibilities, one of which involved going up and spending a bit of time with Clyde Hover Ferries.

My first experience of Clyde Hover Ferries was when I went down to Largs to find their hovercraft. There were a few people on the beach where the craft was going from, but all the seats were being taken out. I stood there and watched as they drove some sheep on board, leaving the passengers behind. They were taking these sheep to Little Cumbrae. The hovercraft came back again, they put the seats in, got the passengers on board and off we went.

I spent a day or three there and I tried driving a hovercraft – very badly as it turned out. There were about three pilots for their one hovercraft. The idea was that we might use this as a training ground for pilots to start building up our own experience. I looked at it and thought, "I don't want anything to do with this." They were very keen of course, but it just wasn't possible.

The previous summer, 1965, Les Colquhoun, who had already been recruited by Swedish Lloyd to run the Hoverlloyd operation (although he hadn't handed in his notice to Vickers at that stage) had paid a visit to Clyde Hover Ferries and shrewdly recruited four of their seven pilots to start his operation. He had simply told them to 'watch for the advert'. It was hardly a poaching exercise as, by the time Les got there, it was clear that the operation was in some difficulty and the pilots were somewhat concerned about their future. The three remaining pilots were not prepared to move south and so they soldiered on until more or less the end.

With no experienced hovercraft pilots available, Tony Brindle then looked at Hovertravel as a possible training facility, but decided that at the end of the day he would be better off carrying out his own training. He came to the same conclusion as Hoverlloyd that the best way of gaining experience and forming a cadre of expertise to eventually run the N4s was to start small:

> *I came to the conclusion that the only practical way to get the hovercraft operation started was to set up our own company and operate an N6. We then had to decide where that company would run to and from. I looked at the Humber, Blackpool, Southport, South Wales and Cowes and decided that the sensible place would be to operate from Cowes to Southampton where we would set up a miniature service. I extrapolated an analysis of weather statistics from the Calshot and Dover Straits lightships. It was a very interesting exercise in the event, but overall, for other reasons too, it made sense to operate on the Solent.*

First Obstacles

But, as with everything else, it was not as simple as that. British Rail had a tacit agreement with Red Funnel ferries that they wouldn't encroach on each other's Isle of Wight services. Red Funnel operated the Cowes to Southampton route and British Rail operated the Portsmouth to Ryde and Lymington to Yarmouth services:

> *Tickets were interchangeable in those days, so I felt the draught from Southampton. We had an agreement and I was upsetting the applecart. I said, "Well, I'm going to do it anyway", and British Rail eventually agreed, explaining to Red Funnel, "We're very sorry, but this is not a ship". So I now had the blessing of British Rail to run from Cowes to Southampton. Now that became an extraordinary story.*
>
> *The Chairman of Red Funnel and the Chairman of the Southampton Harbour Board, as it was then, was the same man, as a result of which the Harbour Board said, "You can't come to Southampton with a hovercraft. We can't have that thing whizzing about at unbelievable speeds". We had some*

very amusing correspondence which continued for a long time, but basically it went along the lines of, "You can't come to Southampton". I said, "I am coming to Southampton." "You can't come to Southampton because you haven't got anywhere to go – you may not enter the port of Southampton." "You can't stop me – Merchant Shipping Act 1894 – port of refuge etc."

But they wouldn't let me in, so I walked from the Itchen all the way to Western Shore looking for somewhere I could land with an N6 and I found this little place which belonged to the Southampton Corporation where they were going to build the Itchen bridge at some future date. I couldn't start a service from Cowes unless I had somewhere to land near Southampton. So I went to see the Town Clerk of Southampton and said, "I want that site". He said, "We're going to build a bridge there", and I said, "You're not building it yet – you're certainly not going to build it before I move on to the N4". So it was agreed that I would rent this site off Southampton Corporation on the understanding that it was totally and absolutely in confidence, especially to Southampton Harbour Board, who only had jurisdiction over the water and not over the land.

When the Harbour Board eventually found out that I now had a landing site their determination to stop me went up a gear, particularly when I showed I would be operating a service every half hour. Their next move was to say, "There is no tonnage measurement for hovercraft; therefore we cannot raise dues upon you per entry; therefore you can't pay us; therefore you cannot come." So I wrote a very silly letter which said, "I've noted what you say, but also note under your rules that, if you have an unmeasured vessel, you may raise dues on the basis of displacement. When I come into Southampton I will be operating on the hover. The volume within the cushion of an N6 – whatever it was – on the basis of 100 cubic feet per ton – maritime practice – means I have a negative displacement of X. Therefore on the basis of your current dues it would appear you will owe me 2s 6d for every entry I make."

It was ridiculous, so the clerk to the Southampton Harbour Board, with whom I later became friends, rang me up and said, "I think we'd better meet". So we met and had a very good lunch at which it was agreed we'd pay 1d (less than ½p) per passenger. Once he agreed that we had a deal it was approved by the Harbour Board.

Seaspeed's Southampton Terminal, near the site of the proposed Itchen Bridge, yet to be built.

Finding a place in Cowes wasn't so difficult because there was the slipway at West Cowes, which was specifically designed for the *Princess* flying boat. Unfortunately it still had the *Princess* that actually flew (ALUN) sitting on it. So, one way or another, Tony was having an interesting battle at both operating ends to secure his sites:

> *I'd rented the place off Westland on which I'd also acquired an unwanted flying boat. I said, "Please will you move your flying boat," and they said, "Sorry, it's not ours. We've sold it to the Ministry of Aviation." So I wrote to the Ministry of Aviation, "Please will you move your flying boat," and they said, "a) we haven't got anywhere to move it to and b) we've sold it anyway, but the chap we've sold it to hasn't paid for it." Well, that wasn't very much help and they also added, "although the chap hasn't paid for it, he's pinched the wheels, so we can't launch it."*
>
> *I got a local chap in who suggested we would roll it into the sea on some pipes, but I couldn't move it because it wasn't mine. So this time I wrote to the Minister of Aviation and said, "Move your aeroplane. Whether you have title or not is a moot point, but move it. It's under your keeping." They didn't, so I wrote*

again and said that if this is not moved within seven days I shall take lien upon it. They didn't and I took lien. So I found myself the proud owner of the unwanted Princess *flying boat.*

Westland, who by now was caught up in the saga said, "We'd better do something about this", so we put it in the sea and it went up the river to Falcon Yard. It was eventually taken to Calshot and broken up.

Seaspeed's West Cowes Terminal where *Princess* flying boat ALUN had been parked.

Finally, with a terminal at each end, permission to operate the route and buildings purchased, Tony was looking at the organisation of the company.

The first person he hired in the staff recruitment process was a chief engineer. For the last three years of his railway life Tony had been based in Plymouth and came to know several naval personnel.

Tony had first met John Lefeaux in 1963 when John was Executive Officer of *HMS Raleigh* at Torpoint, across the Tamar from Devonport and ran into him several times after that, mainly on the London to

Plymouth train. Their paths crossed again in mid 1965. John[4], who was later to become Seaspeed's Managing Director, tells a slightly different story, but Tony Brindle recalls:

> *Following my visits to Cowes and Yeovil, it was clear to me I needed a chief engineer, who ideally was (a) a competent technical aide, because my planning, operational and organisational work load was substantial, and (b) my particular concern about the effects of salt on an aviation structure. I thought that this latter issue might be a problem and I was trying to assess the working life and depreciation rate of the craft.*
>
> *While playing trains in the West Country, I happened to meet several naval officers, some of whom were from the time of my own service. I thought, "Nobody knows more about aircraft structures in a marine environment than 'flying plumbers' (Air Engineer Officers) in aircraft carriers." I knew of two engineer commanders who were coming to their end of service and discussed my concerns with both. Both wanted a job, but one of them was promoted to captain (in Australia), so I told Lefeaux to write to the British Rail Director of Personnel, as I was confident of his help, especially as there were no such skills within British Rail.*
>
> *I certainly knew what qualifications I wanted as I was a qualified engineer myself. When John Lefeaux joined in April 1966 the organisational structure was fully planned and some posts were being advertised. His first job was to make the initial selection of the 'Hover Dollies'! He turned out to be technically very competent indeed and a pleasing working companion. We enjoyed a very good working relationship throughout our time together in Seaspeed.*

Although John was still with the Navy with a release date set for mid 1966, he greatly assisted Tony in discussing how the organisation might be put together.

It turned out that he was required earlier and, with a slight reduction in pension, was released from the Navy and joined British Rail Hovercraft on 14th April as Management Staff Grade 3. Unfortunately, the Staff Department was currently out of copies of the BR rulebook, so initially Lefeaux was not allowed to go on 'the running lines' without a qualified escort! Welcome to the world of British Rail!

---

4 John Lefeaux is the author of *Whatever Happened to the Hovercraft?*

Meanwhile, Tony had decided that he could actually start a service in June that year – 1966:

I said to British Rail, "I am going to start a service in June." "That's fine – so that will give you a year to do everything before the N4." I said, "No, it will give me two years before the N4." They said, "No, no, no you mean June next year." "No," I said, "June this year." "Oh, it's not possible." "Yes it is."

So we recruited people, but British Rail considered that I should only recruit from British Rail. We had lots of discussions about it such as where we would find aeronautical engineers with jet engine experience in British Rail. I made it clear that I'd got to have the right to recruit where I wanted. As far as pilots were concerned I would stable horses with British Rail – I would take half of my pilots from British Rail staff and I would take the other half from wherever I thought I could find them and I'd decided on the sort of people I wanted.

Aviators or Mariners

At this time there was much controversy as to the sort of skills and experience necessary to produce safe and competent hovercraft pilots. The manufacturers were firmly of the opinion that an aircraft background was required and Don Robertson of Hovertravel, being an aviator himself, was in agreement. They naturally gravitated to what they were used too, it would seem.

On the other hand, Peter Kaye, in setting up Clyde Hover Ferries decided to employ mariners, although the reason for the decision is not clear. Only Brindle, who leaned towards mariners, at least had some sort of logical explanation:

I'd made the decision that I would not take anybody unless they had a Master's Foreign-Going Ticket. The reason is very simple. I'm going to be operating the N4 on the busiest waterway in the world and if it breaks down, and please God it never does, but I'm sure it will, we simply become a vessel not under command[5] and at that stage I've got to have a Master Mariner in command if I've got two or three hundred passengers on board. We never recruited people who didn't have a Master's ticket.

---

5 'Not under command' – Nautical parlance for having no means of control.

One of the two Seaspeed SR.N6s approaching the ramp at the Southampton base.

Les Colquhoun, who had several discussions with Tony on this important aspect of hovercraft operation, was less certain. Although the other two pilots at Vickers had been mariners, Les was probably heavily influenced by his own experience and that of Sheepy Lamb, Chief Test Pilot of Westland, as to which qualification was preferable. The fact that Clyde Hover Ferries had a ready supply of trained SR.N6 pilots solved a big problem for Les quickly, but as one of those pilots recruited for Hoverlloyd, George Kennedy, said, "I don't think Les was initially 100% comfortable with the fact that we were mariners".

Although Tony alluded to the fact that the hovercraft was operating in a two-dimensional marine environment, the point that seems to have been missed at that stage was the importance of having mariners for anti-collision navigation, particularly in reduced visibility. Unlike their aviator colleagues, mariners had had a formal training in this art, backed up by several years of experience, more of which will be explained later.

Recruitment

For British Rail Hovercraft, the first pilot to be employed was a Norwegian called Hermod Brenna-Lund, who was persuaded to join by phone and telex, despite the British Rail Board's unhappiness of taking on people 'blind'. Tony had heard about him through Westland. He was a Master Mariner with 11 years sea-going experience in the Norwegian Merchant Navy, in large passenger liners, rising to the rank of 1st officer.

He also had experience in command of hydrofoils and, prior to joining Seaspeed, he was employed as a hovercraft commander by Scanhover, piloting the SR.N6, making what proved to be the longest open sea hovercraft operation anywhere in the world – 44 nautical miles. He had also been employed by the British Hovercraft Corporation to undertake hovercraft ice trials in the Baltic. Tony Brindle:

> *Everybody said this Norwegian chap was such a natural – he could make this thing 'talk'.*

Left to right: Hermod Brenna-Lund, John Syring, Peter Barr and Tony Brindle.

Brenna-Lund became their senior and training captain, and eventually the first captain to take the SR.N4 on a commercial trip across the Channel, but there is one nice little story told by John Lefeaux that is worth including:

> *We moved down to Cowes in early May and for a few days stayed at the Gloster Hotel, but soon moved to the Royal Corinthian Yacht Club as it was out of season and they had cabins to spare and staff able to give us dinner and breakfast. Brenna-Lund joined us there bringing with him his fiancée. She was given a cabin at the club also. Nowadays the fact that the housekeeper sometimes found one bed undisturbed in the morning would not attract much comment but in 1966 things*

were different and action was called for. A special licence was organised and they were married about a fortnight later in the church just behind the Royal Yacht Squadron[6].

Peter Barr and Alan Burns, both serving as deck officers with British Rail ferries, followed Brenna-Lund. Peter's story is an illustration of the misguided concept by some that aviation experience should be a prerequisite to piloting a hovercraft. Prior to applying for a job he had commenced flying lessons with an ultimate aim of taking an air traffic control position in Australia. His total flying experience was six hours when he attended the interview:

As I recall the interview panel consisted of Tony Brindle, John Lefeaux and Sheepy Lamb. One had to have been on the railways to get a job, which was sheer good fortune for me because I'd just got my Master's ticket. We lived on the edge of Portsmouth and, not wanting to go back to sea with the New Zealand Shipping Company, I said to my wife, Sue, that while I was deciding what I was going to do with the rest of my life, I'd join the local Portsmouth ferries, which would bring in enough money pro-temp. The railways owned most of the ferry routes around the coast in those days – 1966.

All vacancies within the railway system have to be advertised internally and it is only when they can't find qualified people are they permitted to go outside. They had literally hundreds of people with Master's tickets, but the only difference was I had six hours flying. I thought six hours flying was pitiful, but Sheepy Lamb thought that put me light years ahead of the other 300 or whatever applicants.

There was a bit of irony to Peter's luck:

Ian Dalziel, John Jackson, Julian Druce (all subsequently taken on by Seaspeed) and I were all on the Portsmouth ferries and in the winter time in those days you used to spend an awful lot of time on standby. We had to standby in a place they used to call 'The Hut' on the end of the pier. Ian was fanatical about hovercraft and he used to drag me down almost every day when we weren't actually on the ferry to watch the Hovertravel

---

[6] One tends to think of the 'sixties' as epitomised by 'flower power' and the 'Summer of Love' in San Francisco, but in fact these were local phenomena and had little influence on society in general until the very end of the decade and into the '70s. The rather staid and disapproving attitude presented here is far more typical of the times.

N6 just the other side of Clarence Pier. That was my only interest in it really. I was far more concerned with this air traffic control idea. As luck would have it, of course, when the interviews came up I got an interview and Ian didn't. He was the enthusiast and I wasn't, but he was about 35 and they were looking for people a bit younger. He got in on the second round because he was persistent. I'd just got Masters at the age of 28 and they were basically looking for people under 30.

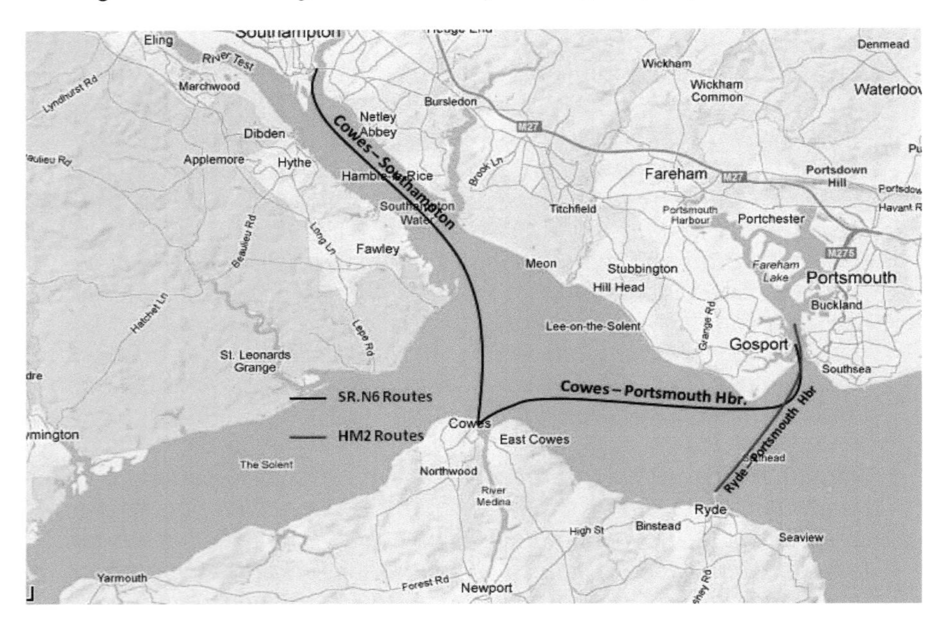

John Syring, who later became the Civil Aviation Authority's hovercraft test pilot in 1974, joined Seaspeed in the second intake of pilots:

I started my seagoing career with Union Castle from Southampton to South Africa. I became discontent, as did a lot of officers with seafaring, because, with a young family, leave was very badly apportioned and liable to be cancelled. I came ashore with a Master's ticket and worked for Marley Tiles. I was living on the banks of the Solent at Warsash and I saw this little thing (an SR.N6) buzzing backwards and forwards.

About a year later, still at Marley Tiles, I said, "Blow this, I can't stand this roof tiling business – it's not for me". I applied for a job on the British Rail ferries going from Ryde to Portsmouth and they said they couldn't get me in till next year. One day I got a phone message saying, "Where are you, you should have started last week!" – typical British Rail – I didn't start. I then applied to Townsend and was interviewed by them. I also applied to British Rail Ferries at Harwich and then I got both jobs. I thought, "Ah, Townsend 1966 – no, there's going to be a tunnel", so I joined BR at Harwich and there I was when I saw the advertisement for hovercraft.

Seaspeed's first SR.N6 – 011 departs from Cowes after acceptance trials from BHC.

My interview was with Sheepy Lamb, Brindle and probably Lefeaux, and it was Brindle of course to whom I owe my job, because he was insisting on Merchant Navy officers. It was an interview requirement that I should know about gas turbines and flight theory, so I went to the Seafarers Education Service to find out more, but all they could give me was a book on steam turbines, so I knew all about those. On flight theory the best they could do was to lend me a book, with biplanes, triplanes and zeppelins in it. At the interview Sheepy, being aviation orientated, asked me what I knew about gas turbines and flight theory. I said, "Well, I know about steam turbines", and then told him what I knew about biplanes, triplanes and zeppelins. Brindle asked me my hobbies and I said, "I sail", so tick from Brindle. Sheepy asked me what I drove. When I said an MGTC, he enthusiastically replied, "Oh great – a sports car", so tick from Sheepy. They were all mad keen drivers, so I got the job and joined in December 1966. It's amazing how one remembers little things on which half your life hangs.

The cabin of a 36-seater SR.N6.

British Hovercraft Corporation (BHC) carried out the initial training. One of the first trainees was Tony Brindle himself:

> *My instructor was Don Ellis who came to Westland in about January 1966 and I think I was his first pupil, but I also had Sheepy and Harry Phillips, depending on who was around. I collided with a tree on the first day.*

Peter Barr recalls his training:

> *Don Ellis from BHC trained me, but Sheepy Lamb did all the checking out initially. Hermod Brenna-Lund only needed updating because fingered skirts had just come in as well as puff ports, which were introduced as we were setting up the service. No matter who you speak to Hermod was the best N6 pilot I've ever seen. He could do most things with an N6 that most of us wouldn't attempt.*
>
> *I guess I had 20 or 30 hours before I went solo. We then did Sheepy Lamb's check out across Browndown Beach, a shingle beach with all the ridges and the bumps. We rattled around on there, with Sheepy cutting the engine on you and that was it – we were launched into carrying passengers on our own. We'd*

actually started to learn on the old craft without fingers and without puff ports, so the new craft with fingers and puff ports was much easier to handle.

Various things happened while we were training. The IHTU was dashing about and they overturned[7] *an N5 outside Cowes within a week of us starting. But word got round that we'd flipped over as there were always people around on the Solent watching what was going on. They saw a craft go over and they knew we were on the Cowes-Southampton route and assumed it was one of ours. I don't remember having any particular incidents in my time driving the N6.*

On 15th June 1966, Seaspeed had taken delivery of their first SR.N6 – 011 – a returned craft from Scanhover, which was the same day as 'Hovershow '66', a mini-Farnborough of the hovercraft industry, opened at Browndown, Lee-on-Solent, where Brenna-Lund took lots of people on pleasure trips for 2s 6d (12½ p).

Cowes – Southampton Underway

On 5th July, with a staff of about 30, Mr. Stanley Raymond, Chairman of the British Rail Board went to Cowes to open the service with the one SR.N6, but according to John Lefeaux he did not seem very interested in what they had to show him. It had taken just 12 weeks to become operational, which is an enormous credit to the entrepreneurial and organisational skills of Tony, ably assisted by John. The following day, Seaspeed was carrying its first scheduled passengers on the 12-mile route in 20 minutes between Cowes and Southampton. The service operated seven days a week over a 12-hour span, with a reduced timetable on Sundays, and a departure every hour from each end.

As all the early SR.N6 operators found, the SR.N6 was subject to excessive finger wear, particularly the early experimental fingers. In addition there were problems associated with the salt spray in relation to the engines and propellers, thereby pushing up the operating costs to almost unacceptable levels. The service was basically set up for training personnel and gaining hovercraft experience prior to the start of the SR.N4 services on a yet unspecified route. With the vast resources of British Rail, backed by the government, a profit was not the main criterion. Even so, with the high staffing levels and equally high operating costs, it is difficult to see how the operation could even have

---

[7] The SR.N5 was fitted with the flexible jetted skirts modified to avoid this type of capsize. It was found that undue wear of components in the region of the jet itself was the cause of the accident. The maintenance schedule timing was reduced to avoid a recurrence of the problem and indeed no further accidents occurred.

covered the basic costs. However, as Tony Brindle points out:

> *We were very much the permanent exhibition for hovercraft operation. We made extra calls at Westland's, East Cowes, almost every day carrying a constant stream of both governmental and commercial VIPs. I was also anxious to establish our own ethos. If I hadn't I might have had a whistle and a green flag wished upon me!*

Earl Mountbatten was a regular traveller to his home in Broadlands and as Governor of the Isle of Wight. The Queen made her only trip in a hovercraft to Cowes for Cowes Week in 1968. Tony Brindle is behind The Queen and Charles, The Prince of Wales, can be seen at the entrance of the hovercraft.

The second SR.N6 – 009 – the second of the Clyde Hover Ferries returned craft, arrived shortly after the first to provide welcome backup to cope with the inevitable loss of service as a result of early day technical failures. John Lefeaux noted:

> *Many components on the craft were unreliable or had short overhaul lives requiring too frequent replacement, which was not only expensive, but also hard and awkward work, often in foul weather.*
>
> *The construction of the craft and its systems and components, mainly followed quite normal aeronautical practice. Sometimes*

this resulted in components being rather more complicated and certainly more expensive than really necessary, but the real problems came from the use of these craft in an environment in which aircraft do not normally operate continually.

Unions were part of daily life from day one and Tony Brindle decided that TASSA (Transport and Salaried Staff Association) should represent everyone except the pilots, who were represented by the MNAOA[8] (Merchant Navy and Airline Officers Association). John Lefeaux commented:

As time went by I found both unions' attitudes to developments selfish and disappointing.

Seaspeed was not left entirely to their own devices and had to deal with several departments within the British Rail structure. Some were helpful and some were not, demanding time-consuming information and statistics and were not best pleased when they did not get the attention to which they felt they were clearly entitled. The Brindle-British Rail partnership was already showing signs of strain:

In the meantime British Rail had some wonderful ideas about building 'St. Pancras Station' at Southampton and I'd upset them by going to the Ideal Home Exhibition at Olympia and buying a terminal.

Dr. Sydney Jones, Chairman of British Rail Hovercraft, showed about as much interest in the project as the chairman of the British Rail Board, and was generally only seen at monthly board meetings. In fact, Jones often hindered what Lefeaux and Brindle were doing. He then introduced Don Bartlett, a civil engineer attached to the main British Rail Board, who had been one of his assistants at the Research Department in Derby, to act as a go between.

Occasionally, people arrived from the British Rail Research Centre at Derby, part of Dr. Jones' domain, with problems to solutions they had heard about, but had not bothered to see what it was all about before coming up with a solution. John Lefeaux recalls:

One chap arrived with samples of a coarse fabric developed for sheeting over open goods wagons, which he hoped would solve the skirt wear problem. He was the first to admit that it would not once he had seen the problem for himself. Another arrived with

---

8 It is interesting to note that the only union in Hoverlloyd was the MNAOA, representing the pilots. The MNAOA was also a form of insurance in that if a pilot or flight crew officer had an incident resulting in dismissal, not only would the MNAOA provide legal representation, but they would also provide a sum of money.

something to paint on the slips at the terminals to reduce skirt wear, but the problem occurred at high speed at sea.

Tony Brindle's story of the hole with a grill is worth regaling here:

When we were operating the N6 at Cowes, Jones came down one day and said, "What happens if you come in too fast and go over the fence?" I said you end up in the street, but we won't. "Ah, but you might." "Well, I don't think so. After all, all I have to do is shut the HP (High Pressure) fuel cock and the skirt collapses." The cushion decay rate on an N6 was very rapid indeed and level. We had learnt on one occasion that if you went over a large open hole, the bow came down hard and structural damage occurred.

The next thing that happened was a party came from the research centre and looked at the problems of a hovercraft having to stop urgently. I said, "There isn't a problem." However, they all went off back to Derby and three weeks later they asked for a meeting because they'd solved the problem. I said, "But I haven't got a problem." They said that I might have and they'd worked it out very carefully.

So they came back to Cowes – all these people – I'm trying to run a hovercraft service and do all sorts of other things – plan terminals, buy buildings, recruit staff, get a fiscal structure working, plus having all these drongos here saying they'd solved the problem. They wanted to dig a big hole and put a grill over it. I said, "All you have to do to stop a hovercraft is take the cushion away. Now, let's take your point. You've now got a hole and my hovercraft can't stand up in it. So what are you going to do to get it operational again? May I suggest that your solution is to fill it with water?" And they said, "Somebody has told you." I said, "Have you done the calculations of how much water is going to be needed?" They said, "No, we'll just get a big pump." "You will need a pump about the size of the one at King George V Dry Dock in Southampton to turn this craft round in five minutes." "Yes." "Where are you getting the water from? At low water it is way out there."

"I don't need this. If I want to stop the hovercraft I turn the engine off – it stops instantaneously then drops to the ground." "Oh."

This had taken five people three weeks – typically British Rail.

So even before the SR.N4 operations got off the ground there appeared to be 'clear blue water' between the free enterprise approach and attitude of Swedish Lloyd's Hoverlloyd operation and British Rail's Seaspeed rather staid and bureaucratic *modus operandi* (despite the best efforts of the capable and charismatic Tony Brindle), which was to endure throughout the time the two companies were independent from each other, and cause a clash of cultures on their eventual merger.

John Syring at the controls of a Seaspeed SR.N6.

John Syring did, however, have a major scare with the N6, which made Tony Bridle's view of stopping an N6 sound over simplistic, when anything, including a grill, might have been welcome:

> *I started training in late 1966 and got a licence for the N6. I had about 70 hours before I went solo with passengers. I remember doing my first solo from a little beach to the west of Cowes off a concrete apron, which was a wartime relic. That dreadful feeling, realising when you went out and you turned and came back you mustn't twitch the throttle to the extent that the engine failed. You knew that if you are accelerating briskly, or you didn't decelerate when coming through zero, the engine would either over speed or overheat and stop. That was the*

biggest fear, but if it did could you remember how to restart it, which is remarkably simple, isn't it?

It was about the 100 hours mark when I inadvertently put my career on the line. I was going from Cowes to Southampton with a full load of passengers and I thought I was pretty damn good as one is at 100 hours. I came up the River Itchen with a slight tail wind and I was faced with a 90 degree left turn to go up the slip. As I started this 90 degree left turn I realised I was going rather faster than I needed to be, or should have been, but it was too late to shut the throttle, as by now I was in a drift. (Had I closed the throttle at that speed with sideways drift the likelihood is that the craft would have turned over.) Under my lee side were the yacht moorings so I couldn't straighten up and go there. I had to continue this drift, continue this drift, continue this drift; more power to try to kill this drift until I went up the slipway completely sideways very briskly. I shot over the top of the slipway on to the level ground, still going sideways and stamping on the rudder until I was now pointing back out to sea but still going backwards at high speed. I put on full forward throttle until it stopped and I put it straight down. I sat there totally drained of everything. I just sat there staring at the windscreen.

The next day I was summonsed to Tony Brindle and Lefeaux's 'Court of Inquiry'. Had I not done what I did I would have gone clean through the terminal. This arrival was the most dreadful mess up you have ever seen.

Unfortunately, Geoff Howitt, the ARB (Air Registration Board) test pilot, who under normal circumstances was a charming man, was on board. He told Brindle that it was the most amazing stall turn he has ever seen, but if that pilot ever does it again he is out of a job. So Brindle said, "Explain yourself." I told him I didn't know what a stall turn was for a start, but the incident had been a complete horlicks from start to finish. I'd made a dreadful mistake and couldn't do it again to recover the situation in a thousand years if I tried.

They let me get away with it. It was a mistake, but it was the 100 hours over-confidence thing. I got away with it, not through any great skill of mine, but by the luck of the gods. It was just bad luck to have 'the man himself' from the ARB on board.

From the start of operations on 6th July 1966 to 10th April 1967, Seaspeed had carried 67,600 passengers, mainly on the Cowes-Southampton route. This was without doubt the most successful of the three routes they operated on. John Lefeaux:

Some VIPs like Earl Mountbatten, who used the service regularly between his home at Broadlands and his duties as Governor of the Isle of Wight, were little trouble, but others expected attention; I recall visits by MPs Jeremy Thorpe and Edward du Cann. However we did not begrudge the time in August 1968 when The Queen used our service to come to Cowes for Cowes Week; the only time she ever travelled by hovercraft, I believe.

Expansion and the HM2

In March 1967 Seaspeed opened a second route between Cowes and Portsmouth Harbour, tying up to a pontoon in the latter as a result of being unable to obtain a hard standing in or near the harbour. John Syring:

Tying up to a pontoon in Portsmouth was rather fun because the N6 wasn't designed to do that. Going in was fairly easy – once you'd put the thing in the water you could bobble in there without difficulty, but coming off was the problem, because you're alongside a jetty with bows into the pontoon with as often as not a beam wind. If you came astern you would drag all the way along the pontoon jetty and drop sideways round the corner probably to where you didn't want to be, resting on the corner as you went. So you either had to pick it up and hover off, which was dreadfully noisy with lots of spray, as a result of which you'd fill the Captain Superintendent's office with water, or you could try to spring off in the good old-fashioned way in a seamanlike fashion by going hard a starboard full ahead and getting the stern out. It was a difficult, messy and ungainly manoeuvre.

That service didn't succeed because people weren't used to going that way. Their whole lives were built around either Portsmouth-Ryde or Cowes-Southampton. You couldn't suddenly persuade enough people to change their habits on a slightly unreliable service.

The service only ran for two and a half years, closing in September 1969, but in April 1968 Seaspeed launched a third service from Ryde Pier to Portsmouth Harbour using the 65-seater Hovermarine HM2, which had rigid sidewalls, a flexible skirt at the bow and the stern and twin underwater propellers. The 51-foot craft was powered by two Cummins diesel engines, which gave a top speed of about 35 knots, cutting the crossing time compared with conventional ferries over the four-mile route by half, to ten minutes.

The HM2 on initial trials for Seaspeed. The non-amphibious craft was not the success it was hoped for.

Although the concept impressed Tony and John Lefeaux, there were many teething problems, including all sorts of difficulties with the skirts and propeller cavitation. The craft was also vulnerable to damage by hitting baulks of timber and other debris in the water. Peter Gray, who joined Seaspeed in 1971, subsequently joining Hoverlloyd in 1972, started on the HM2s, by which time they had been greatly improved. Even so, they still had big disadvantages over the amphibious machines:

> *They were great fun to drive once the craft was in the hover. Handling was good, with the widely-spaced propellers providing excellent directional control and good manoeuvring capability. Although the propulsion engines delivered strong astern power, approaches to the berth in Portsmouth could be a little hairy – 15-20 knots had to be maintained until the last moment to counteract the strong tides running off/onto the pontoon. This wasn't easy as the propulsion clutches required a 7-second hold in neutral while moving from ahead to astern – it often seemed a very long time. Worse still, if boat speed was a little high the water pressure acting on the stationary props would often stall one or both engines as astern power was selected – very embarrassing.*

The passenger cabin of the HM2.

> *En route problems, apart from the wakes from container ships and our own engine system failures, were mainly debris-related. Engine seawater intakes were frequently clogged and the coarse pitch, shallow draft propeller blades made of a bronze material were vulnerable to impact from flotsam. The slightest dent resulted in excessive vibrations and an out-of-service run to the Hovermarine factory on the River Itchen at Southampton for a lift-out and propeller replacement.*

Seaspeed persevered with the HM2 service until September 1972, but it was clear that the traffic wasn't sufficient to justify continuing the service. Over the winter of 1971/72, Seaspeed stretched their two SR.N6s by ten feet to become the Mk.15, which increased the passenger seating to 58, and up-rated the Gnome engines to cope with the extra payload. The SR.N6, in whatever form, was too expensive to operate to make a profit. Seaspeed continued the Cowes-Southampton route until 1976, when in May that year it transferred the ownership to Solent Seaspeed – a joint Hovertravel and BHC venture.

Returning to the last quarter of 1967, having got the initial SR.N6 services under way, attention now turned to preparations for the introduction of the SR.N4. The question was, where would Seaspeed operate from and to? With a heavy responsibility on his shoulders for making a success of the first prototype SR.N4, but with little enthusiasm from his own chairman, Tony Brindle set about his daunting task with the help of John Lefeaux. An article in the April 1967 issue of *Flight International* said:

> *"The staff are very, very confident indeed about the commercial future of hovercraft," says Mr Brindle, "and we are laying down a career pattern of the type to attract young men."*
>
> *How long this enthusiasm, so manifest at Cowes, will continue in the light of a detectable lack of enthusiasm at BRB headquarters remains to be seen.*

One of Seaspeed's stretched SR.N6 Mk.15s, capable of carrying 58 passengers.

Where was the SR.N4 to run?

After getting the Solent services operational, Tony Brindle's next and most important task was to decide on where he would run the SR.N4 to and from. The original idea of running a Solent service initially, so that BHC could keep an eye on developments, was ruled out by Tony as a complete non-starter. With only one unbuilt and untried giant craft to his name, Tony, having taken the decision to run a cross-Channel service, was keen that it should be one on which, or near where, British Rail ships were operating in the event that the one craft service failed for any reason and passenger and cars had to be transferred.

In early October 1966, Tony, John Lefeaux, Brenna-Lund and a couple of engineers set off from Cowes in an SR.N6 along the coast towards Ramsgate, passing Brighton, a possibility, but ruled out as too far to France, before making a refuelling stop at Newhaven, which had the attraction of a British Rail ferry service to Dieppe with rail connections at either end, but dismissed because it was too far from the main traffic flow from France, Belgium and Holland.

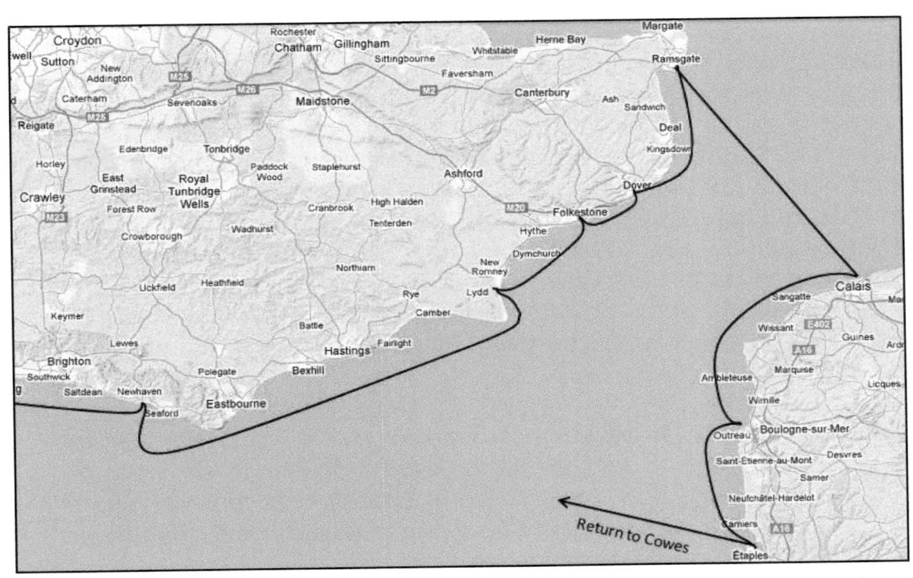

Tony Brindle's search for his SR.N4 operating bases using an SR.N6 with Brenna-Lund, John Lefeaux and 2 engineers starting from Cowes, Isle of Wight. October 1966.

Moving east up the coast, lay Lydd Airport, a mile and a quarter inland behind Dungeness, which was built in 1956 and at that time was used by Silver City Airways for a car and passenger service to Le Touquet operating mainly Bristol Freighter aircraft, carrying four cars at a time with a flight time of 20 minutes. The aircraft were coming to the end of their lives and the service was about to be discontinued. There was therefore an attraction in taking over an existing service as both Lydd and Le Touquet had rail connections.

The full extent of the difficulties in getting a giant SR.N4 from the sea 1¼ miles inland to Lydd Airport are explained in Chapter 11: *The Battle for Pegwell*. In any case, apart from the other obstacles, it was clear from the start that each storm would change the contour of the beach where the hovercraft would make the transition from water to land or vice versa, and with it the possibility that the craft wouldn't be able to get onto or off the beach. The other drawback was there was no ferry backup service.

It became clear that it had to be Dover or Folkestone on the English side and Calais or Boulogne, on the French side as they had the backup from British Rail Ferries and their French partners SNCF, which had now been renamed Sealink as a result of the amalgamation of various British Rail regional shipping services. Having recently named their operation Seaspeed, Tony and John, were flattered and amused by their choice of name and wondered what they might have called it if they hadn't given them a clue.

It was quickly apparent that Folkestone was not suitable because the harbour was too small for a giant hovercraft and too exposed in the southerly gales. At Dover Harbour they parked on the Camber, which Townsend had used for their ill-fated cross-Channel SR.N6 service the year before, but that particular area was too steep and too restricted for an SR.N4, but Dover seemed to offer the greatest potential to meet Tony's criteria. From Dover they travelled round the South Foreland and headed north past Walmer and Deal, with the Goodwin Sands on their starboard side, before arriving at the Hoverlloyd terminal in Ramsgate Harbour where they had arranged to spend a few days. Tony Brindle:

> *Les Colquhoun very much wanted me to join with him at Pegwell Bay. I was perfectly happy for Les to do what he wanted, but I was going to be in operation at least a year in front of him and I couldn't go where I would have to abandon passengers with no alternative.*

The next stage on leaving Ramsgate on 10th October was to investigate the French side. Unfortunately John Lefeaux, much to his annoyance, was called to a meeting in London, which he was obliged to attend, leaving Tony, Brenna-Lund and the two engineers to take the craft across the Channel from Ramsgate to Calais. John planned to meet up with them later that evening by catching an evening ferry.

They landed on the beach to the east of Calais Harbour at about the spot where the Calais Hoverport, capable of handling the SR.N4s was eventually built. Tony had arranged to have talks with both the Calais and Boulogne Chambers of Commerce – arch rivals in the ferry port business – and also acted as the port authority for their respective ports. Tony:

> *Boulogne was very anxious for us to go there and Calais was very convinced that we couldn't go anywhere else but Calais.*

They hadn't been on the beach at Calais long when a uniformed official associated with '*Affaires Maritime*' turned up in his '*deux chevaux*' demanding to see the hovercraft's 'papers', which in this case was the 'Permit to Fly[9]'. Unfortunately, Tony hadn't brought it with him, added to which the 'Permit' only covered the operation between Cowes and Southampton:

The official said, "Where are you going from here?" I told him we were going to Boulogne as we had a presentation to make. "Mais non, monsieur. You may go back to England, but you may not go to another point in France." Brenna-Lund and I left the craft on the beach with two engineers, we got into the official's little deux chevaux *and off we went to a hotel in Calais, where I telephoned Foxy (Lefeaux's nickname) and said, "Foxy, go to the ARB, get me an 'Experimental Permit to Fly' with passengers included, and get down here tonight – I don't care how you do it – charter an aeroplane from Southend and fly over, but I've got to be in Boulogne tomorrow morning". Foxy was amenable to that.*

In the meantime we decided to give this chap a very good lunch. I asked him what he liked to drink and he said, "Whisky". So we bought a bottle of whisky and he hit the bottle. We had a very good lunch and at about half past two he slumped in his chair. Brenna-Lund and I got him out of his chair, one each side, and carried him out of the restaurant, put him in the back of his little deux chevaux *and I drove his car back to the craft on the beach east of Calais, where we left him above the high water mark. We never did go to the authorities in Calais at all. We just took off and went straight to Boulogne.*

We got into Boulogne Harbour at low water where lots of chaps were digging for lugworms in the mud. Suddenly, this hovercraft appeared, but they didn't take the slightest notice. I said to Brenna-Lund, "Where are we going to go?" "There's a long slipway – let's have a bash at that." We went whizzing up it and parked ourselves in the car park. We found a parking meter and jokingly pretended to put a coin in it. Somebody took a picture, which appeared in Paris Match. *Of course this thing made a lot of noise, so by the time we'd been in Boulogne for ten minutes everybody knew we were there.*

Anyway, we had a very good session next day. We shipped everybody on board the craft and went down to have a look at

---

[9] See Appendix 2: Hovercraft Licensing.

the airport at Le Touquet while we were there, going up the river at Étaples and landing on a convenient slipway. Going through the surf was an interesting game. But the best place to operate from seemed to be Le Portel, a stretch of beach just on the south side of Boulogne Harbour where the Boulogne Chamber of Commerce agreed to build a hoverport.

At that point I knew that our route would be Dover to Boulogne.

Although the route length was longer than Dover-Calais there was one major advantage that couldn't be ignored:

As far as the short sea routes were concerned British Rail and SNCF, French Railways, had effectively, a long-term pooling agreement. The French end was directed from Paris by 'Armament Naval', the marine arm of SNCF, whose great white chief was M. Philippe Graf. He in turn appointed a delightful retired French naval aviator to deal with all matters relating to hovercraft, which effectively included the organisations of Seaspeed, the prefecture of Boulogne, the Bertin company, SEDAM[10], and the French Maritime Authority. He was heavily involved with us right from the planning of the N4 services.

I had a lot of discussions with SNCF as British Rail had a very close relationship with them and in those days SNCF operated a higher proportion of the cross-Channel ferry services. It was clear from an early date that SNCF would wish to put their N500 into service alongside Seaspeed's N4. So the British Rail Board had said to me, "You have to work with French Railways whether you like it or not." Our Continental marketing was therefore already in place.

The co-operation I got as a result of my intention to operate to Boulogne was enormous. SNCF was even happy to build a railway line and put a station in at Le Portel right next to the new terminal at their cost, so we had a direct rail link from Le Portel to Paris. This was very much the reason why we chose Boulogne over Calais.

Calais Chamber of Commerce, was certain that Hoverlloyd would operate to Calais, so their arch rival Boulogne was very anxious to ensure that Seaspeed operated to Boulogne, as a result of which they too were extremely helpful.

---

[10] SEDAM, were at the time constructing their N300 craft, the smaller version of their giant N500, yet to be built.

After lunch and a press conference at Le Touquet Airport, the SR.N6 crossed the Channel to Newhaven for an overnight stop before returning to Cowes the next day. And so by mid-October 1966 the decisions of where to operate from and to had been made.

When Tony Brindle decided to base his SR.N4 at Dover, British Rail regarded the decision to operate alongside their ferries as in-house competition, and therein lay much of his management problems. Even so, Tony persuaded the board that it was the right thing to do and made a deal with Dover Harbour Board that they would build a new hoverport in the Eastern Docks at no cost to him.

Now the long process started of drawing up agreements with the respective authorities at Dover and Boulogne and having his decisions ratified by the boards of British Rail and SNCF, but the decisions that had been made were kept a total secret until after the first Pegwell Bay Public Inquiry. Had this information been known before the first inquiry there would not have been the necessity for a second inquiry, and would have avoided much angst in the Hoverlloyd camp. BHC too might have been spared the indignity of their chief test pilot putting up a case as to why Dover was most unsuitable for SR.N4 operations.

But Tony found that he was just as much in a political arena as an operational one and therefore felt fully justified in staying silent until he had all his ducks in a row.

Chapter 9

The SR.N4 Takes Shape

The Final Design

It is interesting to speculate now how the SR.N4 might have fared had it been built in the early '60s, in the sequence in which it was designed. In other words, in its original 'Morphy Richards' manifestation prior to the N5s and N6s. So much was learnt during the development of the smaller craft that eventually benefited the larger, that one can't help feeling that building the N4 then would have been a disaster. For instance, trying to go through all the developments to the skirt configuration on a machine that size would have been an incredibly expensive exercise.

For the SR.N4 to be the success it eventually was, depended solely on what had been learnt in the ensuing years. How a skirtless N4 would have performed doesn't bear thinking about. What is certain is that it would never have crossed the Channel as a commercial venture.

When the structure for 001 was laid down in the autumn of 1966, BHC's designers could feel a lot more confident in what they were trying to achieve.

Without question the development of a viable skirt system changed everything; the final piece of the jigsaw put in place just in time for the N4.

A Breakthrough in Skirt Technology – 'Fingers'

Despite intensive research and development, coupled with the operational experience gained on the smaller machines, the 4-foot flexible skirt, formed as an extension of the annular jet, was far from satisfactory and had a limited life. The flapping of the jet as it passed over water at high speed caused the links and their attachments to excessively wear both themselves and the reinforced rubber walls. Initially the links were reinforced rubber diaphragms, but the longest life was achieved with chains.

The important breakthrough came with an idea provided by an HDL engineer, Denys Bliss. His solution was to form triangular sheets of reinforced rubber into a row of individual scoop shapes attached to the cushion.

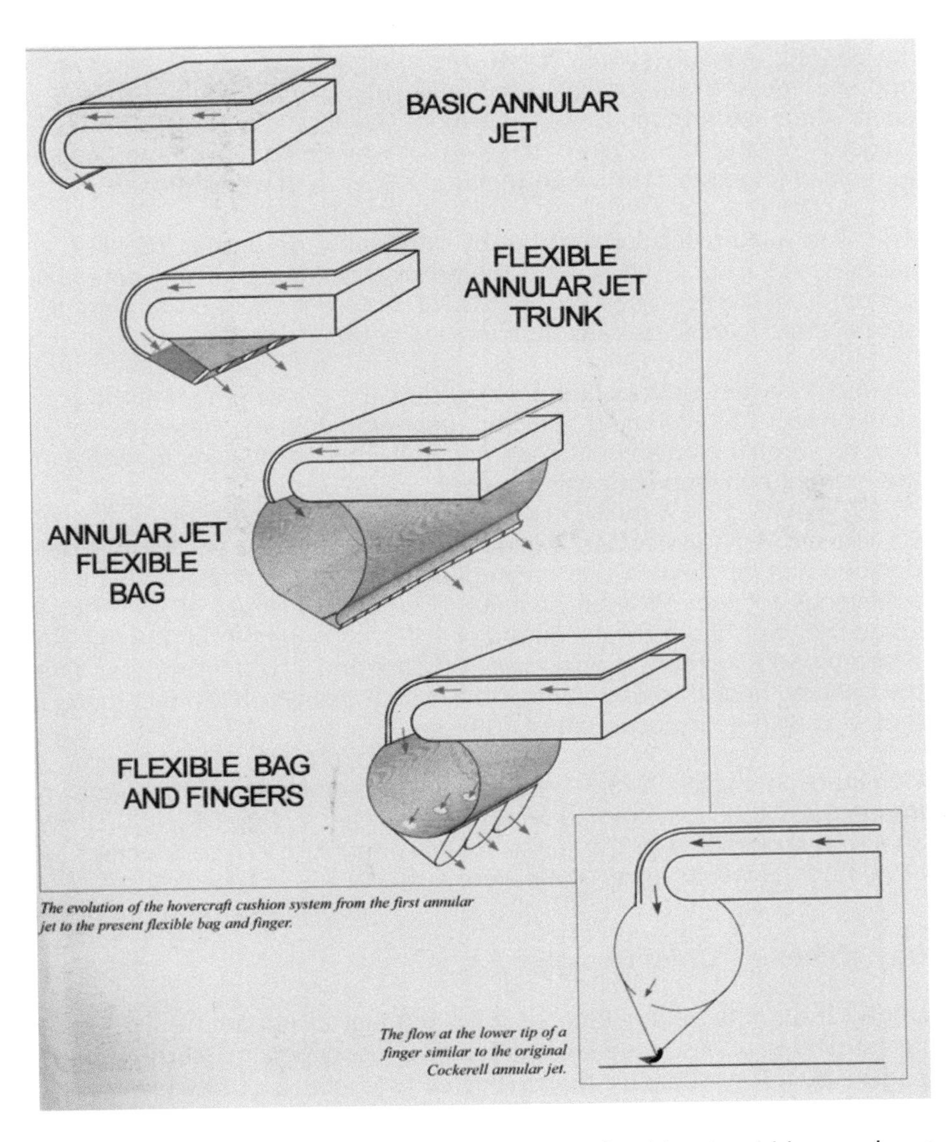

The evolution of the hovercraft cushion system from the first annular jet to the present flexible bag and finger.

The flow at the lower tip of a finger similar to the original Cockerell annular jet.

In fact, originally it was intended that the whole skirt should be made up of these segments, but the company engineers thought it would be better to hang them from the bottom of the jet. These appendages were soon to become known as 'fingers'. The inner and outer skirt were joined together and a series of holes were made to allow the air to escape through to the fingers to be deflected under the craft. The whole system thereby provided a concertina-style flexible interface with the water surface, creating a more efficient seal.

After a series of model tank tests, the modified skirt with its fingers, which represented about 30% of the total depth of the skirt, was fitted to the SR.N6[1] early in 1966. There was a slight increase in performance and sea-keeping capabilities over small waves, and a slight decrease in calm water performance. But the biggest improvement of all was the longer life of the lower part of the skirt. It was much easier and cheaper to replace fingers as and when required rather than repair the jet.

This was a vital step forward but by no means the end of the skirt and finger story; it took many more years of development, and potentially dangerous failures at sea on the SR.N4, before it could be brought to a stage where it could be reasonably relied upon.

Even before the success of the finger idea, the development of the basic skirt system had signalled sufficient improvements to enable designers to considerably modify their ideas generally, but particularly in respect to the amount and type of power required.

By the end of 1964, when design activity on the N4 resumed, it was decided that the flexible skirt conception reduced the power requirement sufficiently for four Proteus engines to be used instead of six. At this stage the craft had a displacement of 165 tons, a length of 128 feet and a beam of 83 feet with a payload of 33 cars and 116 passengers. Over the ensuing year this was refined to the final design of 130 feet long, 78 feet wide with a displacement of 168 tons.

Construction started in mid 1966 and for the remainder of that year and during 1967 this aptly named Mountbatten[2] Class giant SR.N4 001 took shape at Columbine in the same hangar in which 15 years earlier just three huge *Princess* flying boats were built.

Ray Wheeler – Chief Designer

During the same year, 1966, as a by-product of important changes to the structure of the hovercraft industry – described in more detail on page 245 – Ray Wheeler found himself promoted to Chief Designer of the newly formed British Hovercraft Corporation and, in consequence, in command of designing and building the SR.N4. If taking on something never before attempted in aircraft manufacturing, replete with unknown questions requiring previously unheard of answers, wasn't enough, Ray found himself also saddled with a number of other millstones which might have crippled lesser men:

---

[1] By 1967 all new craft were fitted with this type of bag/finger skirt.

[2] Lord Louis Mountbatten was the first person of note to take an interest in Cockerell's invention, but only in January 1968 did the BHC board announce that they had decided to name their hovercraft designs after the Governors of the Isle of Wight. The SR.N5 was named the Warden Class; the SR.N6 the Winchester Class; the SR.N4 the Mountbatten Class; and the BH7 the Wellington Class.

Ray Wheeler, Chief Designer of the SR.N4.

I had grown up with the N4 because when it was in the project office I was helping with the structural design right at the beginning with Derek Hardy working on the 'Morphy Richards'. When the chief structural engineer retired, I was pitch-forked into that. Dick Stanton-Jones was promoted to technical director from chief designer and he then gave me the job. I remember the two things I was told. First Dick said, "You've got to make sure the basic weight of this craft remains not more than 100 tons." Then Lewis Boddington, who was the group technical director, came along and said, "Ray, you are inheriting 700 engineers – you have three years to reduce that to 300 and you can't have any money to do it." Those were my two instructions. There were jobs available, so people moved from one to the next.

The serious stuff started when I took over as Chief Designer. But don't forget, three weeks after I took over as Chief Designer, the Black Arrow satellite launcher was perched on top of me as well, all at the same time. I was Chief Designer of that too. Two years later they had to appoint a new Works Director and, because he was new and not very experienced, they turned to me and said, "You can handle all the training as well".

Anyway, the SR.N4 stayed at a basic weight of 100 tons, so I was pleased with that.

The SR.N4 buoyancy tank of modular construction being built mid-1966. An SR.N5 under construction in the foreground.

The front end of the car deck with the steps leading to the front passenger cabins on either side.

The SR.N4's 19-foot Dowty Rotol propeller – the largest aircraft propeller in the world. 1967.

The platform shape was to be identical to the SR.N5, rectangular with a semi-circular bow and a length to beam ratio of 1:1.66. The platform structure of the craft was made up of 24 watertight compartments, as safety was a prime consideration in the design of the SR.N4. The fuel was contained in FPT Limited flexible bags in the buoyancy tanks in essentially the four corners of the platform. Fuel could be transferred between tanks fore and aft and across the craft in order to give a better trim, depending on the load and wind conditions.

The SR.N4 stern section showing the aperture for the rear car deck doors and the Proteus engine exhaust apertures – two either side. 1967.

The SR.N4 port engine room showing two Rolls Royce Marine Proteus engines in position. 1967.

Continuation of the assembly of the roof and superstructure of the SR.N4 001 with the start of construction of 002 in the foreground. 1967.

A 12-foot diameter centrifugal lift fan.

The lift fan air intake and the pylon mount for the front left transmission system. 1967.

Six independent electro-hydraulic systems, driven from the main gearboxes, powered pylon and fin movements and four further systems powered the variable pitch propeller mechanism.

A self-contained bilging system enabled every compartment of the buoyancy tanks to be drained independently. There was a separate battery-powered[3] system for each engine, which in the event of a total power failure, enabled them to be shut down. Five 21-inch landing pads (known as 'elephant feet') were placed on the underside of the buoyancy tanks in such a way that the craft could rest on three and remain stable. They would also rest on the jacks when lifting the craft for skirt maintenance.

Although safety requirements back then were nothing like the standard they have risen to today, passenger safety was a high priority. Ray

[3] There were four 'lift and push' switches located in front of the engine throttles, which were colloquially known as the 'Jesus Christ' buttons. When activated, an electrical pulse tripped the high pressure (HP) fuel cocks, which immediately shut down the engines. The necessity for them was that if all power was lost the engines could still suck fuel with no means of control. These switches were used just once in an incident described in Chapter 21: *SR.N4 Hoverlloyd – The First Years,* that nearly wrote off the cross-Channel hovercraft.

Wheeler's deputy, Chief Engineer Albert Weeks, said:

> *We designed the craft so that a little old lady in stiletto heels with her bustle on fire would survive the journey whatever happened to the craft.*

Reorganisation

Early in 1966, while all this intense activity was going on at Cowes, a significant reorganisation of the hovercraft industry was underway. Westland and Vickers concluded that the market was not big enough for both of them, so after lengthy discussions they decided to merge their interests and, on 1st March 1966, became The British Hovercraft Corporation (BHC). Westland held 65% of the shares, Vickers 25% and the remaining 10% were taken up by NRDC.

The first board of BHC was rather large, with Westland supplying six directors, Vickers three and two representing the interests of NRDC. At the same time Dick Stanton-Jones was promoted to technical director and Ray Wheeler to chief designer, a position Ray was to hold until BHC ceased trading in 1983.

Since the object of forming BHC was to stop British companies competing with one another for a limited worldwide market, the NRDC, through its wholly-owned subsidiary, HDL, refused to give Vosper a licence to build hovercraft. With two NRDC directors on the BHC board this was clearly a conflict of interest. It was at this point that Christopher Cockerell put himself at loggerheads with the new 'establishment' by vehemently opposing this decision, as he felt that competition was the best way of advancing technology. He told Ray Wheeler:

> *The British Hovercraft Corporation offered Duckworth, Managing Director of NRDC, and Hennessey, who was his deputy and Chairman of HDL, seats on the BHC Board. As a result NRDC refused to give a licence to Vosper to build hovercraft, thus giving BHC a near monopoly. What could I do? The only thing I could do was tell the press.*
>
> *Duckworth and Hennessey came off the BHC board; I got a letter from each of them giving me the sack; and Vosper got their licence.*

What a courageous and resourceful man. Cockerell was to receive some considerable consolation three years later in 1969 when he received a knighthood. No doubt he eventually felt somewhat vindicated because in the autumn of 1972, Westland, who had purchased the Vickers[4] interest in October 1970, acquired NRDC's 10% stake, at which point the British Hovercraft Corporation became a wholly-owned, but autonomous, subsidiary of Westland Aircraft Limited.

Mr. Frank Batterby, the BHC Chargehand, in his splendid bowler hat, which in that era was a 'badge of office'. 1968.

Further Progress

Following the March 1966 merger, hovercraft activity continued unabated. In May, the IHTU demonstrated the SR.N3 in Germany and Denmark and design of the SR.N4 started in earnest.

The first major event in which the new BHC played a major role was between the 15th and 19th June when the industry staged Hovershow '66

[4] As a result of the merger of the Westland and Vickers interests no further Vickers designs were built, although their designers did participate in the modification of the SR.N5 and SR.N6 craft for specific customers, including the Army's 200 Squadron craft. In addition, these craft and the first three BH7 military craft were built at the Vickers Southampton (Itchen) factory. On completion of these craft no more work was available and the Itchen plant was sold.

at Browndown on the Solent. The show was opened by Lord Louis Mountbatten and some 40 exhibitors took stands. In addition to BHC there were craft and models produced by Britten-Norman, Denny's and Hovermarine, the latter two being involved with sidewall hovercraft, as well as several exhibitors displaying equipment associated with the hovercraft. The Ministries of Defence and Technology were also represented and, as already mentioned, British Rail Hovercraft was running pleasure trips from the beach. It was indeed a showcase for the hovercraft industry, ably organised by Douglas Hammett, a Chartered Engineer, who was BP's[5] Hovercraft Co-ordinator, and attracted a large number of British and international visitors, both military and commercial.

Hoverlloyd's 003, *Sure*, far left, nearing completion next to Seaspeed's 004, *The Princess Anne*, with buoyancy tanks laid and the first military BH7 under construction. 1969.

Prior to the event, however, Hansard records an exchange in the House of Commons between Mr. David Crouch MP, and the new Minister of Technology in the Labour government, Peter Shore, in answer to a question:

---

[5] BP was a generous supporter and sponsor of hovercraft activity during the period. As well as being clearly seen on the side of the SR.N4s, the BP shield was also a familiar sight at small amateur meets.

Most of the cost of the Hovercraft Exhibition will be borne by the industry, but Hovercraft Development Ltd. – the subsidiary company of the NRDC (National Research Development Council) – will be spending £4,000 on its exhibit. Several departments are co-operating in the exhibition, but the only expenditure which it has been possible to estimate would not amount to more than £2,000.

Hardly a sum suggesting whole-hearted support for the industry.

It was probably no coincidence therefore, that the Ministry of Technology announced in the same month the intention to purchase two BH7 single-propulsion and lift-system military hovercraft.

BH7 Mk.1 on trials with the Royal Navy. 1970.

The BH7's origins started in a 1965/66 feasibility study by Westland (commissioned by the Ministry of Aviation) of hovercraft in the role as both a fast patrol boat and logistical support craft. After consultation with the Army and Navy, the Weapons Development Committee accepted their proposals and, in 1966, announced their 'intention' to place an order for two 50-ton BH7 single Proteus-engine hovercraft to be built, one for each role. In the final analysis the BH7 was a quarter-size SR.N4, with as many compatible components as possible being built into it. It may be cynical to suggest that the possibility of a government order was in some way connected to its miserly contribution for a major exhibition of Britain's latest technology, but nevertheless there always seemed to be a battle for unqualified support for the hovercraft in government circles.

A few months later, design of the craft commenced but problems with the SR.N4 had delayed the BH7[6]. Eventually the first one was launched in December 1969 and delivered to the IHTU in September 1970. The Iranian Navy ordered and took six BH7s.

In 1975 the IHTU was re-commissioned as the Naval Hovercraft Trials Unit (NHTU) but was disbanded in 1985 on the basis that their exploratory venture had reached a conclusion. In other words the defence budget had been cut yet again and the hovercraft was way down the list of priorities for keeping. The role of the IHTU and the NHTU and what they did and achieved would cover a book on its own, but suffice it to say that it was a big disappointment for BHC, whose parent company, Westland, was in financial trouble at that stage. Inevitably it was another nail in the coffin of the development of the hovercraft.

Even in 1966, despite the huge advances in development, it was clear that there was still some uncertainly both in government and private enterprise circles as to what this new splendid contraption could practically be used for, plus grave concerns about its economics.

In the USA, Bell Aerospace decided to establish a production line for their military version of the SR.N5, which they called the SK5. Three well-armed versions of this craft were to see service in Vietnam with the United States Army, together with three much-modified BHC-built craft, which were in the service of the US Navy.

To What Future?

In October 1967, 001 was wheeled outside on rails in front of the press and dignitaries with one notable exception – Christopher Cockerell. He and his wife had once again been excluded from this historic occasion. On 20th November, 001 made her first engine run. Eleven weeks were spent on extensive ground-running checks where faults were found and remedied. On a chilly 4th February 1968, SR.N4 001 was launched, and again Cockerell was snubbed by the board of BHC. Instead, he watched the fruits of his invention take to the water from the Seaspeed Terminal on the other side of the river.

What would the future hold – would this be the start of a large SR.N4

---

[6] Four mark versions of the BH7 were built.

production line with worldwide orders, or would it go the same way, as some cynics thought, as the *Princess* flying boats? The answer lay somewhere in between.

The SR.N4 having been wheeled out of the hangar on rails, without its skirt, bow or stern doors. October 1967.

PART TWO

Cross-Channel Adventures

Chapter 10

Cross-Channel Adventures

The 'Sixties'

The sixties were not just a time of radical social change, innovations and advances in technology, space exploration and in the arts; it was a time of enthusiastic expansion in other areas too. No less so than in the growth of tourism.

The decade had just seen the inauguration of the 'package tour'. Seemingly overnight the average British family could, for a mere £30 or £40, jump on a plane and spend a week in sunny Spanish resorts like Torremolinos and Benidorm; and before long other brand-new holiday locations rapidly followed. Hotels by the dozen shot up in quick succession and inevitably the new and lucrative market eventuated in a number of crises, driven by a certain amount of over enthusiasm on the part of the proprietors, and exacerbated by a lack of builders to cope with the work. It was not uncommon during this period for some visitors to be moving in as the builders were moving out.

Those were often the lucky ones; the British papers of the time were full of horror stories of hotels minus plumbing, lighting or even worse, foundations! Despite the journalists' field day, the exponential rise in Brits clamouring to get away to lie in the sun blithely continued. Guaranteed sunshine was the thing, a lot better than sitting around in your plastic mac in the drizzle at Blackpool or Margate.

By the end of the sixties, when we hoverers appeared on the scene, the second phase of this UK summer migration was well underway. What the new adventurers travelling to the mysterious 'conteenong' had discovered by this time was that, contrary to the mythology, it wasn't half as 'foreign' as they thought. You could find plenty of people who spoke English, foreign currency wasn't so hard to understand and what's more, beer and fish and chips were available. The latter showing that, when it came to judging their market, the Spanish hoteliers were pretty much on the ball.

The upshot of this new confidence in venturing abroad was the thought occurring to Mr. and Mrs. Average, "Why not take the car next year?" In no time, long lines of vehicles were lining up at the Channel ports, full to the brim with people and luggage.

In subsequent years if French cartoonists wanted to depict *les Rosbifs* on holiday, invariably they were drawn against a Mini Minor, down on its axles, its roof-rack straining under a *Leaning Tower of Pisa* pile of luggage. Usually standing nearby, a couple; he, very seventies, long sideburns, *viva Zapata* moustache, pot-bellied in vest and braces, topped off with a knotted handkerchief; she of 'fuller figure' proportions straining the capabilities of a short sun dress and wearing an over-large straw sunhat.

Funny – but like all good jokes not so far from the truth.

The Cross-Channel Boom

By the beginning of the seventies, it was said that if you rowed a tin-can raft into Dover Harbour, someone would have driven a car onto it. To this day the cross-Channel market from Britain to Europe is the largest in the world, in both volume and value. Where else is there an island of some 60 million reasonably well-off people, all of whom are screaming to get off to the sun each July to September?

It was undoubtedly a lucrative and growing market and two shipping companies, Swedish Lloyd and Swedish America Line were eager to enter it, but there was a small problem.

In 1965, from the Swedes' perspective, the principal cross-Channel port at Dover was already filled to capacity with the British and Continental rail service ferries plus the only private operator Townsend – both of whom were not inclined to allow another aggressive shipping company an easy entry into the market.

As early as March 1965 in a joint venture, Swedish Lloyd and Swedish America Line had started negotiations with Westland Aircraft Limited to enter the cross-Channel market with a new "150-ton hovercraft for an 80-mph cross-Channel service between England and France", so reported *Lloyd's List*. The national press also took up the story with the *Daily Mail* venting its disapproval:

> *In 1964 the Norwegians snatched a lucrative cross-Channel ferry service from British Rail, who said it was 'uneconomic'. That was bad.*
>
> *What is worse is to learn that a Swedish firm plans to run cars and passengers across the Channel by hovercraft. For this is a British invention and we are years ahead of anyone else with it. So we were with vertical takeoff and swing-wing planes, but the*

Americans walked off with both and started large-scale production. Now we have to buy them back from the US.

If this is allowed to happen with hovercraft we shall really begin to despair of the British will to go ahead in the modern world.

In early April the *Financial Times* ran the story following an interview with Swedish Lloyd's Managing Director, Mr. K. Andersén:

Mr. Andersén said that he could see no objections to its operation in Britain, and considered that the potential market for a fast cross-Channel service was extremely large and would continue to grow. So far he was not aware of any other company competing with his own, but no doubt if and when Swedish Lloyd started the service other companies would follow.

Mr. Andersén confirmed that the cost of the 150-ton hovercraft[1] *would be in the region of £1.2 million. Fares would be about the same as charged for a normal crossing by boat, but the journey time would be very much less. However, he did not rule out the possibility that prices could be reduced at a later date.*

He did not think that there would be any early possibility of a hovercraft service linking Britain with Scandinavia. At present the craft was more suitable for comparatively short runs, high density traffic, and provided a very rapid turnaround time.

The significance of this was that, although the Swedes might not have got everything right, they certainly appreciated that the craft was only likely to be commercially viable on high-density short-sea routes, which subsequently proved absolutely correct. From the beginning they had intended to operate from Ramsgate Harbour. The idea would be to start in spring 1966 with two Westland SR.N6 38-seater hovercraft operating between Ramsgate and Calais for route proving and information gathering. These would be followed in 1968 with a giant SR.N4.

The *Daily Telegraph* had got wind of possible competition and noted:

The possibility of a Swedish shipping line starting a hovercraft service between Ramsgate, Kent and the French coast may

---

1 The All Up Weight (AUW) of the SR.N4 Mk.1 was variably quoted, anything from 165 tons to 175 tons. At this juncture in 1966 the most commonly quoted number by the press was 150 tons.

speed a decision by the British Railways Board on whether to operate a hovercraft on one of its well-established cross-Channel routes.

British companies have been reluctant to place an order, even though the National Research and Development Corporation (NRDC) is prepared to assist financially. Southern Region has been discussing the possibility for several months. Recently the matter was sent to the British Railways Board for a decision.

The NRDC was in fact prepared to loan money to Westland to build the hovercraft – they were not prepared to directly subsidise an operator. It would appear that pressure was indeed brought to bear by the government on a reluctant state-owned British Rail to purchase a craft. Some degree of sympathy could be expressed for their reluctance because British Rail was a ferry operator and they could see that with a hovercraft they could end up competing against themselves, as well as with this new Swedish company.

Following the announcement by Swedish Lloyd and Swedish America Line to start a cross-Channel hovercraft service, Townsend Car Ferries, the competitor to British Rail at Dover, also got in on the act with the announcement that they would acquire a £100,000 SR.N6, capable of carrying 38 passengers, for a pioneering Dover-Calais service. If successful, they too would order a giant SR.N4. Westland hailed the announcement as, "Townsend's decision is the breakthrough we have been waiting for."

So a bold step was taken with the formation of Hoverlloyd Limited in November 1965 and recruitment commenced.

Les Colquhoun

Any history of Hoverlloyd as a company cannot ignore the contribution of Les Colquhoun, its first operations manager and managing director. Moving on from what had already been a distinguished career in aviation he oversaw from the very beginning the development of the company with a clear vision of what was needed and the determination to make it happen. What became an enterprise to be intensely proud of, producing a highly-motivated workforce, achieving a level of dedication and a standard of presentation second to none, can be attributed solely to this one remarkable individual. Despite his short tenure as MD, moving on in 1971, Hoverlloyd was indeed lucky to have him during those vital formative years.

LES COLQUHOUN, DFC, GM, DFM

Les Colquhoun, Managing Director of Hoverlloyd from 1966 to 1971, was a pilot with a distinguished career in the RAF during World War Two and later as a test pilot for Vickers-Armstrongs. He was in all respects a most remarkable man, a natural leader, through whose quiet encouragement the company's unique management style was initiated and nurtured.

In 1941, Colquhoun had being flying fighter sweeps over northern France with No. 603 (City of Edinburgh) Auxiliary Air Force Squadron when he was detailed to deliver a photographic reconnaissance Spitfire to Cairo. During a stopover in Malta it became apparent that the situation there was desperate. He and his Spitfire were seconded to No. 69 Squadron and, so hard-pressed were the island's defenders, that for a while the squadron comprised only Colquhoun and one other pilot. For nine months during 1941 and 1942, Colquhoun piloted an unarmed and unescorted pale blue Spitfire out of Malta, flying over Italy and assessing the enemy's position. Travelling at great speed and using considerable guile, he would continuously out-manoeuvre the enemy, before landing back at Luqa with his anxiously-awaited photographs.

Not until May 1942 was the squadron reinforced by Spitfires transported in the aircraft carrier *Eagle*. By then, Colquhoun had survived 154 operational sorties, each of which had been flown mostly over the sea, a risky business in a single-engine Spitfire. Following his exploits in Malta, Colquhoun returned home for Mosquito operational training and in 1943 he joined No. 682, a Mosquito photographic-reconnaissance squadron based in Algeria and operating over Tunisia and Italy, moving to San Severo in Italy in September 1943.

He remained with 682 Squadron until October 1944, during which time he flew 82 operational trips over enemy territory involving 262 hours of operational flying. He was awarded the DFC for his photographic contribution to the successful conclusion of the North African and Italian campaigns and his part in the preparation for the landings in the south of France during the summer of 1944.

In 1945, after a short period as an instructor, Colquhoun was posted to Vickers Supermarine where he joined the team of production test pilots. Here he tested the later marks of Spitfires as they rolled off the production lines. He also displayed a special aptitude for putting new types through their paces, particularly the Attacker, an early carrier-borne naval jet fighter. In May 1950, he was flying an Attacker when the outer tip of his starboard wing folded up. The ailerons locked and Colquhoun began to lose control of the plane. Coolly, declining to eject, he stayed put in the hope of discovering the cause of the fault. Flying by rudder alone, he managed to bring the speed up to more than 200mph. Although this was about twice the Attacker's landing speed, he was able to land at Chilbolton, Hampshire's 1,800-yard runway, with the length of a cricket pitch to spare. By putting his life on the line, Colquhoun, as the subsequent investigation revealed, had enabled the fault to be identified. For that particular exploit, not surprisingly, he was awarded the George Medal.

At the start he was particularly concerned that if we were to run this impressive new concept in cross-Channel transport it should have impressive onshore facilities to match. Moving firmly away from the 'few sheds on the beach' paradigm, he insisted that the hoverport should present itself no different to any other commercial airport carrying passengers overseas.

He tackled the difficult negotiations with the various opposing arguments with determination and diplomacy and at the same time ensured the Swedes did not scrimp on the costs of the infra-structure. The end result was a magnificent hoverport that in presentation and facilities more than rivalled anything available at any domestic airport in Europe at the time. It was an achievement to be proud of.

There was a downside however. Les Colquhoun's vision had not been shared by his Swedish backers and therefore was not included in the initial budget estimates. It is likely that, like other initial purchasers of hovercraft, the directors were taken in by BHC's view of simply running off the beach with the minimum of infrastructure. They probably imagined starting the service with a few demountables and temporary structures at a minimal cost.

The actual expenditure was far from minimal. The unforeseen infrastructure cost of the building – well over £1 million – was a major contributing factor to a massive budget deficit over the first two years of the SR.N4 service. Something like £2 million was rumoured as the total blow-out figure – in today's terms a staggering £22 million! During those first two years of operation we lived with the constant expectation of closure. Every day we drove into the hoverport at Pegwell Bay, heart in mouth, expecting to see chains on the gate and the great adventure at an end.

The budget deficit undoubtedly led to Colquhoun's resignation in 1971. Nonetheless, despite the heavy price paid, Les Colquhoun's bold and courageous vision was eventually justified. Hoverlloyd could not have expanded and prospered through the ensuing years without his early determination to get things working in the right direction.

But we move ahead of ourselves. Long before the arguments concerning the type of building at Pegwell, initially the plain fact was there was no guarantee the company could use Pegwell at all.

The SR.N6 Days 1966 to 1968

The two SR.N6 craft for Hoverlloyd's initial pilot operation were delivered early in 1966. At the same time, the four experienced SR.N6 pilots from the now defunct Clyde Hover Ferries were joined by three new recruits, namely David Wise, Ted Ruckert and Bill Forsyth. After a short SR.N6 technical course at Cowes and about 20 hours practical training on the craft itself at Ramsgate, the three new men were ready to go solo.

A special pad inside Ramsgate Harbour had been built as the operating base and, at Calais, a hard standing was put down on the west side just inside the harbour entrance, but whoever designed the ramp at Ramsgate had little idea of hovercraft operations, because it was shaped like an upturned saucer, so that unless the craft was driven straight up it was liable to slide off.

Hoverlloyd's SRN.6 *Swift* arriving at Ramsgate Harbour from Calais. Hoverlloyd's colour scheme was originally blue. It was only when Seaspeed started at Dover with a similar colour arrangement the decision was made to change to red. It could be said that Seaspeed had done Hoverlloyd a favour as the red and white was a lot more striking. 1966.

The landing pad still not complete when the SR.N6s arrive at Ramsgate. 1966.

Inaugural ceremony of Hoverlloyd's operation from Ramsgate Harbour. April 1966.

The N6, with its single propeller was not nearly as controllable as the N4 later proved to be and the newcomers found it a handful to learn to drive. Apart from the basic difficulties of control, they soon learned that plough-in was a real danger. If nothing else it was at least a pointer to what to expect of the larger and heavier N4. David Wise, who eventually became Associate Director of Operations, remembers it well:

> *The SR.N6 was quite challenging and difficult to control in limiting conditions. We could wind the craft up to about 50 knots in calm seas, but the biggest danger was a high-speed 'plough-in' when the front of the skirt would 'tuck under' making increasing contact with the water. The drag at the front end would tend to lift the craft at the back end during the quick deceleration – an alarming experience when it first happened.*

As David also points out, it was never intended that this much smaller craft would run a regular service to Calais; it was too small to cope with anything but the mildest of Channel weather:

> *We had a 20-knot and 4-foot sea operational limit, and therefore cross-Channel operations were only undertaken during the summer. The problem was that the weather situation could change*

quite quickly and there were occasions when one could get to Calais, but not back again. It was never intended as a commercial operation, and for all its limitations it did help us to build up valuable cross-Channel operating experience in preparation for the arrival of the larger craft.

Welcoming function in the Mayor's Parlour 1966 – 'The Originals'.

Front row left to right: Roy Mortlock, Bill Forsyth, Bill Williamson, Emrys Jones, Les Colquhoun, Mayor of Ramsgate (Mr F.R. Smith), Tom Wilson, Cyril Nurthern, Jim Weston, John Bartlett, Maurice 'Monty' Banks.

Back row left to right: Terry Halfacre, unknown, David Wise, George Kennedy, unknown, Ted Unsted, Ted Ruckert, unknown.

Obtaining the 'Operating Permit'

Before passengers could be carried for 'reward or hire' it was first necessary to obtain a 'Permit to Fly', the issuance of which in the early days was solely the preserve of the Air Registration Board (ARB). Clyde Hover Ferries and Hovertravel had already been through this arduous but necessary exercise and now it was the turn of Hoverlloyd, followed a few months later by Seaspeed for their Solent operation.

Even at that stage it was becoming increasingly clear that special legislation for hovercraft was required, for no better reason than this new form of transport had aviation, maritime and road vehicle characteristics, none of which individually applied to the hovercraft. How the matter was eventually resolved is described in *Hovercraft Licensing (Appendix 2)*.

But at the time, Les Colquhoun and Emrys Jones, his chief engineer, were engaged in a time-consuming exercise, mainly in accordance with ARB regulations, setting out in detail the qualifications of the staff, how they would be supervised and organised, and how their work would be inspected and recorded. In particular, the engineering organisation had to be minutely detailed, but the mitigating factor was that it was a one-off task.

Tom Wilson and Bill Williamson, the two senior SR.N6 captains, who came from Clyde Hover Ferries, and commanded *Swift* and *Sure* on the inaugural press trip to Calais on 6th April 1966.

Swift and *Sure* on the completed pad at Ramsgate Harbour. 1966.

Press boarding *Swift* for the inaugural press trip to Calais, April 1966. Les Colquhoun on the ramp. The craft still had an old skirt with no fingers.

Calais Hoverport just inside the western breakwater of Calais Harbour. 1966.

The Start of Operations

Having obtained the 'Permit to Fly', the first scheduled service to Calais was commenced on 30th April 1966, but prior to that, on 6th April, a publicity run was made by both craft to the new Calais Hoverport, carrying 70 VIPs and the press. The weather was not ideal according to the journalists' admittedly inexperienced accounts. The sea-state was rough on the outward journey but comparatively smooth on the return. Describing herself as the only woman of the party, Kathleen Welsh of the *Daily Telegraph* verged on the poetic:

> *To the roar of engines and in a huge flying curtain of spray, I set out with 69 other pioneers yesterday on a journey across the Channel.*
>
> *On the outward run, from Ramsgate to Calais, it was a bumping ride on a flying switchback as we flew on a cushion of air over waves five to six feet high.*
>
> *On the return trip our craft behaved like a gentle-moving rocking chair, skimming across the sunlit ripples. Such are the moods of the Channel.*

One cannot help wondering if the mood of the passengers was not also aided and abetted by the slap-up luncheon in the Hotel Sauvage before the return.

No official record by the crew now exists, but the crossing times were estimated by the passengers to be one hour and 20 minutes to Calais and a little less than an hour back to Ramsgate. Under the headline, 'Channel Hovercraft make light of heavy going', David Fairhall of *The Guardian*, someone who obviously had a nautical background, talking authoritatively about the dangers of the Goodwin Sands, probably had a more accurate report:

> *For although the Force 3 to 4 winds were comfortably within the licensed limit for this 38-seat Westland N6, steep 6-foot waves piled up by the tide flowing over the sands gave us a distinctly uncomfortable, if exhilarating spell.*

Although this must have been an anxious day for the company, the largely impressed reports from journalists, commenting on how well the N6 dealt with the weather, was a bonus and worth the risk.

During the summers of 1966 to 1968 the two craft, named *Swift* and *Sure,* would generally make cross-Channel trips, but if the weather was unsuitable, short pleasure trips were operated from Ramsgate to the Goodwin Sands and from Southend.

Sometimes 'pleasure trips' were not quite the pleasure they were meant to be, as Ted Ruckert relates:

> *My logbook records that on 25th May 1966, the first year of the SR.N6 operation from Ramsgate Harbour, we set off for Calais with George Kennedy in command and myself as navigator. The weather was not good and near the East Goodwin Lightship a large wave hit the craft and smashed both front windows.*
>
> *I had my head in the radar at the time, and the next moment a huge plate-glass window crashed onto the radar and we were ankle deep in water. The passengers were not amused, but the good old Rolls Royce Gnome gas turbine engine kept running, even though the control panel had been deluged.*
>
> *When we got back to the beach, adjacent to Ramsgate Harbour, it was almost dusk and we were all very cold – teeth chattering etc. An old lady staggered down the beach, stuck her head through one of the broken windows and said, "Are you doing any more pleasure flights today?"*

Despite these occasional mishaps the N6 operation proved to be a popular draw-card for holidaymakers and locals alike. At low water, when large areas of the shallows would have dried out, trips to the Goodwin Sands were also a popular pastime. Even this comparatively safe excursion out to the sands in calm water could have its problems. Ted's following story illustrates that not only were the first employees learning to handle a new concept in travel, they were also only just embarking on an understanding of what the responsibilities of handling large numbers of passengers were all about:

A potential SR.N6 recruit, but the pilots couldn't get her to handle the joystick properly!

I remember the night I took 36 children with their teacher out to the Goodwins at low tide for an 'environmental lesson'. (It wasn't called that then!) The weather was fair but deteriorating and by the time we got back to Ramsgate it was raining and blowing a hooley.

That night I lay in bed listening to the storm, wondering whether a poor soul was wandering around on the sands looking for us as the tide came in. Fortunately not – but there had been no head count and they had scattered in all directions.

I had no sleep that night! Health and Safety today would have had a ball!

Hovercraft channel crossings from

RAMSGATE TO CALAIS

and Hoverlloyd pleasure trips

NOW RUNNING Make sure that you are one of the first to experience this new travel sensation

---

DEPARTURE TIMES FROM RAMSGATE Please note these new times

(All the times shown are approximate and subject to revision without notice. The times underlined indicate the basic regular service. Craft will depart at the other times shown according to passenger bookings)

8.00 a.m.
9.00 a.m.
10.15 a.m.
11.30 a.m.
4.00 p.m.
6.00 p.m.

Pleasure trips (Duration 10-15 minutes) whenever craft are not engaged on the Cross Channel Service they will be available at the Hoverport to run pleasure trips.

INFORMATION FROM THE BOOKING OFFICE, HOVERPORT, RAMSGATE PHONE: THANET 53701

Day trips operated from an allocated patch of beach adjacent to Southend Pier. A 15 minute trip cost 10s (50p).

Standing by the 'ticket office' left to right – Jim Weston, Engineer; Ted Ruckert, Pilot; Pat Lawrence; Office Staff. Summer 1967.

An excursion to the Goodwin Sands at low tide in an SR.N6. 1967.

It wasn't only the flight crew who were getting used to a new set of experiences – the engineers were having their fair share too. One of the first of the engineering staff to join the company was Mike Fuller, who remained as Crew Chief up until the Hoverspeed merger. His experience of spending the night on the Goodwins is the sort of thing that no one would be likely to forget:

During the summer of 1967, Hoverlloyd operated two SR.N6 hovercraft on cross-Channel trips, running four round trips a day.

I reported to work one evening in June at 8 o'clock as usual expecting to see two craft on the pad after the day's operations, ready for the night-time inspections and maintenance. But only Sure *was there as* Swift *had broken down on its way back from Calais and was now sitting on the Goodwin Sands, eight miles off Deal.*

The two crew members, Captains Tom Wilson and Ted Ruckert had reported that the oil pressure in the main gearbox had dropped to zero, and as it was low tide they put the craft down on the sands.

I was bundled onto Sure *with John Bartlett, the other airframe engineer, with a host of pipes, oil drums, spare pumps and other parts and off we went into the gathering dusk. When we arrived, the 36 passengers were wandering around on the Goodwin Sands. We landed alongside, off-loaded all our bits and pieces. The passengers boarded* Sure *and returned to Ramsgate with Captain Bill Williamson at the controls.*

It soon became obvious that the gearbox trouble was terminal as the main lubrication pump had failed and the box was in bits internally. The craft was dead, because the gearbox drove both the lift fan and the propeller so we were going nowhere.

The tide had started to come in so we dug the anchor in and informed base that we needed help quickly! It was now dark, apart from the lights of Deal in the distance, and very large tankers were charging down the Channel behind us. An SR.N6 hovers about three feet off the sea but floats very low in the water and is quite a frail-looking craft to be floating around in among the busiest shipping lanes in the world.

Base radioed back that they had contacted a large Deal-based fishing boat called Sea Symphony *and that she would put to sea at dawn and tow us back to Ramsgate. The tide fairly races in over the Goodwin's, but our puny little anchor did sterling service and we stayed put, although we did start to wallow a bit, making John turn quite green. We then heard on the radio that* Sure *was on its way out to us and it soon arrived with sandwiches and thermos flasks of coffee, which was most welcome, but not for John! We decided that he was not needed and he went back on* Sure, *leaving me with the two captains.*

I had to check the buoyancy tanks every hour to see if we were taking on water, which we were, but not a lot. This involved clambering around the decking on Swift *and entering the plenum chamber with the hand pump. Tom later awarded me the rank of 'Honorary Able Seaman-First Class', which for me was something, from a former officer of a North Atlantic weather ship.*

At 04:30 the Sea Symphony *arrived but the tide was now going out, so the skipper advised us to cut the anchor rope and drift out to deeper water into an area of the sands called the Kellet Gut. This we did and at 06:00 the towrope was made fast and we were on our way back to Ramsgate, about seven miles away.*

A skirt on a floating hovercraft fills with water, estimated on an SR.N6 to be 12 tons worth, and this, plus the raging Goodwin's tide, held our speed back to one knot. This meant it might take about seven hours to reach Ramsgate. So Sure *came back out with some more sandwiches, coffee and warmer coats. We finally made it into port at one o'clock in the afternoon as the tide had turned, and we managed to drag* Swift *some way up the pad with the help of the harbour master's Land Rover. I was given the next day off, and so I joined the very small band of people who have spent a night on the Goodwin Sands and survived.*

Being an engineer, and an aircraft one at that, Mike can be excused for thinking the tankers transiting the Straits were a danger to the craft. Even in ballast these vessels would likely have been drawing at least eight metres, however close and menacing they may have looked, but with the hovercraft well up on the sands they could not have come much closer than three or four miles. Perhaps even further if the *Sea Symphony's* instruction to cut the anchor and drift on the ebb into the Kellet Gut is correct. It would suggest *Sure* was north of the Kellet and therefore well away from the big ship traffic.

Apart from a job well done this other story told by Bill Williamson illustrates how, by 1967, confidence in the N6's capabilities had grown. Force 6 from the north west – right on the nose – would have been the absolute limit for the small machine:

One day we ran Les Colquhoun across for a meeting in Calais. On the trip across the weather was north-westerly Force 4 to 5, but coming back it had increased to Force 5 to 6, which is pretty unpleasant when coming out of Calais. We were just getting settled in when I heard a shout from Bill Forsyth, "There's something over there!" whereupon we saw the white overturned hull of a little dinghy and on the back where the rudder was there were two people hanging on to it. They were about a mile off Calais Harbour. I dropped the craft down, more by fluke than anything else, right beside them. We opened the hatch and Bill and Les and two others used a heaving line to get the first one up. She was a small thin lady with a lifejacket on and a greasy oilskin over it, and we got her up without too much trouble, but then it came to the bloke. He was a big heavy man also wearing a lifejacket and an old style oilskin, but we couldn't get him up over the bow. We had to secure him with the rope and drag him along the side of the craft back to Calais, which didn't take long downwind.

In recognition of a heroic effort of using a difficult to control vehicle, not designed for the task, Calais Chamber of Commerce presented Bill with a Certificate of Commendation.

Arrival at Ramsgate Harbour hoverport with the craft in the background up on 'jacks' for maintenance. 1966. Courtesy Pat Lawrence.

Engineer Dave Parr (left) and Assistant Chief Engineer Monty Banks (centre) in the maintenance area. 1966.

Sir Robert Menzies, former Australian Prime Minister, and family with Les Colquhoun far right and Katie Colquhoun, 4th from left, during a visit to Ramsgate while staying at his official residence at Walmer Castle in his capacity as Lord Warden of the Cinque Ports. 1966.

The Hoverlloyd team at Calais Hoverport. Left to right front: Tom Wilson, Ted Ruckert, Bill Williamson, Les Colquhoun, David Wise, Terry Halfacre.
Back: Yves Lefebvre, Calais Chamber of Commerce, Ted Unsted, Jean Demeulener, a broker. 1966.

Dave Wise at Cowes

In the meantime in BHC's East Cowes hangar, the first 165-ton SR.N4, designated 001, was taking shape.

Because of David Wise's extensive experience in the aircraft construction business it was natural for the management to seek his advice. In order to get better feedback as to progress at Cowes, David suggested that Hoverlloyd should have a representative on site, pointing out that it was usual for someone from a purchasing airline to be in attendance at the factory where the aircraft was being built. David was appointed to the task and duly dispatched to Cowes to act as liaison between BHC and Hoverlloyd.

At that point the buoyancy structure and some of the superstructure of 001 had been built. Viewed with a little suspicion at first, he managed to prove he was there in a constructive capacity and was made welcome – the start of an excellent and productive relationship between manufacturer and operator, which lasted throughout the company's history.

Hoverlloyd had been the first to sign purchase orders and therefore could reasonably expect to receive delivery of the first craft built, 001 and 002. Inexplicably today, the British contingent had no inkling that the Swedish directors had already declined – quite rightly of course – to take the prototype, leaving it open for British Rail to purchase it if they so wished.

With no apparent knowledge of these decisions David's assessment was not unnaturally the same. His advice to Les Colquhoun and Emrys Jones, the technical director, was that Hoverlloyd should in fact acquiesce to taking the second and third craft, leaving the first craft to Seaspeed to complete development work and trials.

This was the right course of action; if Hoverlloyd had taken what, admittedly, was still a prototype machine in 1968, and experienced the total unreliability that proved to be the case in the first abandoned season at Dover, it is doubtful that the Swedes would have continued to persevere. It is more than likely they would have cut their losses and written off the whole exercise. It is also possible then that Seaspeed, reluctant starters in this new form of cross-Channel transport, would have abandoned their proposed operation from Dover. The British Railways Board would have breathed a sigh of relief and gone back to running their ferries without the distraction of the unwanted and commercially-dubious hovercraft.

David Wise

David Wise, ex-Merchant Navy officer with Shell Tankers, left the sea and managed to secure a job at Vickers in the Flight Test Technical Department. David had become a flight test observer on the Vickers Vanguard before moving to the BAC 1-11 project. (In late 1963 one BAC 1-11 was lost during stalling trials. It crashed after getting into a 'deep stall' situation, killing all on board. David had been flying on that aircraft the previous week, but luckily for him he was on a week's leave when the accident happened.)

By 1965, the BAC 1-11 programme was coming to an end and the future of further commercial aircraft being designed and built in the UK, apart from Concord (later Concorde) in collaboration with the French, was looking highly unlikely. At the same time David was aware that a new and exciting amphibious craft, called a 'hovercraft', was being developed by both Vickers and Saunders-Roe. He thought his marine and aviation experience would lend itself to this new form of transport and decided this was an opportunity for a career change. David knew Les through his connection with Vickers and applied to join him at Hoverlloyd. His application was successful and David joined in the second set of pilots taken on early in 1966.

He trained on the SR.N6 for the early cross-Channel and pleasure trip services from Ramsgate Harbour, before being based at Cowes to represent Hoverlloyd's interest in the building of their first SR.N4. When the craft arrived at Pegwell Bay he was one of the first three captains to take command.

David became Hoverlloyd's Operations Director in 1972 and, following the merger with Seaspeed, was highly regarded by both ex-Hoverlloyd and ex-Seaspeed personnel alike. He stayed with Hoverspeed until his retirement in 1993.

Other Duties

Apart from the two principal objectives of the initial years at Ramsgate; to evaluate the feasibility of a cross-Channel route to Calais and to ascertain whether Ramsgate Harbour was a suitable base, the first flight crew recruits found themselves involved in a number of other activities, not all directly connected with the task in hand.

Considering the concentration of operational experience in Hoverlloyd at the time it was natural for other organisations and enterprises intending to use the new technology to look to the company for advice and training.

One of the major contracts during the first years was the training of army personnel from 200 Squadron of the Royal Corps of Transport. A number of officers and NCOs of the regiment were trained on the N6 at Ramsgate, eventually seeing service with the hovercraft in the Far East.

During this time a new route was proposed in Italy, between Capri and the mainland. The company Aeronave took delivery of an SR.N6 in August 1967 and Bill Forsyth, much to the disgust of his colleagues, all of whom fancied a holiday in the Mediterranean, was picked to train the Italians on the route.

Bill Williamson with Officers and Men of 200 Squadron.

A number of studies for other potential routes were also commissioned by the Ministry of Transport, few of which came to fruition. The idea of a hovercraft service from Heysham to Douglas, Isle of Man was quietly shelved when weather data from the nearby Morecombe Bay light vessel revealed recorded sea heights in excess of 25 feet. That would have been too much for even the N4 to handle.

A similar study concerning problems of weather was conducted on the Goodwin Sands, investigating the effects of 'pluming', the resulting effect of waves coming from two directions, producing a punching force[2]. The study was principally targeted at the problems of North Sea oil exploration, in full flow at the time, but was obviously of interest to Hoverlloyd too.

So, although winter weather kept the craft tucked up in Ramsgate, the embryo flight crew had enough with which to occupy themselves. Perhaps an unforeseen outcome of what was probably originally viewed as convenient use of people with time on their hands, such involvement

[2] A force which most likely was the cause of the front windscreens being smashed in Ted Ruckert's earlier recollection.

with company's affairs had the effect of establishing the flight crew as not just hovercraft drivers, but part of the management team.

Without doubt, what today would be termed a sense of 'ownership' contributed to their high morale and dedication to the company's affairs; a system and mindset which continued on when the N4 service began. Flight crew continued to concern themselves with the efficient operation of the service. Everything from 'Duty Captains' acting as day-to-day operations supervisors, overseeing in close cooperation with traffic officers, the juggling of craft and crews to achieve maximum efficiency of departures and load factors, down to more mundane duties such as compiling crewing rosters. The flight crew, whether requested or not, took an interest in it all.

Going back to the first formative years in Ramsgate, the various studies and contracts, while interesting and adding to the income, were not directly concerned with the company's future wellbeing. What was of vital importance to the future was to ensure that the looming SR.N4 operation had the best possible start, and that was very quickly understood to mean the best possible base.

Viewing what was available, or more correctly, what was not available on the East Kent coast, the only solution was Pegwell Bay, but first the government and the public had to be convinced.

Chapter 11

The Battle for Pegwell

Uncertainties

Despite the continued good progress made by the SR.N6 pilot programme through the second year of operation, 1967 must have been a year full of nail-biting uncertainty for the new company. Two public inquiries had to be dealt with, both of which had the potential to put an end to plans to operate the larger SR.N4 and therefore put an end to the company altogether.

The small 36-seat SR.N6 about to enter the narrow entrance through the staggered breakwaters of Ramsgate Harbour with Pegwell Bay and Richborough Power Station in the background.

It had become clear right from the start, despite Hoverlloyd signing a 49-year lease with Ramsgate District Council, that Ramsgate Harbour was impossible as a base for the larger machine. The harbour was in many ways too small for the N6, let alone something seven times larger. Another location was needed, preferably not a harbour at all, as entering and leaving through stone-walled entrances in strong crosswinds and sea was not the ideal scenario for any hovercraft, and certainly not for something as large as the N4. What was really needed was a wide open area, one that could provide a large clear space to manoeuvre and park and, just as importantly, with possibilities for expansion.

Another necessary feature of any possible site for such a fuel-hungry craft was a reasonably short route distance to Calais. The only location that filled all of those criteria, all of them essential if the new project was to be successful, was Pegwell Bay. There simply was no alternative.

Not in My Back Yard!

Artist's impression of the proposed Pegwell Bay Hoverport.

Once Hoverlloyd had announced its intention to build a hoverport at Pegwell it was a foregone conclusion that there would be objections.

Although the extensive coastline south of Ramsgate is not the most alluring of beauty spots – in fact part of the bay prior to Hoverlloyd's short occupation, was being used as a corporation rubbish tip – it was nevertheless well known as a bird sanctuary. A virtual catalogue of feathered creatures attracted bird watchers at all times of the year and another not so attractive animal, the common lugworm, was dug out of the mud at low tide by scores of 'bait diggers' who made a good business out of selling them to fishermen.

So apart from the coastal residents of Cliffsend, who would find themselves hard up against a busy, noisy and possibly smelly hoverport, and quite naturally not keen on the idea, there were plenty of others to support them.

Pegwell Bay extends south from Ramsgate about one mile and a half to the mouth of the River Stour, north of the town of Sandwich. The Stour running north west from the old town is all that remains of the Wansum Channel, which in Roman times made today's Isle of Thanet region a real island. The bay is bounded by a wide sweep of cliffs from the eastern edge of Ramsgate Harbour, decreasing in height from the north east until they finally merge into the saltings on the north bank of the Stour within half a mile to the south west of the proposed hoverport site.

The Ideal Spot

At the foot of the cliffs, and south westward towards the river estuary, was a generally narrow, untidy and apparently uncared for, sandy beach, largely overgrown with grass. Above the cliff, edged by the long stretch of the Viking Green[1], was the small Ramsgate suburb of Cliffsend.

As it appeared at the time, the beach was hardly the most desirable place to take the children on a summer afternoon. The bay itself was extremely shallow and the mud flats at low water reached out almost to Ramsgate in the north and well past the Stour in the south. In contrast, as a prospective hoverport it was ideal, despite the fact that the hovercraft would often have to cover as much as a mile and a quarter of mud between the landing pad and the sea[2].

The First Inquiry

As soon as Hoverlloyd announced in mid 1966 their intention to apply for planning permission to build on the flat below the Viking ship, it was to be expected that Cliffsend residents, only a few hundred metres away, would object. They were strongly supported by various environmental groups and many other interests in the area. Nevertheless there were many supporters for the proposal, of which, quite naturally, Ramsgate Borough Council, keenly egged on by the local media, was the most enthusiastic. To hear all sides of the argument, a public inquiry took place in Albion House in Ramsgate and lasted seven days from the 3rd to 12th January 1967.

---

1 Named after the Viking Ship *Hugin,* which still stands as a monument on the grassy area above the old hoverport site. The vessel was rowed into Pegwell from Denmark in 1949 to mark the 1,500th anniversary of the legendary landing of the brothers Hengist and Horsa.

2 The mud turned out to be the only minor drawback to an almost perfect location, which is explained in more detail later.

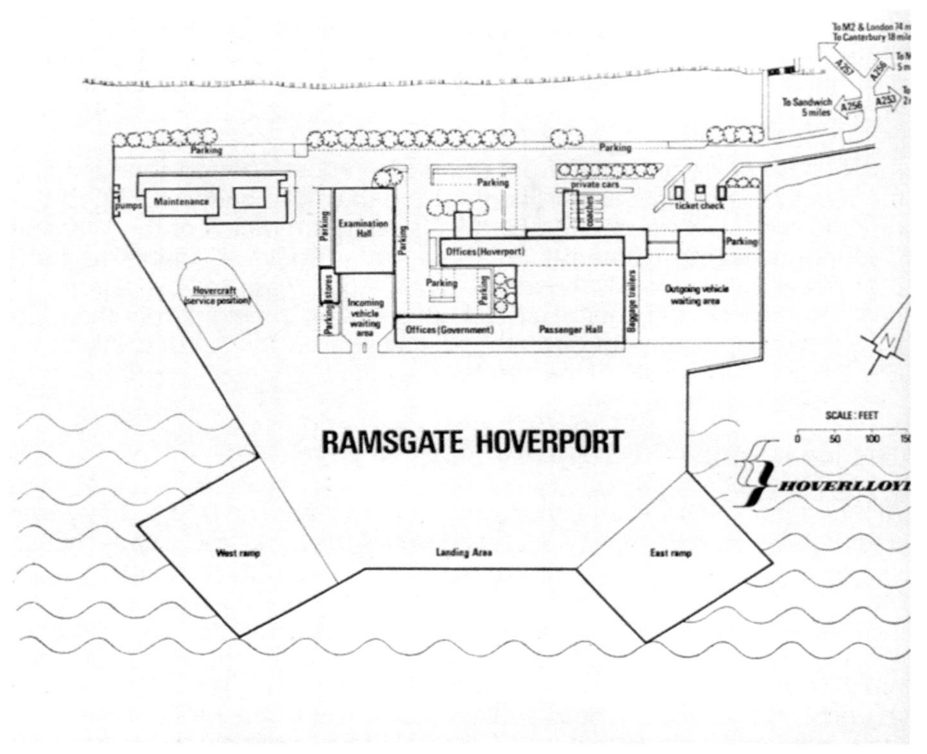

Plan view of the new Pegwell Bay Hoverport.

The whole of the Isle of Thanet region stands out to this day as the poor cousin of Britain's South East, generally supposed to be its most affluent region. In contrast, Thanet, with its poor connections to London at the time, placing it well beyond the stockbroker belt, had a demonstrably poor economy. By 1967, its heyday as an exclusive beach resort for the Victorian gentry had long since gone and, although over the ensuing years it survived the slide downmarket, switching its image to cater for London's East-ender day trippers and coach holidays for northern pensioners, the new trend for package holidays was starting to affect that market too. Whelks and jellied eels were no longer appealing to holidaymakers who were already starting to acquire a taste for paella on the beaches of Ibiza.

If the severe downturn in the holiday trade was not enough, the closure of the USAF base at nearby Manston airfield in the early sixties was a further economic blow. Reading the local papers now, one gets the decided impression that Thanet's politicians were desperate and, other than hopeless fantasies for perhaps designing a bigger and better whelk, were totally bereft of new ideas.

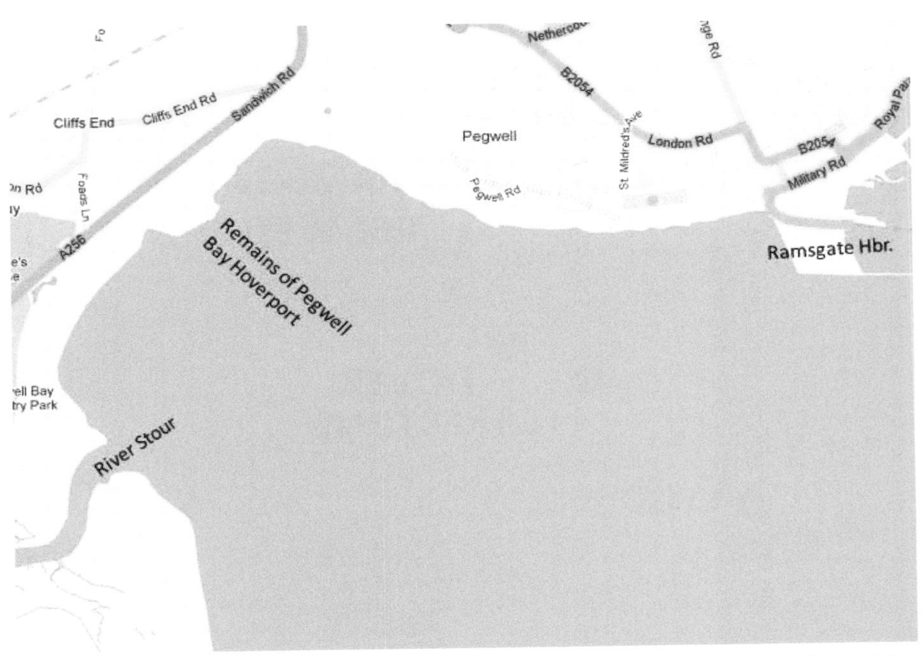

Pegwell Bay today. In the top left hand corner can still be seen the outline of the concrete pad, all that remains of the hoverport.

The only employment prospect of any note in the Thanet area was the Pfizer drugs complex at Sandwich.

Ramsgate businessmen saw the new multi-million pound scheme as manna from heaven and so the battle lines were drawn.

From today's perspective perhaps the most unexpected issue was that the major objector was Kent County Council (KCC). The council's principal concern was the prospect of creating another cross-Channel focal point with all that meant in terms of road construction – and possibly rail – to cope with many thousands of cars, foot passengers and the concomitant infrastructure, but they presented their objection as:

> *The proposed hoverport would be contrary to the county council's coastal policy and detrimental to amenity. It would restrict access to the coast, add to the traffic problems and the noise would intrude.*

They did, however, declare that if it could be proved that no alternative site was available to Pegwell Bay then the county council would concede that the national interest must override planning objections.

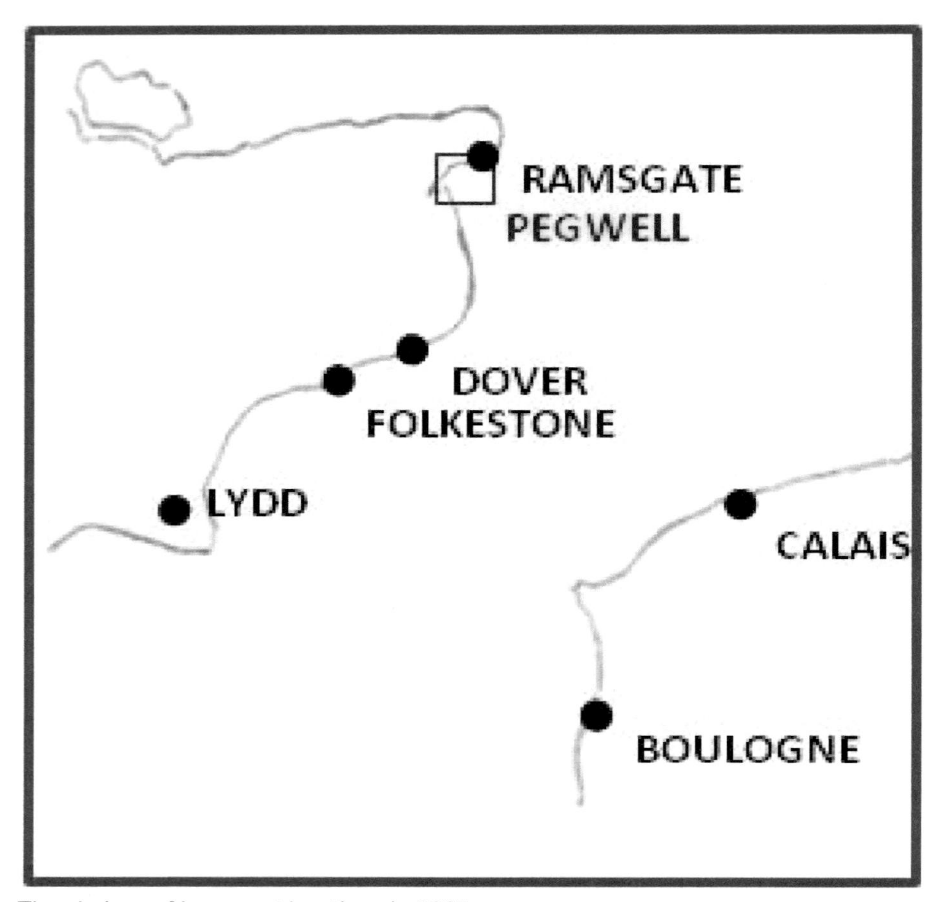

The choices of hoverport locations in 1967.

The council preferred to concentrate on Dover as the existing major gateway, which itself was badly in need of better access. They also had at the back of their minds the long-term prospect of a Channel Tunnel; the location of London's third airport, which had yet to be decided upon; and the large-scale development of Ashford, which would require huge funding for infrastructure improvements. Although major highway schemes were already underway it would be a decade or so before they would be completed.

In retrospect, the KCC submission looks like a huge amount of fence sitting. Although broadly against the Pegwell site, and all for the prospect of using Dover as the designated port for the Channel, they were at the same time curiously unprepared to commit themselves fully. Phrases like, "not able to assess fully the regional considerations", and "not in a position to put forward positive alternative sites", reek of ambivalence. If the county council was unable to make a judgement about its own area of responsibility, or come up with a reasonable assessment of sites in the county, then who was?

Dover Harbour – not the most sheltered of harbours. This picture is of the modern day harbour.

As to the reliability argument that 'early cross-Channel hoverports' should be sited close to existing ferry terminals – for the latter read Dover – was the KCC suggesting that, if reliability was proved, a company could up sticks and move to somewhere else more suitable? The KCC's obvious response to any new application would be, "well, you are running a perfectly good and reliable service out of Dover, there's no need for you to move". In any case if the N4 operation was to be that unsatisfactory that it needed the constant presence of conventional ferries to take the cancelled loads there wasn't an awful lot of point running the service at all.

Mr. K. Davis, General Manager of Dover Harbour Board, gave evidence. One has to wonder whose side Mr. Davis was really on. His description of the difficulties the large conventional ferries experienced in Dover Harbour was damning evidence against its use as a hovercraft terminal. If large displacement vessels couldn't enter at times, then what was the lighter, smaller, wind-prone hovercraft going to do?

The fact that Mr. Davis also claimed that Dover had superior facilities to the otherwise 'smaller facilities' proposed at Pegwell Bay revealed his misunderstanding of what was planned. Certainly Dover was the largest ferry port in South East England but the facilities proposed by Hoverlloyd – surely there must have been a plan of the site on view at the proceedings – were in effect the size of a large city domestic airport and a modern one at that. Dover didn't have the available space at that time to build anything half as big.

Although Mr. Davis' description of the harbour control system at Dover was probably meant to emphasise the ease in which the SR.N4 could be accommodated, it only served to highlight a vision of exactly what Hoverlloyd didn't want. The picture presented was that of an already busy and crowded port that needed a traffic control system to deal with the congestion, forcing a high-speed craft, which relied on a fast turn-around to maintain profitability, to stand in the queue waiting for permission to move.

After all, in a small community with few cars and clear streets, who needs traffic lights? Although there was a control facility at Pegwell – and at Calais – the name given to the 'Radio Room' revealed its main function. It was there to maintain communication with the craft at sea from a safety and company management point of view. Any control element was reduced to nominating landing positions on the pad and an update of weather conditions.

Maybe Mr. Davis, although professing a neutral attitude to the question, revealed the true position. Dover did not want the competition.

Hoverlloyd's Case

From the company's point of view Pegwell Bay was not just the best choice of location, it was the *only* choice. To the two criteria already mentioned – clear unobstructed access and a location providing a short route to Calais – could be added the need to find a route direction that provided the most advantageous angle to the prevailing weather from the south-west. The almost south-easterly course from Ramsgate to Calais was perfect and, if the natural 'breakwater' of the Goodwin Sands is taken into account, the route could almost be considered to be custom designed.

Such reasoning cut no ice with the opposing camps. The local residents at Cliffsend simply didn't want it on their doorstep at any price and Kent County Council was fixated on Dover, despite Pegwell Bay's obvious advantages.

Apart from the argument that Dover was the nearest location to the French coast, a mere 21 nautical miles to Calais and was historically the 'brand name' for cross-Channel travellers, from a hovercraft perspective the angle of the route across the prevailing weather was not the most advantageous. The return journey in a westerly gale was almost straight into the wind.

Putting aside the suitability of the route, Dover Harbour itself proved not to be the most sheltered of harbours. Situated as it was on the edge of the promontory of South Foreland, the cliffs on this corner have a natural acceleration effect on any strong wind from the west. There are days when large ferries can no longer safely negotiate the harbour entrances let alone a light, wind-affected hovercraft.

Another factor to be considered was the traffic density in the port. It was already the busiest harbour in Southern England in the mid-sixties and was to expand enormously during the ensuing decade. By 1967, traffic was already sufficient to require a radar traffic control system to maintain safety of movement in and out of the harbour. To expect a lightly constructed free-moving craft like the N4 to pass through those busy harbour entrances was something akin to building walls on either side of the main runway of a large commercial airport and allowing road transport to have equal access – the risk of catastrophe was obvious.

As alternatives to Dover other harbours were proposed as far afield as Newhaven, more than 50 nautical miles further west. But perhaps the most bizarre suggestion was that of Lydd Airfield, well over a mile inland from the nearest beach access.

The Inquiry report describes in detail what the reality of such a proposal would mean:

> *The land between the foreshore and the airfield consisted generally of undulating shingle, traversed by roads, railways, telephone and power cables, and containing private housing, hotels and holiday camps. The area was very heavily populated during summer months.*
>
> *Crossing controls would be required at the coast road, minor access roads and at the Romney, Hythe and Dymchurch Railway's and British Railway's lines. Protective barriers against flying shingle, dust and noise would be required where the flight path passed near houses. The flight path over the shingle would have to be levelled for a length of about 1¼ miles and to a width of about 200 feet. It would also have to be adequately fenced and lighted. Telephone, power and other services would require modification to pass under the flight path.*

Whoever suggested Lydd in the first instance never thought their crazy idea through until the shortcomings were pointed out, but it also serves to highlight the nature of the proceedings as a whole. Literally no one attending had much idea as to what this new invention would entail in terms of safe and efficient operation. Not even the manufacturer, the British Hovercraft Corporation. The fact was that the first machine, 001, was still on the building blocks, still many months away from its first test flights.

Sheepy Lamb, although possessing a fund of experience on the single-engine SR.N5 and N6, found it difficult to imagine how a four-engine, controllable pitched vehicle might handle. Much of his evidence talked about the aerodynamic control of the N6 and the very real difficulties that presented, not realising that the control system of four swivelling pylons would make aerodynamics largely irrelevant. To this end he more or less wrote off Dover by stating:

> *No harbour could be recommended as suitable for hovercraft operation, but Dover Harbour was especially unsuitable.*

This was hardly helpful to BHC, as in their 1964 paper *Hovercraft Channel Link Proposal*, in response to the Ministry of Transport white paper of September 1963, *Proposals for a Fixed Channel Link (Bridge or Tunnel)*, they had suggested that a minimum of 16 SR.N4s would be required to run a service from Dover to Calais. There was even a plan of Dover Harbour in the report. The other problem here is that at this point Sheepy's employers, BHC, was already in talks with British Railways to sell an SR.N4 for cross-Channel operation, most likely out of Dover, but as yet unconfirmed. BHC's vested interest lay in ensuring both cross-Channel companies' operations were started successfully, not promote one to the detriment of the other.

The fact that Sheepy failed to pull off the balancing act leads one to speculate whether there was a reason why his deputy test pilot, Harry Phillips, was called as a witness for the BHC case in the second inquiry.

After seven days of deliberation, the chairman's report, all 190 pages of it, was delivered to the minister on 4th March 1967. The answer was simple. The report agreed with Hoverlloyd: Pegwell Bay was the only sensible place to operate a hovercraft service for the type of service they proposed.

With what must have been a great sigh of relief – although now a year behind – the project could go ahead. Unfortunately this was not the end of the controversy after all; there were moves afoot to inaugurate another operator of cross-Channel hovercraft.

Back to Square One

Only a few months after the successful conclusion of the inquiry, in the early summer of 1967, British Rail Hovercraft announced that they had come to an arrangement with Dover Harbour Board to operate their SR.N4 from a hoverport to be constructed within the Dover Eastern Docks.

This was by no means a surprise to Hoverlloyd. The fact that British Rail was considering starting a service from Dover was an open secret during the inquiry. What did come as a very nasty surprise was that the move raised a number of questions, not the least of which was – was there a need for two cross-Channel hoverports so close to each other? Further, if this other company was quite happy to consider operating out of Dover and, what is more, thought it could do so in perfect safety, had Hoverlloyd's previous submission been totally wrong and misleading?

The government of the day felt it had no choice but to reopen the inquiry to provide some answers. To some extent it is likely the administration may have felt that this was also a way of getting the somewhat jingoistic tabloid press off its back. As we have seen, the British newspapers had had a field day, castigating the government for allowing a foreign company to be the first to purchase what was a British invention and worse, operate it from British soil on a lucrative high-density route heretofore the reserve of the British and French.

As already mentioned, the fact was that the first machine had always been available to a British enterprise after the Swedes had declined to purchase the prototype. Why this fact was kept secret, especially from the British Hoverlloyd management, remains a mystery to this day.

The Second Inquiry

On 5th June, Mr. Greenwood, the Minister for Housing and Local Government, decided to reopen the inquiry into Pegwell Bay because, "a commercial enterprise had decided, for reasons of its own, to operate from another place", namely Dover Harbour.

The scope of the re-opened inquiry – again conducted by Mr. Hilton, which lasted 3½ days from 12th to 15th September 1967 – was limited to hearing evidence and submissions on the relative advantages of using Dover Harbour as compared with Pegwell Bay. So, although Hoverlloyd continued planning their operation, this setback resulted in more delay, expense and uncertainty for both them and the British Hovercraft Corporation and even raised questions in parliament.

So, like it or not, in that year, Hoverlloyd found itself dragged back into a second tiresome inquiry, having to justify once more perfectly reasonable conclusions that had taken two years of careful work to accumulate.

The first question to be answered – had Hoverlloyd been remiss in misrepresenting their previous arguments concerning the suitability of

Dover? – was not so easily dismissed. Mr. Jupp, QC, counsel representing Seaspeed, went on the attack:

> *These were matters which should have been properly investigated before the original inquiry and Hoverlloyd only had themselves to blame for the present reopened inquiry. The minister had not delayed matters unnecessarily. If the application were not granted it would be after the full investigation which Hoverlloyd had failed to carry out in January last.*

These were hard words indeed and must have been particularly galling to Hoverlloyd after all the hard work they had put in. To be accused of misleading the original inquiry, and now being held up as the villain of the piece after making a perfectly reasoned decision to choose Pegwell as a site in preference to Dover, was much worse.

Finding themselves dragged into an argument initially created by the British government's xenophobia on one hand, and BHC's desire to make two sales rather than one on the other, and now to be told it was their fault for the trouble and cost of a second inquiry, must have been especially hard to bear.

Not content, Mr. Jupp twisted the knife further:

> *What was now revealed was that Dover Harbour was suitable and safe for use as a hoverport. That was the manufacturer's view – a view shared, in fact, by all the witnesses. The only doubt was the precise extent of the operational limitations, which would have to be accepted because of weather conditions. The suggestion that there should be only one hoverport did not come from Seaspeed, but if there was to be only one then the evidence was overwhelmingly in favour of it being Dover.*

Contrary to Dr. Sydney Jones' position, which was to object strongly to Pegwell Bay, and contrary to his orders, Tony Brindle, Seaspeed's General Manager, instructed Mr. Jupp to point out that they had no wish to become involved in the Pegwell Bay controversy. They would say nothing at all against the proposal to establish a hoverport at Pegwell Bay unless Dover Harbour was attacked. Pegwell Bay was obviously an excellent site, but for commercial reasons Pegwell Bay was not a feasible proposition for them. They preferred Dover Harbour and would operate there in any event.

In the final analysis the rival companies, accepting that they were operating the same hardware, had different objectives and requirements.

Hoverlloyd wanted to run a totally independent service at Pegwell, taking advantage of the open, sheltered site and better weather routing and accepted the possibility that at times of extreme weather and other operational difficulties it would necessitate transferring their passengers to the ferries at Dover. Seaspeed had the opposite approach. They realised that they were prone to more weather cancellations at Dover but calculated that quick transfers to the nearby ferries was a reasonable compensation. Hoverlloyd wished to provide a maximum reliability service in all conditions, right up to the most severe weather, which would have been a risky policy within the stone-walled confines of Dover Harbour. Seaspeed had no intention of going that far. With the ferries close at hand they would transfer at the first sign of risk.

Incidentally, although the discussion centred on whether Hoverlloyd's verdict on the unsuitability of Dover was a valid one, for Seaspeed the vagaries of the weather as it affected location and route were not the real issues at all. Although Seaspeed saw itself as an independent service to the British Rail ferries, they were nevertheless subject to the overall core focus of British Rail, the railway link to the continent. The ferries were there to connect to the railway on both sides of the Channel, and as far as the British Rail hierarchy was concerned so were the hovercraft, *ergo* that required using the accepted railheads at Dover and Boulogne and no argument.

Two Services or One?

As before, BHC was in the difficult position of not wanting to 'rob Peter to pay Paul'. Somehow they had to play the game that resulted in two winners, not one. Support Hoverlloyd's decision to base their operation at Pegwell, but also suggest that the different style of the British Rail enterprise would fit effectively into their ferry operation at Dover.

Harry Phillips, BHC's deputy chief test pilot, trod the middle road, but had to contradict Sheepy's statement as to Dover's unsuitability at the first inquiry:

> *The SR.N4 was designed to operate in and out of harbours and was being sold for such operations. It would be able to operate satisfactorily and safely from Dover Harbour if certain limiting weather conditions were accepted. Only in extreme conditions would there be any need for concern. In the national interest, however, hoverports at both Dover Harbour and Pegwell Bay should be proceeded with.*

The question of whether two services were a feasible commercial option appears to have been decided in the end, not so much on the possible traffic volume or any other business consideration, but on the overriding issue of 'the national interest'.

How the 'battle' for Pegwell Bay was won

BY ROBERT BRIMMELL

IT WAS a wintry January 10, 1968, with a thin carpet of snow on the ground when the East Kent Times scooped the world Press with the news that Government permission had been given for the hoverport at Pegwell Bay, Ramsgate.

The news, headlined in a special edition, was greeted with the popping of champagne corks and glasses raised in celebration because victory for Hoverlloyd had come after a long and arduous battle.

It was the climax to months of cliff-hanging tension that followed two mammoth public inquiries while the future of Thanet's most important development of the century hung in the balance.

When Whitehall finally gave the go-ahead for the world's first international hoverport and confounded sceptics who suspected a political torpedo the result could not have been better.

Because in the end Ramsgate scooped the hovercraft pool with Government insistence that

WE BROKE THE NEWS WITH A 'SPECIAL'

hoverports being built on the South East Coast.

The battle for the hoverport will long be remembered in Ramsgate as an issue which aroused passionate arguments from a whole range of protagonists both in favour and against it.

First round

Round 1 of the 'hover' inquiry began on January 3, 1967, in Ramsgate Council Chamber and immediately

Ministry inspector Mr. Charles Hilton alighting from the SR.N6 'Sure' to view the site in 1967.

planned flight-path over the mud.

He peered through

'CONTRIBUTION TO THE NATIONAL

But when the resumed

At this point BHC was anxious to see confirmed sales on its books, and this could only be achieved if the first three N4 machines were sold for operation in England and proved themselves to the world to be the new viable form of transport they promised to be.

Les Colquhoun, Hoverlloyd's Managing Director, didn't mince his words:

> *If the planning permission applied for was refused, then a death-blow would be struck, not only for Hoverlloyd, but also for the British hovercraft industry. If this did happen Hoverlloyd reserved the right to reconsider its position. It would be better for them to break their contract with the British Hovercraft Corporation, and pay damages thereby incurred, than to embark upon an operation, which for them, would be commercially unacceptable.*

It was recognised by all the parties involved that if the Swedes backed out – and they would have had every justification for doing so – if their requests for the most advantageous site were ignored, then the whole large hovercraft concept might be set back years, or more than likely, not get off the ground at all.

Mr. Charles Hilton, who chaired both inquiries summed up:

> *If, having considered the relative merits of hoverports at Dover Harbour and Pegwell Bay, the British Hovercraft Corporation and the National Research Development Council have reached the conclusion that both are necessary and it is not simply a question in choosing one or the other, then, in my opinion, they are entitled to put that view forward.*
>
> *The British hovercraft industry at present has a lead in the design and production of hovercraft – the SR. N4 is the first of its size and type in the world, and therefore has a chance to capture the world market. It is clearly in the national interest that it should. There is a danger, however, that operating experience with the SR.N4, vitally necessary if the industry is to consolidate and maintain its world lead, may be lost or delayed by the failure to make available, as a matter of urgency, the best possible facilities. Moreover, if their application is refused, Hoverlloyd may well abandon their order for two SR.N4s. This itself would be a serious blow to the British Hovercraft Corporation and damaging to the prestige of the industry.*

And so, on 15th November 1967, the argument ended as to whether or not a hoverport should be established at Pegwell Bay, but the decision was not announced by the government until 10th January 1968. It had been a bruising and worrying 11 months for both Hoverlloyd and the British Hovercraft Corporation. Without doubt, had Mr. Hilton's recommendation ruled against Pegwell Bay, it is very likely that the SR.N4 would have gone the same way as the *Princess* flying boat – one would have been produced and flown, but never entered proper commercial service, and two more would have been constructed and mothballed.

This is because, had Hoverlloyd been refused Pegwell Bay and decided to throw in the towel, it is quite conceivable that Seaspeed, without any giant hovercraft competition, would have eventually abandoned the SR.N4 project, and with it the hopes of BHC of selling these large hovercraft to other potential operators. In fact, according to Tony Brindle, that would have indeed been the case, as Dr. Sydney Jones, British Rail Ferries and Seaspeed's chairman regarded the hovercraft as a thorn in his side and a distraction from the business of running profitable conventional ferries. Whether the Ministry of Technology, having ordered British Rail to operate an SR.N4, would have allowed this to happen is a matter for conjecture.

Tony Brindle:

> *Les Colquhoun and I had a very friendly relationship and when it came to the final inquiry at Pegwell Bay, about Pegwell Bay, I*

was asked to appear. I went and was asked, "Do you have an objection?" and I said, "No, I have no objection at all. My job is to see that hovercraft are developing." If an operator wants to come and take his chance at Pegwell Bay, that's a commercial decision. But that was fatal as far as Dr. Sydney Jones, the chairman, and I were concerned, because he wanted me to see Hoverlloyd off the premises and didn't care how. He just made life absolutely impossible, but once the decision had been made about Pegwell Bay there was nothing he could do.

But the important issue at the time was that the battle for Pegwell was over and Hoverlloyd had won, but Seaspeed had also got what they wanted. Now the battle for the cross-Channel hovercraft market was to commence. A battle, which sadly, for all their commitment and demonstrable superiority and through no fault of their own, Hoverlloyd would eventually lose.

Chapter 12

The SR.N4 Prototype Trials

Pre-Launch

While the building process with the SR.N4 was in full swing, both Hoverlloyd and Seaspeed kept a careful eye on developments. Hoverlloyd had David Wise *in situ* staying in the Fountain Hotel in Cowes, but with the Seaspeed SR.N6 operation on the doorstep it meant that Tony Brindle, John Lefeaux and the senior pilots were able to hop across from their Cowes terminal to BHC at will to keep an eye on progress. As Seaspeed was to take on the responsibility for putting a prototype hovercraft, the largest ever built by some margin and ahead of its time at that, into service on the showcase cross-Channel route, they had a vested interest in trying to find out what they were in for at an early stage.

Tony expressed his concerns about dealing with BHC:

> *The great problem in dealing with Westland was that their background was virtually wholly military – all their engineers, marketing people, everybody was militarily orientated and that was very well reflected in the N6. The gangway and seat widths and so on were not designed for the broad bottoms of ladies. If you're building a military aircraft you're generally building it for fit slim soldiers. You're not trying to cater for blue rinse ladies from Detroit. So a lot of the finish, a lot of their ideas and design structures had a military background as a result of which one of the great difficulties in dealing with BHC was that their knowledge of civil passenger operations was minimal. I knew about civil passengers, as opposed to military personnel, but BHC didn't comprehend the pressures of running a commercial operation. It's a bit like comparing operating a passenger ship with operating a warship. The passenger ship has got to sail because it's got a ship full of people. Now, if it's military and it's inoperable you simply put off the operation until the problem is fixed, whenever that might be.*
>
> *I had a running battle with BHC's service department, which meant that we were permanently at war. If something went wrong I wanted it fixed now, not in half an hour's time – now. This was the scenario with the N6s, but I knew it was going to be like this with the N4.*

While Dick Stanton-Jones and I were never particular friends, his fellow Joint Managing Director, Lewis Boddington, and I had a very good rapport. He was a delightful Welshman, a highly competent engineer, a former Director of Naval Aviation, and the inventor of the Angled Flight Deck in aircraft carriers. He was very bright and had a good understanding of what I was looking for in constructing a commercial operation.

The control cabin/flight deck of the SR.N4 with its 'initially bewildering array of 110 dials and numerous switches'.

The Seaspeed pilots, in between operating the Solent services, attended lectures on the SR.N4 at BHC given by the very capable and thorough Peter Habens, who was up to speed with every nook, cranny and piece of equipment on the craft. They also attended a course on the Proteus engine at the Rolls Royce works at Ansty[1], near Coventry.

---

[1] Peter Barr recalls, "We did the Proteus course in Coventry at the same time as David Wise, George Kennedy and Bill Williamson from Hoverlloyd. They stayed in a very posh hotel up the road and we were scratching around in some scruffy place, as British Rail expenses were pretty meagre."

John Syring remembers one early day extra-curricular learning attempt in the mock-up of the control cabin at BHC:

> *There was a full-size wood mock-up of the control cabin of the SR.N4 in the factory at Cowes. So in between lectures, or whatever, Peter Barr and the rest of us used to go into this mock-up control cabin. It was just plywood and it had a couple of stools in it together with photographs of all the instruments and switches. So we decided one day that we should set about 'getting this thing going'. We got the manual out and said, "Ah, the first thing we need to do is to start an APU[2], so you tell me what to do and I'll do it." "Such and such and such switch." "Um, ah – oh yes here it is." "Start second APU." We eventually got two APUs going and then we thought it was time to go for coffee.*
>
> *We came back from coffee and by lunchtime we had one of the main engines going and then we realised it was going to take the whole day to start this machine, at which point somebody said, "This can't be commercially viable!"*
>
> *You'll recall how slick it actually was – about a minute and a half to get the thing going. It was so amazing how we stumbled through this bewildering array of 110 dials and numerous switches – something like a* Princess *flying boat.*

First Flights

Finally, on Sunday 4th February 1968 at 2.00pm, with Sheepy Lamb at the controls, and Harry Phillips in the right-hand seat, SR.N4 001, set off on its first test flight with the stern falling over the edge of the Columbine slipway, an irreversible operation as the ramp from the main assembly area was only a few feet wider than the craft. Sheepy's report recorded the wind as southerly Force 5-6 and went on to describe a 'hairy' departure.

At this stage, what eventuated as the perfect control system, the rudder pedals producing a turning moment and the yoke providing lateral control, was one option among several, and in fact not chosen for the first flight. Instead, a mode where the front pylons were controlled by the rudder and rear pylons controlled by the yoke, was selected.

---

[2] Auxilliary Power Units provided the electrical power for the hovercraft, including the ability to start the engines.

SR.N4 001 leaving the Columbine slipway on 4th February 1968.

Sheepy's report noted:

> *The craft came into the hover at approximately 5,500rpm turbine on each system. In this condition small trimming adjustments were made on visual instructions from Mr. Worner, who was on the hard standing. Once the craft was approximately level with pitch settings +12° on the front -10° on the rear, turbines were advanced fairly rapidly to 8,500rpm and pitches on the rear system were advanced to +12°. The front pylons were offset 10° to port while the rear pylons were held central. As the craft advanced down the slipway the stern, as anticipated, began to swing to starboard. About 15° angle of yoke was applied to port to check this swing, but this seemed fairly ineffective at this stage as the pitches were going through zero. The yoke was pushed forward to advance the pitch angles all round and increase the forward speed down the slipway. There was very little lateral movement as the craft went over the edge and carried on down the slipway in a port turn. Once in the water the craft continued a 90° turn to port through the wind and finished up facing the pier.*

SR.N4 001's departure from Columbine viewed from beside the Pilots Office.

SR.N4 001 immediately turning back to face the ramp through lack of control.

We, as SR.N4 captains looking back now with the benefit of a few thousand hours of experience with a fully developed control system, can appreciate that Sheepy and his team were dealing with the virtually impossible and downright dangerous. It was a wonder that manoeuvring in such a tight situation with a minimal level of control, the day did not end in disaster. If the swing of the stern to starboard had continued there was a definite chance of a collision with the hangar. On the other hand if the craft had initially swung stern to port, the craft may have continued stern first off the wall causing who knows what damage to the rear end. At least swinging the way it did Sheepy was left with the option of gunning it forward down the ramp. That he took that option without hesitation is a tribute to his fortitude and presence of mind.

SR.N4 001 being carried by the wind out of the harbour.

It has to be said that test pilots are a breed apart, but most of their work is with aircraft, which, although of new design, still follow the usual format of wings and a tail-plane, with the expectation that the controls will be much the same in structure and function. It needed a special kind of man to handle this totally new machine, unpredictable in its response to the elements and a control system that no one had experienced before. That Sheepy Lamb achieved this not only admirably but with a fair degree of aplomb and sense of humour is to his everlasting credit.

The report continues:

> *Attempts to steer the craft on the front pylons only in these conditions were proving abortive and full differential applications of pitches were made with positive to port and negative to starboard to turn the craft through 180°. The craft turned through 90°, but due to the sternway, which was now being made towards the front at East Cowes, the negative pitches on the starboard side had to be reduced whereupon the craft turned back in the direction of the true wind. The wind at this time was carrying the craft towards the breakwater and it was decided to take advantage of this and maintain a heading in the direction of West Cowes whilst the wind did the work of getting the craft out of the harbour.*

SR.N4 001 on trials. February 1968.

Once clear there was a ten-minute juggling act to get the craft on a heading towards Calshot Spit. It was quite clear that Mode 2 pylon control was not going to work, despite it being logical that it should, and any further trials in this mode were abandoned. On the first outing the craft went on to record air speeds of up to 60 knots and water speeds of up to 50 knots in seas of about 4 feet.

Towing trials were also part of the first day's event. A rendezvous was made with the tug *Ower* off Prince Consort Buoy with both craft stemming the full tide. Sheepy's report also noted:

> *All in all, the problems of taking the tow to the tug went extremely well considering the self imposed difficulties of an unmade-up bridle, which is essential for this form of operation. Once shackled on, the tug made headway into the ebb tide and achieved an estimated towing speed of three knots, which was a fraction above that of the tide. Whilst this was going on, the Harbour Master aboard Thomas' Bros. launch astern, was assessing the slipstream from the propellers and jet blast from the exhausts in an idling condition with various propeller pitch settings. A loud 'crump' from the back doors, which resounded throughout the craft, denoted a certain amount of over enthusiasm from Mr. Thomas and his brother, who rammed us from astern. An assessment of the damage was passed by Mr. Harris to control, which, though serious, did not warrant any interference with the trials programme.*

Once the craft had been moored the engines were washed. A very cold and wet trials team, who had given their utmost under the very difficult and adverse conditions, were then disembarked by two launches alongside.

Although the 2 hour 30 minute trial was judged an overall success, one of the biggest problems to emerge was that throughout the trial the craft was longitudinally trimmed ½° bow down and a rectification of trim by at least 1¼° to 1¾° was considered desirable[3]. As was subsequently shown there was no accurate way of measuring trim underway. The picture on her first trial on the previous page indicates, if anything, the craft was bow up.

During Monday 5th and Tuesday 6th February, 001 went out on three further trials, departing from the buoy. There were the inevitable delays and breakdowns, but progress was being made and experience gained in handling the craft. Sheepy's report continued:

> *On the Tuesday morning the craft moved away from its buoy in heavy rain showers, but lighter wind conditions, and proceeded west towards the Needles. After passing Totland, some rough*

---

[3] Far from being bow down, the picture on the previous page of the N4 underway shows a distinct stern trim. It is not clear what trim measurement system the trials team were using. If it was the installed instrument then it was of minimal use and very questionable as to accuracy. As described later, Hoverlloyd replaced it with a spirit level system.

water could be seen breaking on the Shingles towards the Needles. It was decided to run the SR.N4 along the edge of the bank over this broken water to assess role and pitch responses. Over the first few combers the craft rode extremely well at a water speed of approximately 35 knots. These combers were mainly on the port beam and had all the appearance of surf as the south-westerly swell across the channel broke on the shallow bank. Towards the end of the bank there is an area where the two swells meet from the West and the South and build up against each other. The craft was steered in this direction where the water was seething in a fairly high wall. Estimated height of these seas varied from a minimum in the cockpit of 6 feet to a maximum at car deck level of 14 feet. My impression was 8 feet, but from an elevated bird's eye view I would concede another 2 feet to give a maximum of 10 feet.

As the craft approached this edge of the bank, she was rolling 2° either side of centre line with very little pitch until the first comber pushed the bow up by some 2° to 3°. As the bow descended the second comber hit the port bow fair and square throwing green water as high as the cockpit windscreens. The impact, which was unexpected after the previous three minutes of roughish water trials, tossed people out of their seats towards the deck head, and those standing over the forward area into the air. The craft, however, continued on above hump speed with the bow paying off 20° to starboard. An immediate check was made of all trials team from which it was unfortunately learned that Mr. Newbury[4] had descended with one foot onto a piece of angle iron and broken his leg.

The craft headed back towards Cowes where Jim Newbury was taken off by launch to a waiting ambulance at the slipway. And so ended the morning's trials on a rather unfortunate note.

The afternoon's trials were focused on the effects of a 'plough-in' whereby the front of the skirt progressively catches the water to the point where the contact is such that it drags the bow down, with the result that the craft comes to a shuddering stop. The end of that day's trials necessitated the craft returning to the hard standing at Falcon Yard, some way up the harbour. Although Sheepy wanted to take the craft up under its own steam, the Harbour Commissioners insisted that a tug was attached, and at 4.15pm tug *Ower* was made fast. Sheepy wanted a slack tow to prove his point, but for various reasons this was not possible. Later honour was satisfied and it was agreed that for future sorties up the Medina a tug only had to be in attendance and not attached.

---

[4] Jim Newbury later joined Hoverlloyd as a hydraulic and systems engineer.

It must be stressed, however, that these manoeuvres up and down the Medina are particularly hazardous in such a confined waterway and not recommended other than in the winter months and under optimum conditions. If we are to continue with the production of SR.N4s further consideration must be given to a base outside the harbour.

Once the stern of the craft was clear of the floating bridge the tow was cast off and the craft turned to port for the ascent at Falcon Yard. Here again one is juggling a quart into a pint pot and though, after much trial and error and trimming, the landing pads were got into the reinforced concrete areas, it was not without a lot of cold sweat in the near zero cockpit temperatures.

SR.N4 001 floating on its buoyancy tanks after returning to the harbour.

The above report shows just how uncertain Sheepy was, not unsurprisingly, of the control of the SR.N4. As time went on and techniques and expertise were developed, the craft proved very controllable in confined spaces. Although there were incidents in later years where the craft did get out of control, these were due to inexperience or inattention.

The length of the second trial on Monday 5th was 2 hours 55 minutes, Tuesday morning's was 2 hours 5 minutes, and the afternoon 2 hours 30 minutes. So, with possibly just ten hours handling experience on a still very untried and untested SR.N4, Sheepy's exceptional skills as a test pilot were very much in evidence.

SR.N4 001 about to undergo towing trials. Note the towing lines in the bow and the SR.N5 alongside.

Ray Wheeler as BHC's chief designer was very pleased too with the initial trials. Even so there was much work to be done on modifications before 001 could be considered commercially viable. A big problem that became apparent soon after the craft went to sea was the engine air filtration, which also became one of the hardest issues to resolve. In order to reduce the level of salt ingestion to an acceptable level, very significant changes had to be made to the air intake system. The most serious problem, however, was when checking the effect of an instantaneous engine cut of one of the forward systems (resulting in taking away the power to one of the forward lift fans and propellers), under a full load with a 35-knot following wind, basically simulating a 'plough-in'. The longitudinal deceleration exceeded 0.25g, which was considered unacceptable. As a result, a speed limitation of 50 knots was placed on the craft combined with a limitation of the forward centre of gravity and the maximum car load. Work on a revised skirt system started immediately.

As the trials continued Seaspeed pilots were invited to participate as observers. John Syring:

There was then great competition to stand on board when it was doing its sea trials at Cowes, but there wasn't much room on the flight deck, so we might just be able to elbow in to see what was happening, but we were really just getting in the way. We spent a bit of time on board during the trials, but not as much as we would have liked. My learning to drive didn't start until I got to Dover and then I was instructed by both Sheepy Lamb and Harry Phillips.

Rubber bags were lashed to the car deck in strategic places and filled with the appropriate amount of water to simulate loads.

Sometime later David Wise was invited out on a test flight and a few more thereafter. David recalls a memorable incident:

On one particular trial with Sheepy in the left-hand seat and Geoff Howitt from the ARB (Air Registration Board) in the right-hand seat, all was going well for about half an hour when the craft encountered hydraulic problems, which necessitated shutting down two engines, resulting in the craft having to limp back to its buoy. On arrival, Sheepy, never known to be down hearted and still looking very pleased with himself, turned to Geoff and said, "Well, Geoff, what do you think of it?" "It's very good, Sheepy," replied Geoff, "but it doesn't work!"

There were some who were even less kind. On seeing the gleaming all-white craft moored at its buoy, someone was heard to query, "Another white elephant from Saunders-Roe?" – A clear reference to the *Princess* flying boat.

Chapter 13

1968 – 001 at Dover

Seaspeed takes 001

Trials continued during the spring and early summer of 1968 at Cowes, in between which the cabins were fitted out with other equipment to Seaspeed's specification. After a day's trials the craft was either moored on her buoy outside the East Cowes breakwater or went upstream to BHC's Falcon Yard, just past the chain ferry, to carry out maintenance on the skirt[1].

SR.N4 001 leaving the new Dover Hoverport in the Eastern Docks on trials. July 1968.

One particular test that interested John Lefeaux was the rather mundane but nevertheless vitally important onboard service, namely the toilet discharge system – the first to be fitted to any hovercraft. As Lefeaux explains in his book:

[1] Unlike the hoverports at Dover and Pegwell, BHC did not install a permanent jacking system at Cowes. Instead they used the designed arrangement where portable jacks were lowered through manholes in the car deck. Although infrequently used, the portable system proved invaluable on occasions when major skirt damage occurred, requiring repair away from the home ports.

Provision had been made to empty the soil tanks at sea but it was far from clear whether this could happen without the craft anointing itself all over with effluent mixed in the spray from under the skirt. It was important to know this, because if this happened alternative arrangements would have to be made ashore involving tankers to suck the tanks empty and then getting somewhere to empty them. The trial was carried out with paper and a simulated coloured additive to represent the real thing and showed that there was no problem. So a routine was established to empty the tanks on passage. Nowadays I expect that someone would raise their hands in horror at the thought of this but the quantities involved were not large and cannot have done much harm to the environment.

The cabins were provided with standard aircraft-seat floor mountings, but finding suitable seats seemed to pose more of a problem than first anticipated. The concern was that passengers needed to have some protection in the event of a 'plough-in'. Aircraft seat belts were ruled out on the grounds that they would prove unpopular and be a nuisance, as well as being almost impossible for cabin staff to control.

SR.N4 001 arriving at Dover Hoverport looking towards the Eastern Entrance. July 1968.

Instead, it was decided to make the seat backs strong enough so that if people were thrown against them they would not collapse on the people in front. This proved difficult for the seat manufacturer to achieve as the load would be on the top of the seat back, as opposed to seat level in the case of seats with seat belts. It took some time for them to come up with a seat that would meet the ARB test loading without failing.

SR.N4 001 on the pad at Dover Hoverport with a Townsend ferry on Berth No. 3. Dover Castle is on the hill in the background. July 1968.

Although, strictly speaking, prototype trials for 001 were conducted off the Isle of Wight prior to delivery to Seaspeed at Dover, the still very experimental nature of the SR.N4 programme suggests the first season in service with passengers can only be viewed in retrospect as a continuation of the trials period. Tony Brindle had been keen to have the SR.N4 operational as soon as possible:

There was no dedicated funding. Ideally we would have had a real pre-service trial and development period, but I had to get the craft into service because until I did BHC would not be paid by British Rail.

I kept my office in Cowes – I didn't move to Dover as I was heavily involved with Dick Stanton-Jones in the planning. I'd got a start date and I'd booked the 'Royals' (Princess Margaret and her husband Lord Snowdon) to open the service so I put the pressure on BHC to get the N4 ready in order to operate from Dover in June. I actually signed the cheque at 9.30pm on the eve of the service launch day and the BHC accountant was there to collect it. The first time I had signed a cheque for over one million.

SR.N4 001 positioned in the jacking area. The terminal building is in front of the craft and, directly behind the craft and between the ferry tied up alongside the Eastern Arm, are the World War Two submarine pens, which were eventually demolished and the land reclaimed.

Tony had been determined right from the start to get rid of the 'soot and serge' image with which anything associated with British Rail in those days was tainted. To achieve his aim he appointed early on Mrs. Marcelle Connell as Chief Purserette, a delightful Belgian lady who set very high standards and was to lead her team by example in smartness and courtesy. She remained with the company to become one of the longest-serving cabin crew staff.

On 11th June, 001 went up to Dover where trials, training and fitting out continued. There were plenty of teething problems for both Seaspeed and BHC engineers to resolve. Seaspeed had installed an all-important jacking system on the pad at Dover, but unfortunately it malfunctioned on its first lift by raising the front end of the craft at an alarming angle so, as described by Lefeaux, "...it looked like Concorde on take-off". It took Les Thyer[2], who was appointed by Lefeaux to run the day-to-day

---

[2] Les Thyer was an experienced test engineer from the *Princess* flying boat days. He was also a pilot and had worked on the SR.N2.

engineering operation at Dover, all night to get the craft down safely. One other mishap occurred during training while taking avoiding action when the craft was 'ditched' beam on, extensively damaging the side structure, but it was able to return the 20 nautical miles to Dover. This apparently serious damage was repaired within a week.

Car and coach loading trials. July 1968.

Seaspeed's Dover to Boulogne Service Starts

On 25th July, Michael Donne, Air Correspondent for the *Financial Times* summed up Seaspeed's forthcoming challenge in an article, in which he wrote:

> *Next Thursday, subject to the final certification clearances being obtained, the giant 160-ton SR.N4 £1½m hovercraft in the colours of British Rail Hovercraft's Seaspeed service, will leave Dover Harbour on its first-ever fare-paying passenger trip across the Channel to Boulogne (Le Portel). Capable of carrying about 260 passengers and 30 cars, it is expected to have a full load, for there has been an encouraging public response to the start of the service. In its first week of operation, the SR.N4 will make three round trips daily between Dover and Boulogne. After August 7th, it will make six round-*

trips daily at two hourly intervals between 08.20 and 18.20 from Dover and between 9.20 and 19.20 from Boulogne. In the autumn, the number of round-trips will be reduced for the winter period.

Captain Hermod Brenna-Lund being introduced to Princess Margaret by Tony Brindle, with her husband Lord Snowdon behind her, prior to the trip to Boulogne for lunch. 31st July 1968.

Inevitably, the pioneer of the new cross-Channel SR.N4 hovercraft service, British Rail, has to bear the brunt of the publicity, and of the criticism should anything go wrong. In fact, everyone connected with the hovercraft industry in this country, including the manufacturers of other types of craft, such as Hovermarine, Vosper-Thornycroft and Cushioncraft are hoping that the British Rail service proves successful. For upon that success depends not only the long-term future of the SR.N4 itself – and by implication the long-term future of the British Hovercraft Corporation too – but also the success of the entire hovercraft principle applied to large overwater craft.

Nowhere else in the world but Britain has anyone attempted to build such a large craft as the SR.N4, and thus the start of services this week logs another world technological 'first' for the

U.K. As with all such ventures, the rewards can be great but the price for failure high. If the SR.N4 is a success financially and technically, then the U.K., through the BHC, may expect to log orders for as many as 100 SR.N4s over the next decade, worth around £150m. This is quite apart from further orders for smaller craft such as the SR.N5s and SR.N6s.

Nobody expects that the SR.N4 operation across the Channel will entirely escape teething troubles, although reliability is the principal objective of both operator and manufacturer; for without it there cannot be profitability. The success or failure of the operation will be measured not over days or weeks, but months. By the end of this year, it is hoped that a sufficient body of technical and cost information will have been gathered to convince the many would-be operators now watching from the sidelines.

Around the early months of 1969, the orders should start to flow in. It is at that point that everyone should know whether or not the SR.N4, and the entire concept of large overwater hovercraft, is a sound one.

Princess Margaret and Lord Snowdon seated in the front row of the starboard forward cabin on the newly named 001 as *The Princess Margaret.* 31st July 1968.

Third from left: Francesca Greenwood; 5th from left: Jackie Vannobel; 6th from left: Terry Critchley. Other girls: Monica Craven, Sylvia Smith and Susan Jones.

On 30th July a press day was organised by British Rail and a large contingent of national and local press were taken to Boulogne, with Brenna-Lund at the controls, on a day trip for a very wet lunch. Lefeaux records that Chapman Pincher of the *Daily Express*, to whom he was assigned to look after, was very hard work and seemed quite uninterested in the craft and the fact that he was being taken across the Channel in the fastest public service surface craft in the world. The next day, however, Seaspeed received excellent publicity with big favourable write-ups from most of the press.

The press day provided a good dress rehearsal for the next day when Princess Margaret arrived at Dover with her husband, Lord Snowdon, to name the SR.N4 001 *The Princess Margaret*, following which a large crowd of VIPs from Richard Marsh, Minister of Transport, Sydney Jones, Chairman of Seaspeed, MinTech, Board of Trade, ARB, British Rail Board, Dover Harbour Board, Dover itself and Kent were taken across to Boulogne for a rather more dignified lunch. Tony Brindle recalls that all Richard Marsh and Sydney Jones wanted to talk about was the Channel Tunnel.

The following day, the 1st August, the first fare-paying passengers and cars were carried by the newly named *The Princess Margaret* across the Channel to Boulogne, but it was only three days later before the problems set in, resulting in cancellations due to major structural damage, putting the craft out of action until 8th August and necessitating the transfer of passengers back to the British Rail ferries.

The Princess Margaret making the transition from land to water on the start of the service to Boulogne. Note the vulnerability of the bow to wave impact damage as a result of it being unprotected. 1st August 1968.

The Princess Margaret passing through Dover Harbour's Eastern Entrance.

The Princess Margaret mid-Channel on her maiden voyage. 1st August 1968.

The first three Seaspeed SR.N4 commanders in the 1968 season were Hermod Brenna-Lund, Peter Barr and John Syring, who recalled the situation at the time:

> *With a large number of cancellations there was a lot of coming and going between Dover and Cowes. Some of us were living on the Isle of Wight, so we'd go to Dover for a few days, do a bit of N4 driving, and go home again. It was a job with a lot of time off – it always was. There were these long breakdowns and teams would have to come up from Cowes to sort it out.*

At the start of the service only Brenna-Lund and Peter Barr were checked out as captains because John Syring, for his sins, had been appointed safety captain and was casually told to set up the evacuation routine in his spare time, as a result of which his own personal training went by the board. Peter Barr:

> *I had less than 20 hours when I went solo with passengers on the N4. We'd done a lot of standing around in the control cabin with Sheepy Lamb driving during trials and you pick up quite a lot from that. We all at that time had handled hovercraft, so the principles of hovering were not new to us and of course the machine was a wonderful thing to handle once you got the hang of it. It was very manoeuvrable.*

If you got from Dover to Boulogne without a problem you almost noted it in your logbook. Boulogne, of course, was another issue. Usually within ten minutes of leaving Dover one pylon would go hard over and stay there; an APU would chop somewhere along the line – it really kept you on the edge of your seat. All this was fairly amusing in good weather, but when it was nasty it wasn't much fun.

In those days the skirt was under the bow and we didn't have that sort of bulbous protection at the front, so if you took a sea head-on you were liable to do quite a bit of structural damage. And the other thing was it was bloody fast. It was much faster in the early days than subsequently. I did a trip across in that first season with 10,700rpm on the compressor (11,300rpm normal cruise) all the way in about 30-odd minutes to Boulogne. We absolutely rocketed along and if you weren't careful you could do a lovely plough-in, particularly going downwind. Without the bulbous bag[3] we were hugely susceptible to plough-in. Within a few days or weeks of starting the service, BHC announced they were taking the craft back to Cowes at the end of the season and the skirt configuration was going to be changed.

John Syring relates his early experience with the N4:

My learning to drive didn't start until I got to Dover and then I was instructed by both Sheepy Lamb and Harry Phillips. Because of the safety training I didn't get as much exposure to this thing as I might have liked. Eventually it shot off on its maiden voyage with Brenna-Lund as captain and then it became Buggins' turn to do a passenger run to Boulogne. I went to the management and said, "I don't want to give up my chance of commanding the N4 across the Channel, but you realise what I've been doing, don't you?" "Yes." "Well, I've only been to Boulogne once before and I've only got 11 hours on this machine. Is this really right?" They said, "It's up to you whether you do it or not", so I asked Harry Phillips to sit on the jump seat and off we jolly well went. It was safe, but I was naturally pretty overwrought about it.

Brenna-Lund probably only had about 20 or 30 hours experience. But I did have one bonus, because of the way it all panned out. I suppose I was more of a bookworm than a hovercraft pilot and I managed to get the first licence – I've got licence 001 on the SR.N4. It was BHC's decision, but it put a lot of noses out of joint.

---

[3] The 'anti-plough bag' was fitted to the craft for the second season.

Popularity wise I never really recovered from it. The reason was that we had a written exam for engineering and systems, plus all the things about speed boundaries, operating limits etc. and I got more marks than the others.

The three 1st officers for the season were Ian Dalziel, Martin Godfrey and Alan Burns, all of whom became captains the following year.

The Princess Margaret arriving at Boulogne with her first fare paying passengers. 1st August 1968. (Photo courtesy of Pat Lawrence.)

Brian Laverick-Smith, who had come from the Townsend SR.N6, was one of the two original SR.N4 2nd officers responsible for navigation and anti-collision:

I was one of the first navigators on the N4 in 1968 with Peter Henderson, but some of the other captains and 1st officers would sit in the back seat to have a 'look see', but we couldn't sit anywhere else. We didn't do many services in 1968.

Our navigator training consisted of, "There it is – off you go." That was it. So Peter and I got together and we obtained some information from an aeronautical guy. He said that what we wanted to do as we were going so fast, as we hadn't got time to plot, was to use an aircraft system whereby you put a dot with a chinagraph pencil on your departure point and a dot on your destination, use vectors of 10 degrees and arcs of two miles, and just go bearing and distance from your departure point.

We used the radar in relative motion, so as far as anti-collision was concerned it was just a question of giving everything a wide berth – a very wide berth. I remember the rubber visor on the Decca 629 radar was very abrasive and you'd come off looking like a panda. We didn't have curtains around the radars in the first year.

Peter Barr:

With only two 2nd officers we rotated in such a way that one of us would have been on the radar some days. You had to be prepared to do it all.

With the navigator being the only person to have sight of the radar, and being more or less in control, when the fog came down, particularly coming in and out of Dover, if you didn't have full confidence in the guy in the back, it was a nightmare. We gave them what training, such as they had, on the job and it wasn't adequate. Looking back it simply wasn't adequate. We did have a bit of a fog restriction – we didn't go in total zero. If you had someone in the back who you had complete confidence in you would set off in near zero, but if not you wouldn't. That's no way to run a system. Latterly you could leave and arrive without seeing the ground, but in the early days we wouldn't do that – something like half a cable (100 yards) was our controlling factor.

John Syring's take on navigators:

I can't remember having much fog during the 1968 summer, but fog navigation was an interesting concept because here you had the junior member of the crew, who might be a seasonal in later years, who was learning the ropes, but when it became foggy he was effectively in charge. You'd go through the entrance at Dover, which you had to do at about 20 knots to have reasonable control – over hump speed certainly, whatever that happened to be depending on the load – decelerate and do a 90 degree right turn (to the original hoverport in the Eastern Docks) and be lined up for the slipway pretty smartly without any drift. It wasn't a wonderfully wide slipway, but the good navigators did an amazing, amazing job.

When the craft was serviceable it went well and was popular with customers in fair weather, but when the weather was bad the popularity fell off. The two inner cabins each side of the craft, holding 40 passengers each, were much disliked because there was no impression of speed or hovering and many thought they were claustrophobic.

Boulogne Hoverport with the rail link that had been specially built so that passengers could embark and disembark at the terminal. This picture was taken in 1969.

The unreliability of the service due to weather, technical problems and skirt tears, resulting in 30% of the services being cancelled, was attracting much unwanted bad publicity from the press. The skirt was based on a scaled-up version of the SR.N6 one made of heftier material, but it simply wasn't up to the task of dealing with the sea conditions of the Channel. The initial problem was that fingers were being lost altogether, but there were other issues with it too. As has been mentioned before, the skirt became one of the biggest problems of hovercraft operation to solve and was always a huge operational cost, not least because of the number of 'skirties' it took to maintain it.

Peter Barr:

> *In 1968 we had a 2m or 2.5 m wave height restriction, which in the Strait doesn't take much weather to get to that height. The hydraulics were hopeless and we all went round with these little screwdrivers where you could change the filters at the base of the pylon, virtually every turnaround. The APUs were always breaking down too.*

Even so there were some lighter moments in the early days, as experienced by Brian Laverick-Smith:

> *A week after the start of the service we landed in Boulogne and one of the purserettes phoned up and said she had a stroppy*

passenger who wouldn't leave the cabin and could we send someone down to give the girls some assistance in getting her ashore. So as 2nd officer I was sent down. I put my hat on looking all official and went into the port forward cabin and said to the girls, "Yes, what can I do – where's this difficult passenger?"

"Up front by the window; she's not leaving – everyone's gone. You'll have to get rid of her."

"Right." I walked up and there was this little old lady with her bag on her lap.

She said, "I'm not leaving."

I said, "Why's that my love?"

"I've got my ticket to Paris." She wouldn't leave and said, "The next stop is Paris".

I said, "No it's not. The craft can't take you to Paris. We've got a train waiting for you." Eventually she got off and went on the train.

It was the uniqueness of it all. People couldn't understand what the hovercraft was like, or believe what it would do. That memory always comes back to me because the public didn't know what the craft would do. Some of the captains didn't know either!

For that first 1968 season Les Thyer and his team of engineers had a tough time trying to cope with all the problems this new piece of underdeveloped machinery threw at them on a daily basis. It can be argued that 001 had entered service before it was ready, but the reality of the situation was that BHC had no money to properly develop a prototype over a period before putting a production model into service. In order to finance the future they had no choice but to sell straight off the production line.

The only other alternative was to develop the first craft in the environment in which it was designed to operate, but inevitably at risk to future sales if things went wrong, which in fact they did. Perhaps not catastrophically, but nonetheless sufficiently enough to put a question mark against the whole project.

The Seaspeed operation was suspended after just three months and the craft was returned to Cowes for major modifications, the most fundamental of which was a new design of skirt.

Discharging through the stern doors.

Tony Brindle:

> *We withdrew the craft in October. You've got to remember that the original N4 didn't have that deep skirt at the bow – the anti-plough bag as it was eventually called. One day we lost most of the skirt and clearly there was a need by September that there had to be a major, major change in skirt configuration and the decision was made by me to pack it up. I didn't want to, but I had to.*
>
> *There was also another argument running in parallel, because at that time I was daggers drawn with the British Railways Board saying I cannot run a service with a 'one bus garage'. I've got to have a second craft and please can I send them an order. Their answer was, "No."*

This supposed showpiece for Britain's latest technology clearly didn't inspire as there was not exactly a queue of operators rushing to place orders for the SR.N4. The Swedish Lloyd and Swedish America Line consortium remained the only other operator to have two SR.N4s on order and keep the faith. British Rail announced to the press, on the instructions of Dr. Sydney Jones, that they still had faith in the SR.N4, but would not be placing an order for a second one – a less than reassuring statement to potential BHC customers and one that infuriated Tony Brindle:

The announcement by Sydney Jones that BR would not be buying a second N4 came at a time when it could not have been more damaging to the industry. To say I was infuriated was an understatement. I was incandescent. At that juncture we had had much publicised problems. There were some real possibilities for at least two significant overseas sales of N4s. One was to the Canadian government to operate services in the St. Lawrence Seaway to the islands and the other was a service from Gibraltar to Morocco, and I was involved in both.

What was needed at that moment in the N4's history to really light the sales candle was a clear statement of confidence, but it was negated – sadly it would never recur.

That day the real future of giant civil hovercraft was lost.

I gave my own press conference with the principal transport correspondents, but the damage had been done.

The case for the giant hovercraft as a new form of commercial transport had not been proven and the early promise of worldwide orders was becoming a distant dream. Michael Donne had been right in his article when he said, "*As with all such ventures, the rewards can be great but the price for failure high*", and at this point in time the expectation had far exceeded the reality.

Management Changes at Seaspeed

Despite the problems, Tony Brindle's faith in the product held firm, but by early September he realised that even with major modifications to the craft, running just one SR.N4 would never make for a viable operation – he needed another. David McKenna, Chairman and General Manager of the Southern Region, and Tony's great ally on the British Rail Board, had been moved to another job, leaving Tony totally at the mercy of Sydney Jones, who was never really in favour of the hovercraft project in the first place and was in no mood to order a second SR.N4 on Tony's say so, despite his sound arguments. Tony reminded the board that they'd, "accepted the responsibility to go down the road of aiding, encouraging and developing hovercraft", but even with so much at stake there was a reluctance to exercise that responsibility. Tony's solution was to go back to the Minister of Technology, who had agreed his appointment in the first instance, and propose that Seaspeed became independent of British Rail.

Tony explains his reasoning and the consequences:

Within the British Rail structure each Division/Region had a chief executive and I was much the junior. When the Investment Committee met we were all after the same money. So, for example, the London Midland was saying they wanted money to electrify to Glasgow; the Eastern was wanting five new locomotives and a new signal box at York; the Southern Railway was wanting to do away with a three line system, or whatever; the Great Western wanted to electrify to Cardiff.

I wanted another N4, but I was not in the railway business. I was in the hovercraft business and it was ludicrous that I was fighting for money against a signal box or locomotive. It did not make sense, so I said, "I suggest, gentlemen, that I should go back to the minister and say please can I become an independent corporation exactly as Railway Air Services, which was established pre-war by the Southern and Great Western Railway Companies, becoming British European Airways in 1947.

I thought there was no alternative, but one thing I hadn't thought of was that Sydney Jones was saying, "I know another answer – let's get rid of Brindle".

It was very traumatic I might tell you, because Jones didn't tell me – he told the press. I got a call from Michael Bailey, who was transport correspondent of The Times *saying, "Tony, what the hell's happening in the hovercraft world?" So I was suspended, but there was never a reason given to me, but I knew that it was because of my arguments with Jones.*

Tony wasn't in fact sacked. He was offered another job within British Rail, but by that time he'd had enough and resigned. After Christopher Cockerell, Tony Brindle became the second high-profile figure to leave the hovercraft industry. Following his departure the British Rail Board announced that they would order a second hovercraft on the orders of Dr. Sydney Jones.

Tony's departure was a shock to his colleagues, especially as no reason was officially given. Lefeaux concluded that Tony's go-ahead methods and approach to problems was too much for their chairman, Dr. Jones, and some of the British Rail Board, who were not used to dealing with such a rapidly developing subsidiary and a charismatic entrepreneurial general manager. John Syring noted that:

Brindle had a great flexibility of attitude and approach to things, which after he went, was transformed into a more rigid British Rail approach.

Brian Laverick-Smith's impression was:

> *We were a satellite of British Rail governed by a different regime, which was Brindle, totally Brindle. He masterminded it all – he was in charge and it was Brindle. We weren't British Railways. Brindle was the driving force of it all. When he left it started to sink back into the British Rail bureaucracy. The then managers and directors wouldn't have had the strength or individuality to go their own way. They felt the power of the British Railways influence and accepted it – Lefeaux did – we had a whole series of managing directors. Brindle was the only one who stood out amongst them as an individual. The rest of the guys kowtowed to the administration.*

There is one little anecdote to Tony's departure worth reciting. It seems that the higher one climbed in British Rail the harder one fell, as Tony recalls in his early days with the organisation:

> *Dr. Beeching, Chairman of British Rail, was followed by the chap who had been chairman of the Western, who was my direct boss, and I got on very well with him. I was at a meeting at the Railways Board when his secretary came in and said, "Excuse me chairman, the minister wants to see you right away," so he got up and left. He came back an hour later and said, "It's nice to have known you, I've just been fired." – By Barbara Castle.*

Following Tony Brindle's departure, Dr. Jones appointed Don Bartlett as the new managing director, and with it came a number of changes. Don established his headquarters in London, just ten minutes walk from the Sealink headquarters (formerly British Rail Ferries) and about 20 minutes by tube from British Rail's head office, which was a sensible move because, whether he liked it or not, he realised he was more involved in politics than running a hovercraft operation, and he could keep in touch with people who were sympathetic to the cause. Peter Barr's summary of the situation was:

> *British Rail was desperate to slot us into the railway system. There was no question of saying this is a hovercraft, it's completely outside any railway regulation. It's got to be run by the guys in British Rail as best they can, with the appropriate budget and so on. They were forever trying to drag us back to the railways mentality.*

Tony Brindle had been his own operations manager as well as general manager, having responsibility to both the BoT and ARB for the safe management and running of the service. Don did not have the right experience and none of the captains wanted to take on the role, being quite content to just pilot the SR.N4, so Lefeaux was asked to take on the job in addition to being chief engineer. Don Bartlett remained as managing director until 1973 when more changes took place.

Following Tony Brindle's departure, there were serious issues to sort out to make the SR.N4 a commercially viable machine. The situation was extremely grave for the future of the SR.N4, but to give credit where due, BHC, Seaspeed and Hoverlloyd worked hard and together to resolve the many issues that would enable the continuation of cross-Channel services.

Meanwhile, by November 1968, one of the two Hoverlloyd craft was nearing completion, with the second still in the early stages of construction in BHC's Columbine hangar, while further up river in the Falcon Yard, Seaspeed's *The Princess Margaret* was undergoing major modifications, which would be built into the Hoverlloyd machines before they even saw the light of day. Seaspeed was to see the two Hoverlloyd craft enter service on the Ramsgate-Calais route before their own modified craft resumed operations in mid 1969. But many of the problems encountered in the first trial season were far from solved and another fraught summer for BHC and their two SR.N4 customers was to follow.

PART THREE

Hoverlloyd's Hovernauts

Chapter 14

Cadet to Hovercraft Pilot

Flight Crew 1971. Left to right: Bob Adams, Tom Wilson and John Lloyd – Master Mariners one and all.

The Big Question

Who was to fly this amazing new invention? Strictly speaking, could the term 'flying' a hovercraft be used at all?

At the very start, what should be the prerequisites for those personnel wishing to become 'hovernauts'?[1] Did they need flying experience and qualifications because it was, after all, an aircraft minus the wings? Or did they need experience of the sea because the machine was going to deal with the vagaries of wind and tide on the surface of the sea amongst maritime traffic?

---

[1] 'Hovernauts' appeared in the press in the very early days, obviously borrowed from that other contemporaneous topic of fascination, the NASA space programme. While 'astronaut' gained instant legitimacy, 'hovernaut' was always somewhat tongue in cheek, never meant or accepted seriously.

Up until the Hovercraft Act of 1968 came into force, Air Cushion Vehicles (ACV), as hovercraft were also known, operated under a 'Permit to Fly' issued by the Air Registration Board (ARB). This essentially covered the craft itself, but not the qualifications to 'fly' it.

In those early days some commercial services were therefore being commanded by men who had no specific Board of Trade licence to do so. In this respect they differed from the captains and officers of ships and aircraft. These hovercraft commanders were in this situation largely because, as usual, events preceded legislation. The public, however, was safeguarded by two factors. Firstly, a hovercraft-operating company had to have a verbal understanding with the BoT that its commanders were sufficiently qualified (according to the route being operated and the type of craft) along lines which had already been discussed at the BoT Working Party on ACV personnel licensing; secondly, the operator had to see that his commanders had the necessary ACV piloting training – usually given by the manufacturers of the ACV concerned.

An indication of the quandary this caused was that the very first companies to run *bona fide* passenger services, starting within weeks of each other in 1965, had diametric ideas.

The first, Clyde Hover Ferries in Scotland, chose mariners, albeit with a preference for those who had flying qualifications in addition. Hard on its heels the second company, Hovertravel, operating with the same type of SR.N6 craft on the Solent, exclusively chose men from a military aviation background.

Those enterprises already involved in shipping tended naturally towards mariners and *vice versa*. It was a foregone conclusion that British Rail, already a cross-Channel ferry operator and the Swedish Lloyd-Swedish America partnership, long established in passenger shipping, should view the operations of Seaspeed and Hoverlloyd as maritime and recruit their personnel accordingly.

In the same context, it doesn't take too much of an educated guess to fathom where the British Hovercraft Corporation, with its continued involvement with aircraft manufacturing, stood on the issue. BHC's chief test pilot, Sheepy Lamb, was adamant in his opinion that airmen, not mariners, should be hovercraft pilots. His view expressed to a Board of Trade meeting, latterly convened to discuss manning and operating standards was:

> *They should not be mariners, because mariners don't have the manual dexterity to handle a hovercraft.*

Luckily for us all, his suggestion was politely ignored by the assembled sea captains of the board committee[2] – hardly surprising really.

Not only was the choice of personnel left to the individual operators but the level of prior experience and prerequisite qualifications was also initially left up to the managements to decide. This is not to say the ministry did not maintain a close eye on safety right from the start. It was apparent from the very beginning that the ministry had chosen what proved to be an enlightened and effective policy of keeping a loose hand on the reigns and allowing the operators, within a broad set of rules, a high degree of self regulation.

This most unbureaucratic and praiseworthy of approaches had the effect, whether foreseen or not, of fostering within the operating companies a much higher sense of responsibility than might otherwise have been the case under a more pedantic and prescriptive set of rules imposed from above.

This was by no means a policy of idly sitting back and allowing things to develop on their own. The extent of the ministry's careful surveillance is amply illustrated by the additional policy of putting their own marine surveyors through the hovercraft training programmes, with particular emphasis on the entry-level navigational tuition. All those appointed as marine surveyors with special responsibility for the hovercraft were, prior to commencement, expected to go through the full training programme, and what is more, 'check out' like any other recruit at the end of it[3].

The end result of this inspired regime was the mutual respect it engendered between company personnel and the marine surveyors. Certainly, throughout Hoverlloyd's brief life the 'men from the ministry' were looked on as helpful friends rather than bureaucratic enemies.

---

[2] Even when the enabling Hovercraft Act of 1968 came into being, it was not until 12th July 1972 that it took full effect, including the results of deliberations at the Board of Trade concerning minimum entry qualifications for hovercraft captains. These deliberations had continued through 1968-71 culminating in the official requirement of a First Mate's FG Certificate as a prerequisite to command a hovercraft. This was clearly the lowest common denominator position, as Hoverlloyd and Seaspeed continued to require Master Mariners.

[3] Marine surveyors were qualified Master Mariners working for the Mercantile Marine Office. In addition, back then, they were usually expected to hold an Extra Master's Certificate, a degree-level qualification, which because it was still considered a 'Certificate of Competency', required a 70% pass mark to qualify. A pass level roughly the equivalent of the average honours degree. These guys were no slouches!

Hoverlloyd's Choice

So, despite Sheepy's sage advice, Hoverlloyd chose to recruit mariners and then went one step further by stipulating that anyone requiring a permanent flight crew position in the company should also possess a Master Mariner's FG certificate.

In the maritime terms of the day this could have been considered unnecessarily over qualified, the FG stood for 'foreign going', meaning the holder possessed skills to command a vessel of any size anywhere in the world. Such skills included knowledge of celestial and coastal navigation, magnetic compass adjustment, ship-handling in a variety of situations including tropical storms, not forgetting the commercial knowledge necessary to handle charter party and bills of lading disputes; all part of a captain's daily routine, ensuring his ship was run efficiently and profitably for his company.

In comparison, regardless of its frequent description as a 'giant hovercraft', the N4 was tiny, even by the shipping standards of the day. Its small size and its short route across the Channel should have put it firmly into the Board of Trade's definition of coastal craft, the qualification for command of which was the much lowlier certification of Master Home Trade.

Of course 'manual dexterity' aside, this was something very different, a hundred times more sophisticated and complex than any coaster, or cross-Channel ferry for that matter. The calibre of person to command such a machine required the highest level of marine skill and qualification in educational terms alone. However, whether it was considered at the time or not, in retrospect the real essential prerequisite knowledge for the ultimate safe operation of a fast craft across the Dover Strait was a thorough and comfortable knowledge of radar operation as an anti-collision device in all visibilities.

If there is one attribute which stands out when reviewing the comparative skills of airmen and mariners, it is the whole business of avoiding collision at sea and the role of radar in that pursuit.

Aircraft are fitted with radar but their function is strictly for weather detection and has nothing to do with collision avoidance, which is the responsibility of air traffic control and, in a three-dimensional environment, planes can be stacked at height intervals in complete safety. The two-dimensional world of the sea surface offers no such luxury. Ships cannot be stacked at thousand foot intervals and at times are confined to narrow corridors in places like the Singapore and Dover Straits, which inevitably leads to overcrowding of the sea lanes and an increased danger of colliding.

Collision at sea is a hard fact of life and, even with the advent of radar and its subsequent development, still continues to be an all too common occurrence[4]. The understanding of the maritime 'Rules of the Road', all 32 of them, and the skill to interpret a radar picture in all weathers, plus the ability to recognise the movement of other vessels and objects on the screen, was an essential requirement for any ship's officer and remains so to this day.

In contrast, the aircraft pilot, however well qualified and experienced in the air routes of the world, knows nothing of this and in consequence could be a dangerous liability in the maritime environment. This would have been the case in the average 15-knot merchant ship; in a 60-knot hovercraft it would have been lethal.

The First Radar Generation

Another fact worth considering is purely coincidental. All the young men joining Hoverlloyd in 1969 could be described as the 'first radar generation', those who had joined their first ships in the mid to late fifties and thus had grown up with radar from its inception. They differed markedly from their predecessors, those of a generation used to a less technical world without new-fangled gadgets. Almost to a man, these 'old hands' looked upon radar as either totally incomprehensible and best avoided, or even more catastrophically, approached this new means of seeing in the dark and through fog with great enthusiasm but with very little understanding.

In retrospect, what is difficult to understand today is why, despite the obvious widespread ignorance, very little official consideration was given to formal training in the use of radar. Neither the Board of Trade nor the shipping companies, who had most to lose as damage and loss of vessels escalated, thought to train seafarers in the correct interpretation of what they saw on the radar screen.

In 1956, the well-publicised instance of radar misuse, the catastrophic *Andrea Doria/Stockholm* collision in the approaches to New York, resulting in 53 deaths and the sinking of the *Andrea Doria*, did prompt the British government into changing its policy. This, however, can only be described as an extremely half-hearted and limited measure, which failed totally to address the real problems.

---

[4] In fact the first 15 years after radar was introduced to merchant ships, the incidence of collision actually increased. The operation of radar as an anti-collision device and the era of what became known as 'radar assisted collisions' is covered in more detail in the following two chapters.

In 1957 the Board of Trade introduced the Radar Observer's Certificate as a mandatory prerequisite for candidates for Certificates of Competency at all levels. The syllabus required a broad knowledge of radar functionality, an awareness of its limitations and, even more essential, a thorough understanding of how to determine and interpret target movement on the screen. Manual plotting was encouraged as the only sure way of determining another vessel's course and speed.

In the first few years, nautical colleges found themselves ill-equipped to deliver the course at any sort of effective level. Plotting could be taught as a paper exercise, but interpretation of the screen itself required an actual radar. Most colleges acquired facilities on the waterfront and installed radars looking out to sea. This was a step forward, but not entirely satisfactory either.

The instructors did their best, but the most crucial aspect of radar operation at sea, the ability to understand relative motion on the radar screen, could not be demonstrated in a static environment. Much like expecting a learner driver on the road to remain stationary while the rest of the traffic flows past. This inadequate method of training remained for another five years, by which time the first, again very limited, simulators were introduced[5]. Also, by the early '60s, some colleges were starting to use small vessels, which were the ideal dynamic solution, but not necessarily the most economic. Most establishments could not afford them.

Limited or not, such training was better than nothing, but what remains totally inexplicable to this day, is the fact that the already qualified Masters were not required to take the course. Some more enlightened companies voluntarily sent their Masters on the courses, but regrettably the vast majority stuck to their usual response in such situations – "If it is not mandatory we are not paying for it." The senior people in charge, the very persons who needed to be familiar with radar operation for the safe conduct of their vessels, were left in total ignorance.

The 1964 inquiry into a less publicised, but in many ways, more significant collision in the English Channel between the British registered ships, *Crystal Jewel* and the *British Aviator*, clearly identified that both Masters had totally erroneous impressions of what they were seeing on the radar screen. The inquiry's report further suggested that if two well-regarded and supposedly competent senior captains could make such mistakes it was highly likely that most others at sea also laboured under the same misapprehensions.

---

[5] Two establishments responded amazingly quickly, several years ahead of their rival colleges elsewhere. Liverpool Nautical College acquired an Ultra Electronics simulator driving two Kelvin Hughes radars in 1957. Closely followed in 1959 by Sir John Cass College in London, which installed a Redifon system using three Marconi radars.

This also failed to move the Board of Trade to action and it is an almost unbelievable fact that nothing was done seriously about retraining until the International Maritime Organisation (IMO) decreed in 1995 that all member governments should take steps to ensure maritime personnel under their jurisdiction revalidated their certificates, to bring them up to date with the developments in electronic navigation systems that occurred in the final quarter of the 20th century. Such innovations as satellite navigation, electronically displayed charts (ECDIS) and, at last, the inclusion of radar plotting systems, generally known as ARPA.

For the first time since the introduction of radar in the 1950s, after a mere half century of waiting, the maritime world could confidently expect the whole bridge team on any ship to have a sound knowledge of radar and its safe operation. Some would say, not before time!

Going back to the initial years after the introduction of the Radar Observer's Certificate, a bizarre situation existed where the only person on the bridge of a ship qualified to operate and interpret the radar screen was the most junior officer, the 3rd officer. And, because of his lowly rank in the hierarchy, the odds were his advice was rarely asked for, or if given, invariably dismissed. Robin had the ludicrous experience on a Cunard passenger vessel, as Junior Third Officer with a 1st Mate's Certificate and the obligatory Radar Certificate, of being the only officer of the bridge team not expected to view the radar screen:

> *Cunard had a policy of having two watch-keeping officers on each watch with Master Mariner Certificates on their passenger ships. During the mid sixties, Cunard was desperately short of Master Mariners, so they were taking the 2nd officers off the cargo ships and putting them on the passenger ships for a period of time as junior watch-keeping officers. This was not popular with some because one only received two thirds watch-keeping time towards the time necessary to sit the Masters exam. The compensation was outstanding food and service with a seat at the officers' table in the first class dining saloon.*
>
> *My turn duly arrived and I was appointed as Junior Third to the* RMS Carmania, *an elegant 21,637-ton liner with a green hull (originally the Saxonia, built in 1954) on the Southampton to Montreal run – a route prone to the North Atlantic fog for much of the time when it was not blowing a gale. The senior watch keeper on the 8-12 with me was the Junior First Officer, otherwise known as 'Sir'.*
>
> *When the fog did arrive, I was utterly amazed at how the bridge operated. The captain appeared, at which point the senior watch keeper was banished to the wing of the bridge to listen for the fog horns of other ships as we charged along at 20 plus*

knots. There were two quartermasters on each watch, one of whom was on the wheel[6]*. To my utter surprise the other one manned the radar. His job was to inform the captain if a target appeared on the screen.*

Although Cunard had sent some captains on a radar course, I was the only officer on the bridge of the Carmania *at the time with a Radar Certificate and yet all I did was to stand around waiting for the watch to end. I had absolutely no role to play at all.*

Apart from the absurdity of not using the only qualified person on the bridge to man the radar, 'charging along at over 20 knots', also highlights an all too prevalent fault of that era. The senior officers may have had a minimal understanding of the radar picture but paradoxically, at the same time, a touching faith in its capabilities. The common view was that now there was an instrument that could see in the dark and through the densest fog there was no need to slow down – an attitude which can be identified as the root cause of most 'radar assisted' collisions at the time.

Unlike Robin, who attended the elite pre-sea school Pangbourne Nautical College, Roger went straight to sea from school. Consequently he was unsure whether to be surprised or not at his first reception. In total contrast to Robin's experience, the senior officers on his first ship were all too eager to pass the responsibility to someone else, however junior, and in this case, someone even more in the dark – theoretically – than they were:

I joined my first ship in 1955, straight from school totally ignorant of ships and seafaring in general. It turned out I was not the only new introduction to the vessel. They were just finishing installing the ship's first radar on the bridge.

Both the captain and chief officer were obviously flummoxed by the contraption. The captain called me in and said, "Here you are lad here's the manual, go away and read it, and when you've learned how to use it, you can be the radar officer until we clear the Channel." One more problem solved.

Interesting to speculate now, if Christopher Cockerell had come up with his invention some 15 years earlier and cross-Channel operations had attempted to start in the mid '50s, the likelihood is that it would have been a complete failure. The capability to operate unhindered at night or in fog – a frequent occurrence in that area – would have been

---

[6] Quarter masters were able seaman assigned to specific duties such as steering and gangway watch in port. They were a cut above the average crewmen, but not petty officers.

impossible. Radar, although developed through the Second World War years, was still unreliable and the 'old school' mariners would not have had the skill and experience to operate it safely.

This era of radar-induced death and destruction continued on and only started to tail off when our generation were starting to take on senior positions. The experience related by Jon Morris, who started his career a decade later and joined Hoverlloyd in 1977, suggests that dubious reliability and dangerous attitudes remained a lot longer than one might have expected:

> *I first went to sea in the mid '60s when radar was all the rage. The captains I sailed with though, didn't really understand it. I sailed with the complete range, from one who insisted that it was never turned off, "Keep it warm and dry", to one who said, "Save it for when you need it". I have to say that the 'warm and dry' instruction was the best for the radar. The 'save it until you need it' option invariably meant that when you did need it, it didn't work!*

Navigating Apprentice to Master Mariner

As Hoverlloyd opted for the Master Mariner's Foreign Going Certificate as a pre-requisite for employment, it is worth relating by what road we came to such a qualification.

Unlike most other professional and academic qualifications – except possibly the mining industry – those pertaining to the handling of ships, at whatever level, are known as 'Certificates of Competency'. In those days, such certificates were issued by the British government through the Mercantile Marine Office of the Board of Trade. All aspects of standards, training, examination and qualification were handled by that department.

The system originated in the 19th century when the government of the day had come to the conclusion that if standards of safety in shipping were to be improved, the qualification of mariners needed to be formalised.

Safety is the keyword and, although plenty of book learning and practical experience is involved, this has never been an academic qualification so much as an assurance of safety of operation. The consequence being that pass marks in written examinations were extremely high, never less than 70%, followed up by a searching oral examination, which was designed to ensure the candidate was fully conversant with all aspects of the tasks the certificate qualifies him[7] to undertake.

---

[7] Although female Master Mariners were not entirely unknown in the British Merchant Service prior to the 1970s, it was only about that time that women in any significant numbers started to take an interest in seagoing careers.

Typical of the cargo ships ploughing the seas in the mid to late1950s was British India Steam Navigation Company's *SS Woodarra*, 8,753 gross tons, steam turbine 12 passenger/cargo ship with a complement of just under 100 officers and Indian and Pakistani lascars, built for the UK/Australia wool run. This compares with the modern day Maersk 'Super E' container ships, several times the size, which are expected to be allowed to operate with a crew of just 13 when they come into service.

The Master Mariner's qualification was not the only examination, but the culmination of roughly a ten-year progression starting from as young as 16 years of age. The standard pattern was a four-year period as a 'Navigating Apprentice', immediately followed by the first step on the ladder, the examination for a Second Mate's Certificate, which entitled the successful candidate to go back to sea as a watch-keeping officer.

The next step up was the First Mate's Certificate, generally regarded as the easy one. It required additional subjects such as cargo handling and ship stability, the responsibility for loading and discharging cargo being the principal care of the chief (1st) officer on any ship, regardless of its size or trade.

Thus, after another period gaining experience at sea, the ultimate goal of Master Mariner loomed on the horizon. By this time the age of the average seagoing officer was 26, although there was in fact a minimum age stipulation of 25.

Why so long a training period? In theory all the subjects contained in the three qualifications could be comfortably packaged into a two-year college course. The answer lies in one word – 'experience'. The whole regime revolved around 'sea time'. For instance, the four-year apprenticeship required four fifths of that time to be spent actually at sea. There were a number of exemptions to this basic requirement if the young man attended, from the age of 13, one of the elite training schools, the likes of *HMS Conway*, *HMS Worcester* or Pangbourne Nautical College, establishments that to all intents and purposes were public (private) schools with a strong nautical flavour, then, as much as a quarter of the stipulated sea time could be waived.

On the Dole

Many shipping companies continued to pay their employees 'study leave' whilst attending a nautical college prior to examination for their tickets. Shipping concerns have never been renowned for their philanthropy but in this case there was a method in the madness, an ulterior motive. In some cases the all-important remuneration was promised only if the officer agreed to sign a contract – commonly two years – to return to the company's employ once the new certificate had been obtained. Thus, some return for the money expended could be recouped, especially on the training of cadets and apprentices.

However, never blind to the chance of saving a shekel or two, since during the study period we were technically unemployed, the study leave pay was docked by a sum equivalent to the unemployment benefit.

The result, long lines of 'students' turning up at the local Social Security (Dole) office twice a week. First on Wednesday to 'sign on'; basically a statement to the effect that we were still out of work and requiring some assistance, then again on Friday to draw our 'dole'.

This, of course, played havoc with the routine at college. The classrooms on Wednesday and Friday mornings would have been virtually empty. In fact, on many Fridays, the whole day was written off, invariably once we had drawn our money for the week the usual procedure was straight to the pubs or clubs for end of week celebrations.

We were young and out to enjoy ourselves to the maximum, most of us treated our study time as one long holiday amongst the booze and the girls – a welcome change from the monastic drudgery of months at sea.

One sad and embarrassing aspect of this rather odd arrangement was the downcast lines of the genuinely out of work. Mostly middle-aged men, who watched these cheerful youngsters, many up for First Mate's and Masters and therefore earning good money, obviously comparatively well-to-do, turning up in their sports cars, happily discussing last night's or tomorrow's party, queuing up for, what was clearly to them, merely pocket money. What they must have thought I can only guess – at times I know it made me extremely uncomfortable.

Roger Syms

The period at sea between the 2nd Mate's ticket and the 1st Mate's examination required another 12 months of time actually at sea, and from there, to applying to be examined for Master, took another two and half years.

During the time we are talking about, the '50s and '60s, the usual procedure when contemplating any of these hurdles was to book oneself into one of the dozens of 'nautical schools' dotted around the country. Although respectable teaching establishments on the surface, in reality they were not much more than cramming institutions, which ran a continuous and seamless conveyor belt of tuition throughout the year. A young hopeful simply left his ship, enrolled at the school, sat down at his allotted desk, worked through the tutorials and listened to the lectures. Some months later there would eventually come a day when he would realise he had heard that lecture before and decide it was clearly time to 'put his papers in' for examination.

There were Mercantile Marine examination centres in all of the major British ports, which worked much the same system. The qualifying exams, usually a three-day affair followed by two days allotted to the oral exams were, depending on demand, almost on a weekly basis.

If this should appear to the modern eye an amazingly frenetic, beehive-like activity, it most certainly was, but nevertheless it was necessary at that time, simply to keep up with the demands for manpower from Britain's huge merchant shipping fleet.

In 1957 there were 322 passenger ships, 1,145 cargo ships and 575 tankers flying the British red ensign, making a total of 2,042 – the largest merchant fleet in the world. The great shipping companies such as P&O, Cunard, Blue Funnel, Port Line, British India, Shaw Savill, Clan Line, New Zealand Shipping Company with passenger and cargo ships; BP, Shell and Esso with tankers, to name a few, were symbols of a proud maritime nation with an imperial past. Since that peak position, numbers have steadily declined[8].

---

[8] The *Daily Telegraph* reported in 2007, "Since 1975, the number of UK-owned and registered vessels with a capacity of 500 gross tons or more has slumped from 1,600 to less than 300, while the number of British seamen serving on them has dropped from 90,000 to just 16,000." Current statistics for 2011 list 527 ships registered under the British flag and 275, although British owned, sailing under foreign flags. The latter statistic highlights the phenomenon of 'flags of convenience' (FOC), offered by any country as a means of revenue. The 'convenience' lies in the lower standards for registration, which results in cheaper-to-run vessels. The numbers of vessels flagged in this manner have grown almost exponentially in the last 60 years to become the dominant sector in the maritime world. The two major FOC countries, Panama and Liberia, have 8,891 registered vessels between them, and the tiny island of Malta boasts 1,571 in its register.

To keep up with post-war reconstruction and development, the tanker fleets were especially numerous. As an example, by the end of the 1950s, British Petroleum's tanker fleet numbered 146 large seagoing vessels, at the time claimed to be the largest fleet under one company flag in the world.

At this high point in Britain's mercantile fortunes the country desperately needed mariners and Master Mariners in particular. One problematical fact of seafaring life was, and still remains, the enormous loss of expensively-trained people from their late twenties onwards. Just when a young seafarer was about to become fully qualified, invariably his thoughts strayed back to shore.

In European cultures, men tend to marry in the latter half of their twenties and by the time they are in their early thirties children appear on the scene. To a young married man with a family, spending long months away at sea starts to rapidly lose its appeal. As a result of the considerable turnover of personnel, especially at the senior levels, the teaching and qualifying machinery was continually at full stretch, simply trying to keep up with the shortfall.

Lord Kelvin and his Balls

Seafaring has always been known as a conservative profession. The tendency to look upon anything new with great suspicion has always been a notable idiosyncrasy, but especially marked in the latter half of the 20th century. In a time of phenomenal technical innovation, the attitude, 'if it was good enough for Nelson', still prevailed in the maritime corridors of power with hardly a word of dissent.

A prime example in our day was the gyro compass. By the 1960s the gyro compass was a fully-developed and reliable system, having been in existence for half a century and, by that time, carried on the majority of ocean-going ships. However, from the hierarchy's perspective it was still far too new to be trusted, unlike the magnetic compass which could boast at least a thousand years of development since the Chinese invented it.

Although gyro compass principles and operation was taught at 1st Mate's level, a far greater emphasis was attributed to the magnetic compass. 'Magnetism and the Magnetic Compass', was a major part of the syllabus for the Master's examination.

Paradoxically, Nelson wouldn't have known a lot about it as problems with the ship's compass didn't really surface until iron ships were built in any numbers later on in the 19th century. Although it can be assumed that the long lines of guns would have had some influence on the compass, navigation at that time was still rough and ready enough for a few degrees error on the compass to be largely ignored. An all metal ship was a different kettle of fish entirely, because, in an all ferrous metal environment, a magnetic compass could be next to useless.

The ship's binnacle, unchanged in 130 years.

During the latter half of the 19th century the science of the magnetic compass was perfected, and since then the ship's binnacle has remained, totally unchanged in any respect, since Lord Kelvin patented the final improvements in the 1880s. It is still installed by law on every ship today, looking startlingly antique alongside 21st century radar, electronic charts, and global positioning systems.

So, despite its limited use from a practical point of view, a major part of our study for our Master's ticket was the magnetic compass[9]. The idea was to calculate the effect of the ship's iron structure on the compass and, with the use of strategically placed magnets and pieces of iron installed in the binnacle, neutralise the ship's magnetic effects so all that remained was the influence of the earth's magnetic field. Incidentally, one of Lord Kelvin's final patents was the use of the cannon-ball sized soft iron spheres attached on either side. Inevitably in the seafaring world, it wasn't long before they were generally referred to as 'Kelvin's Balls'.

Examination Nerves

Regardless of the subjects involved, examinations are always a time of stress for any student and it was especially so for the young seafarer. The 'official' nature of the proceedings, run by a seemingly indifferent and unsympathetic public service department, demanding a high level of pass, added to the tension.

And there were those terrifying orals!

As mentioned before, the 1st Mate's examination was a relative stroll in the park – it was the first and last that were the real ordeals.

The 2nd Mate's exam, which could be taken at a minimum age of 20, consisted of six written papers, an oral exam, where the principle subject was the 32 Articles of the Rules of Road, and a signals exam consisting of a test in taking down messages transmitted by both a Morse lamp and semaphore flags. One could fail any one of the three but would only have to re-sit the part failed. The written exams required an overall pass mark of 70%, with a minimum pass in certain subjects.

As Roger says, the signals exams at Southampton were not a lot to worry about:

> *The signals examinations were invariably conducted by ex-Royal Navy Yeomen of Signals. It was usual for these somewhat fatherly veterans to offer the young gentlemen a final practice session before the following day's exam. It was always well worth the few shillings to attend these classes, as it was surprising how often the messages practised the night before appeared to closely resemble those used in the exam. I*

---

[9] This part of our training was never needed at sea. If there was any trouble with the compass a professional compass adjustor was called in. When it came to the SR.N4 it turned out to be totally irrelevant. Although it used a magnetic compass, with a hull construction of aluminium, fibre glass and rubber we were back to Nelson's day with not a lot to worry about, except in an incident related in Chapter 23.

am sure the authorities were well aware of this 'nice little earner', but nobody seemed to object, and we the students certainly didn't!

On obtaining the 2nd Mate's Certificate, the fledgling ship's officer would proudly present it, after much celebration, usually to the shipping company with whom he had served his apprenticeship, and be appointed as 3rd officer on one of their ships. A further year of sea time as an officer of the watch was required before one could go back to college for two months to prepare for the 1st Mate's Certificate, which again consisted of a series of written papers, an oral exam and the signals exam to ensure one's Morse and semaphore skills were still up to scratch.

One might think that by the time the young hopeful had reached the rarefied levels of the examination for Master Mariner, a certain amount of self confidence had accumulated and the whole process could be approached, if not with complacency, at least with something like calm assurance. None of it – if anything, the Master's examination was the pinnacle of terror.

The Deviascope – another instrument of torture.

The Master's exam, like its two predecessors, consisted of the ritual written papers, oral and signals test. The three-hour written paper on magnetism was a 'must pass' and the oral exam was in two parts, one of which was 'swinging the compass' usually on a rickety old revolving table, known as the Deviascope, marked out as a ship with some distant church steeple out of the exam window as a reference point.

An hour was generally allowed for 'swinging the compass' before returning for an hour or more of grilling on general matters. Roger's four-hour grilling at Southampton was by no means unusual:

> *'Grilling' was hardly the word. The oral examination for any of the certificates was terrifying and particularly so, the one for Masters. The examiners were retired Master Mariners, and holders of Extra Master's Certificates and many, after some years of interviewing shaking, inarticulate candidates, week in week out, had developed eccentricities, to put it mildly.*
>
> *My Master's oral examination took four hours, an hour of which was fiddling around with the Deviascope. It was conducted by an examiner who had refined a technique that the CIA would be proud of. It was simple in execution but devastating in effect. He would ask a question which I did my best to answer, I thought, correctly. His response was to stare at me for what seemed like an age without giving any indication that the answer was right or wrong, or that he might want me to elaborate a little more. Ultimately he would conclude this impassive but disconcerting stare and ask the next question.*
>
> *It wasn't too bad when the answers were simple things like, "How many fathoms in a shackle of anchor cable?" However, when it came to procedural or protocol matters it was murder. He told me my ship, a tanker had been in collision in the English Channel and a fire had resulted in one of the forward tanks, what was I to do? I listed all the things I could think of; getting crew organised to fight the fire, look to the continued safe navigation of the ship, maintain contact with the other vessel, assess damage, contact my company and keep other authorities informed etc. etc. At the end of my list the examiner sat and stared. I thought of a few other things I might do, still the impassive stare. I thought of something else, no end to the stare. It was excruciating, I even started thinking maybe I should make sure the ship's cat was OK. I thought better of the cat idea and sat dumb for a few minutes, knees knocking with the strain, after which without any comment whatsoever he asked the next question.*

After three hours of this torture he told me I had passed my oral exam and, having already passed the written exams, I was at last a Master Mariner. Of course I was glad to have obtained my ultimate qualification, but I was even more relieved at the thought of never, ever, having to go through that sort of ordeal again.

As Roger says, the strain on the examiners themselves resulted in a number of cases of what can only be tactfully described as personality disorder. Jon Morris recalls that one examiner had obviously got to such a point of hatred of his job that he wasn't inclined to pass anybody:

My orals examiner was a little ginger-haired short guy; I don't remember his name now. Anyway he failed the first 14 people of which I was number 12. He asked me at the end whether I thought I had passed, to which I replied, "Yes", because I genuinely thought I had. "Well you haven't," he said, "you've failed!" The college complained about it this time, because he had also failed the best and brightest student in the class. This guy knew more about the Master's syllabus than the instructors. From then on though, the little shit passed everybody!

Master Mariner to Hovercraft Pilot

Considering the average age of the original flight crew, prior to and including 1969, most had made the decision to depart from seafaring almost immediately, or at least, a very short time after obtaining their Master's Certificate. What was the attraction of the hovercraft business? Or was it simply that qualified people were needed and anything was better than returning to sea?

At that time seafaring could be a hard business, not so much the liner trades which ran their set routes like clockwork and officers knew exactly when they would be in the home port for a considerable time ahead, but general cargo and tanker vessels could find themselves wandering the world at the whim of charterers, negotiating on a daily basis in London, which could keep them away from home for as much as a year.

The next best thing to settling down in a 'shore job' was to join a vessel that didn't spend so much time away. There was a strong interest in the ferry companies, which were starting to expand at that time. It was still going to sea and still well-paid with the attractive bonus of being home every other day. Hovercraft operations promised to be even better and, because of the noise issue, it was unlikely any service would operate over night, so this would be seafaring with every night in your own bed, plus the benefit of not having to 'keep ship'. This had to be a worthwhile consideration.

Probably there was a spectrum of motivation from straight out opportunism to genuine enthusiasm for a brand new concept.

Robin, an enthusiast for life in general, was captivated by the whole idea and just had to be part of it:

> *I was with Cunard at the time of the HoverShow, July 1966, keeping ship in the London Docks during the seaman's strike, the first national strike since 1911, which lasted for two months and severely damaged the British economy. I went to Pangbourne Nautical College aged 13 and then after 'O' levels joined British India where I served my time as a cadet and then 3rd officer out on Eastern Service.*
>
> *I joined Cunard with a 1st Mate's Certificate and now had just a few months more sea time to complete before being able to study for my Master's Certificate and, as a married man, my thoughts were already turning to a new career post Master's. I came away from the show armed with piles of literature and totally fired up about a career in the hovercraft industry. I had a Private Pilot's Licence, which I thought might help, as there were both mariners and aviators 'flying' what hovercraft were around, but the raw truth was that in 1966 there were not many opportunities to be had in this new industry. The answer seemed to be to follow developments closely, obtain my Master's Certificate and then start pursuing every possible avenue, which eventually led to Hoverlloyd.*
>
> *It was now 1967, and I was chief officer, aged 25, with a brand new Master Mariner's Foreign Going Certificate on Townsend's roll-on roll-off cross-Channel ferry service from Dover,* Free Enterprise III*. It was just at the time Hoverlloyd was thinking about recruiting additional flight crew for their SR.N4 service from Ramsgate to Calais, which was due to start in the summer of 1968, but was delayed. I managed to obtain an interview at which it was explained that it was a very risky venture and had every chance of failing. Never being averse to risk, I considered this an exciting challenge. Furthermore I was invited to take up a current offer they had available of (I think) 15 hours training on the SR.N6 for the princely sum of £100. I believe there were only two of us who took up the offer (the other being 'Big' John Lloyd[10]) but I was confident that this would assure me a place in the Hoverlloyd team to pioneer this great British invention, which it duly did and I joined on 2nd January 1969.*

---

[10] There were two John Lloyds in the company, easily distinguishable by their height.

Opportunism is the art of not only being in the right place at the right time but recognising the opportunity enough to grab it with both hands. Bill Williamson, as it happened, was in the right pub at the right time and didn't hesitate. After getting his Master's ticket, Bill spent a year with a salvage tug company, but one fateful day, as the novels put it:

I was at home in Glasgow and, being winter, there was no one else home so I used to go into Glasgow, to my mate's bar in the Gorbals, sit and have a couple of beers and read the paper.

I was down there when this guy came in and said, "Hovercraft are going to set up in the Clyde." I said, "Do you know the company's name?" and he said, "Come in tomorrow and I'll give you a name and address." And that's how it started with me.

For Bill it couldn't have been more opportune, being the first to apply he became senior pilot of Clyde Hover Ferries and progressed from there to become Hoverlloyd's Operations Manager in 1969.

Roger can claim neither enthusiasm – that came later – nor opportunity, but like many others eventually found himself in a new exciting career almost by accident. It was the result of a negative motivation rather than a positive movement towards any clear ambition:

I had turned my back on my deep sea career in 1966 and enrolled in the Nautical Degree at Plymouth. I cannot say that I had any burning desire to continue my education, but it was just one more option to get away from ships and avail my transition ashore.

I also remember, probably as a result of the amazing advances in technology revealed by the NASA space programme, particularly the use of computers on board the spacecraft, having a very distinct premonition that such advances would also come to affect my future career. My feeling then was that if computers were to control my livelihood, I wanted to be on the right side of the computer. In other words, I wanted to be in a position to control the technology, not have it control me.

By 1969 when Hoverlloyd's N4 service commenced, I was going through a bit of a crisis. I was beginning to realise that I was studying for a qualification which rendered me suitable for the very thing I didn't want to do, most of which would involve working in a London office, just the sort of employment I had gone to sea to avoid. To buy myself time I applied for a seasonal job as 2nd officer with Hoverlloyd.

I joined in May – I had to take time off in June to return to Plymouth for my final exams. Halfway through that first season, because of pressures of completing the MinTech Contract[11]*, I was offered permanency and with no other job prospect on the horizon, I accepted. Something I have never regretted.*

Although held together by a common qualification we came to Hoverlloyd by different pathways and via many different experiences. Our colleagues in other departments came by other routes and with many other different stories on the way. What bound us all together in the end was a growing mutual respect and the sheer excitement of being involved.

[11] Ministry of Technology. The contract was a study to evaluate the viability of hovercraft travel. It covered everything from the operating economics to passenger acceptance.

Chapter 15

Driving the SR.N4

The Prince of Wales, the only craft built as a Mk.2.

The Dover Strait in 1969

As previously explained, we had been trained as seafarers, not airline pilots, holders of Master's (Foreign Going) Certificates, entitling us to command anything on the high seas from the *Queen Mary* down to a sludge barge. This was scant preparation for the reality of driving across the Dover Strait in the equivalent of a jumbo jet at a previously unheard of speed of 60 knots, being at least four times faster than any other conventional vessel at the time.

In those days, before the advent of the Channel Tunnel, ferry traffic between the Channel ports was roughly half the amount of through traffic in the Strait – around 400 transits and 150 crossings a day. The Strait was not a seagoing thoroughfare but a teeming crossroads, the marine traffic equivalent of a major motorway junction.

This was not all; add to the conventional maritime traffic the huge number of yachts and other recreational craft which proliferated in the summer, not to mention the annual Channel swimmers and the odd publicity seekers rowing tin baths across the water, and you begin to appreciate the complexity of the collision problem.

In fact the holiday season fleet, which increased over the years, provided a far greater element of risk; few of these small craft carried radar reflectors and few of the people on board understood the 'Rules of the Road', or had even heard of it. Imagine negotiating that in thick fog without the benefit of flyovers, road markings or signs, or indeed any directions whatsoever.

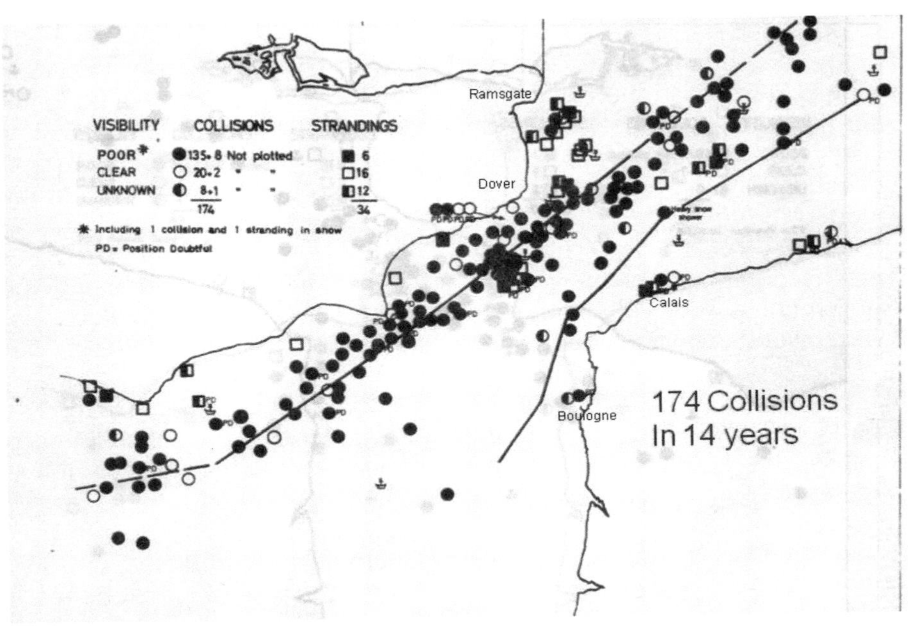

Collisions in the Dover Strait 1957-1971.
(National Physical Laboratory Report Mar Sci October 1972.)

Added to this scene of chaos was a further complication. In the previous decade, leading up to the start of SR.N4 operations in 1969, the critical state of maritime safety in general, brought on by the introduction of radar, was highlighted particularly by the casualty statistics of the very place in which we were going to operate at previously inconceivable speeds – one of the traffic hotspots of the world – the Dover Strait.

Considered now in retrospect, having spent our early careers coping with these sorts of stresses on conventional vessels, it is a wonder we were prepared to consider the idea of driving at high speed across the Strait at all. In fact, the Inter-governmental Maritime Consultative Organisation (IMCO) was concerned enough to seriously consider putting a speed limit on cross-Channel hovercraft when crossing the major areas of traffic density[1].

---

[1] Luckily for us, IMCO dropped the idea of speed limits and concentrated instead on creating traffic separation lanes for the transit vessels.

The National Physical Laboratory diagram on page 351 shows the disturbing statistics of that time, an almost unbelievable average of one collision a month over a 14-year period. In actual fact the average figure is misleading. Owing to various measures to improve the situation, there was a downward trend, starting from the frightening number of 28 casualties in 1961 to a mere 2 or 3 towards the end of the '70s. The year we started, 1969, with a recorded 13 collisions, marked the midpoint.

A large proportion of these casualties could be attributed to misuse of radar. However, in the initial years, another strongly contributing factor was the complete absence of traffic control in the area, the result of which was almost all the maritime traffic of Europe passing in both directions through a gap barely more than two miles wide, south of the Goodwin Sands.

This accounts for the vast number of incidents occurring along the English coast, with very little on the French side. In such confined and overcrowded conditions collisions were inevitable.

As previously explained, the introduction of the Radar Observer's Course in 1957 started to have some effect after the dreadful 1961 peak year, but the series of control measures in the area instituted by IMCO had the most influence.

The Dover Strait Traffic Separation scheme was introduced in 1967, routing inbound vessels – those arriving from the western approaches of the English Channel – on the French side and outbound vessels on the English side south of the Goodwins. At first this dual carriageway system was not declared compulsory, but nevertheless it was immediately followed by the large majority of ship's masters who welcomed the reduced likelihood of head-on confrontations. The scheme also provided an additional benefit; it helped to thin out the traffic in the Straits by providing a larger area in which to manoeuvre.

By the time hovercraft operations commenced in 1969 the situation had improved a little, but was still far from satisfactory. Although UK government legislation in 1967 made it compulsory for all British registered vessels to follow the separation lanes, it was not until 1977 that new collision avoidance regulations (Colregs) made adhering to all IMCO mandated routing schemes compulsory for all vessels worldwide. It was only then that it could be said the final chapter had been written in a sorry tale that had claimed many hundreds of lives and thousands of tons of shipping and cargo.

Despite being well aware of the risks we were taking, we remained by and large undeterred. But of course, being young – in 1969 the majority of the flight crew were in their twenties – we probably relished the risk, accompanied as ever, with those other dangerous cocktail ingredients; youthful confidence and naivety. Nothing bad could happen to us; we were too vibrant, too good-looking and far too clever. It was, let us confess, just one helluva lot of fun!

The Flight Deck

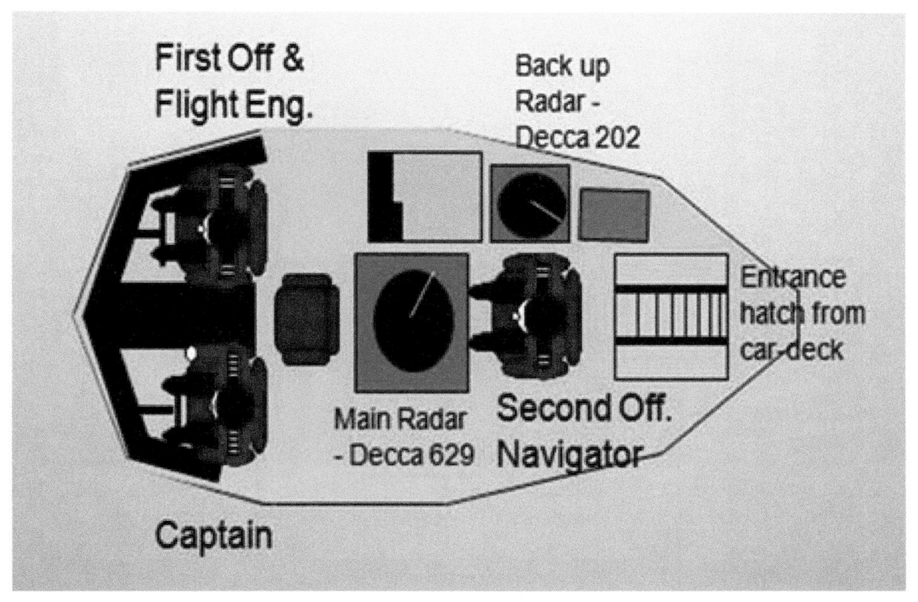

Plan of the SR.N4 flight deck. 1969.

So here we were at the bottom of the learning curve, getting used to the idea of flying an aircraft at sea. After walking the spacious areas provided by the bridges of large ocean-going merchant ships we found ourselves in a new and disturbingly unfamiliar workplace, strapped into seats inside a comparatively small cockpit area, hardly big enough to swing a cat.

It wasn't just a new concept and a new vehicle we had to get used to, it was a whole new working environment.

There were probably good reasons from a design point of view why the control cabin[2] looked a lot like the average aircraft flight deck, but if anything declared unequivocally that this vehicle was built by an aircraft manufacturer it was this.

---

[2] Both descriptions 'control cabin' and 'flight deck' were used, although the former was BHC's official term.

With some sixty-five feet of thwart-ship roof space to play with, the six or seven feet wide cabin was plonked plumb in the middle, resembling something akin to a small limpet on a fair sized rock.

The SR.N4 flight deck with captain's seat on left and 1st officer's on the right. The throttles and propeller pitch levers are on a consol between the seats. 1969.

Into this tiny space, in addition to the pilot and co-pilot controls, was shoehorned all the necessary marine equipment for negotiating the Channel: two radars, navigation aids, radios, even a chart table. Space for the three crew seemed to us almost to be an afterthought.

In contrast, any boat builder would have settled for the usual ship's bridge, spread the structure from side to side, and possibly fitted it with bridge wings. Not so farfetched when you consider the fast catamarans that succeeded the hovercraft were designed in just such a manner.

Not that a marine-style bridge would have been appropriate either, as Mike Rowland-Hill quite rightly points out, because it is likely it would have interfered with the airflow:

> *Of course the control cabin could have been a bit bigger but I thought the design had a lot going for it, as against a thwartship 'walk about' structure. Weight-saving, streamlining and, of course, airflow to the props and fans, which with a 'conventional' bridge would have been compromised.*

The limited space meant that the radar could only be positioned behind the captain and 1st officer, becoming by default the sole responsibility of the 2nd officer, thus settling any potential argument as to whether the captain should have a radar presentation as well.

Left to right: Tom Wilson (captain); David Ward, 1st officer (flight engineer); and Bob Adams, 2nd officer (navigator).

There are obviously arguments for and against such an arrangement. It certainly was an unusual and somewhat unsettling experience for the captain in poor visibility to have to rely totally on someone else's view of the situation and, to a large extent, someone else's judgment as well.

But the fact was that the captain did not have the time to handle the controls, monitor the instrumentation, interpret the radar screen *and* look out of the window, all at the same time. The allocation of responsibilities where trustworthy navigators, well trained in high speed operation, who had sole access to the radar, worked without incident for over 30 years of the SR.N4's existence. It's likely that had the captain had access to the radar it would have resulted in a dangerous overload. Further, there would have been the tendency to engender discussion concerning the radar picture between the captain and navigator, which would have dangerously delayed any decision on how to deal with a particular collision threat.

Roger Syms' later experience with catamarans tends to confirm the hypothesis:

Since my hovercraft days I have had many opportunities to observe how high-speed catamarans operate, where the radar is visible to both the captain and the navigator.

What happens too often on the cats is the captain tries to do everything; look out, look at the radar and drive, relegating the navigator to an advisory position at best. This situation led to at least one serious accident that I know of. On the much faster N4, I firmly believe it would have been disastrous. At that speed having radars available to both the captain and navigator would have engendered discussion not decision and there simply wasn't time for that.

The Flight Crew

In 1969 there were 16 of us altogether, including Bill Williamson as captain and operations manager, making up five crews to operate the first summer schedule on a six-day per week basis.

We had agreed on the one-day-off a week schedule, knowing as we did the parlous financial state the operation was in. It may have been all very altruistic and typical of Hoverlloyd industrial relations, but all of us after two or three months of this regime were wrung out and starting to make mistakes. From a safety point of view it was not a supportable idea and the Air Registration Board (latterly the Civil Aviation Authority) eventually stepped in and set down working hours far more conducive to safety, which remained unaltered right to the last Hoverspeed flight in 2000.

The Blind leading the Blind

In the first years, the experimental 'are we going to make it?' nature of the enterprise was not the only aspect of which the trusting passengers were blissfully unaware. We, the intrepid mariners, although not entirely ignorant, were entering into the unknown too. The painful fact was that no one had ever traversed the Channel before at such a fast speed, or for that matter, even by the route we took on our way to Calais – straight over the Goodwin Sands. So who was going to teach us?

Apart from Hoverlloyd's own original flight crew, who had spent three summers testing the concept on the much smaller SR.N6, when it came to the SR.N4, it was a question of starting again from scratch. There *was* no one else, anywhere, who had any remotely related experience.

We were on our own, learning on the job, and inventing things as we went along. As one example, Robin Paine set the foundation for the necessary navigation expertise by constructing a quick reference chart:

The service started on 2nd April 1969 with Swift. *With just one hovercraft in operation, three of the original SR.N6 intake were captains, the other three were 1st officers (flight engineers) and the new intake of 2nd January, of which I was one, were the 2nd officers and navigators. As we built up to this momentous day, after our initial training at BHC Cowes and Rolls Royce, everyone concentrated on their own particular job.*

Although there was a captain initially in charge of navigation, he had enough trouble learning to drive the hovercraft without other distractions, and everyone struggled to get 30 hours of operational experience in their respective seats before the start of the service. With regard to this high-speed navigation, with moving and fixed targets having the same 'tail' length on the radar, it became quite clear to me that unless we had a more practical and informative 'chart' for the Goodwin Sands and the approaches to Pegwell Bay, we could be in for a heap of confusion and trouble in fog. So for myself I drew up a scale 'chart' with lines between all the fixed objects showing the distances and bearings between each, which I had laminated for longevity. This was subsequently purloined by everyone else and proved an invaluable navigation aid, particularly for new recruits, to the end of operations from Pegwell Bay.

One might say a good initial solution to the problem; nevertheless these remained only 'first guess' answers to previously unknown questions, made by people with no previous experience. As Roger discovered when he joined the company, five or six months after Robin, in the first few months there was very much an air of the blind leading the blind:

When I took time off from my degree studies in early 1969 to commence my navigator training, I was trained to use radar at speed by a guy who I thought was a red hot expert; that is until I found out later he had checked out as navigator himself only three weeks before!

It's difficult to comprehend now how totally unknown travelling at sea at 60 knots (about 70mph) was at that time. The normal type of ships we were all used to averaged a quarter of that speed. In spite of our previous experience and, having grown up with radar, nothing we had encountered thus far prepared us for the problems of speed, where beyond 30 knots was difficult enough, but 50 knots and above made any currently held techniques for plotting target movement totally redundant[3]. At the higher speeds any sort of plotting method was all too slow.

[3] A full description of navigation methodology on the SR.N4 follows this chapter.

How were we going to use radar safely was a major concern; could we maintain our 60 knots cruise speed in zero visibility? In the North Sea and English Channel areas, severely restricted visibility is a common occurrence, with the added inconvenient fact that fogs invariably coincided in summer with days of high pressure, conditions which also meant a flat calm sea. So a craft which could almost be said to have been designed to operate on a billiard table, would be achieving its highest speeds in glassy sea conditions and minimal visibility.

Such a prospect was worrying enough, but, of greater concern – already mentioned above – was the way the control cabin was configured. The narrow cockpit area only allowed for the navigator and his two radars to be squeezed in behind the captain and flight engineer, so the only flight crew member who could see the radar at any time was the most junior member of flight crew team, the 2nd officer/navigator.

When the visibility dropped to nothing the captain had to rely solely on what the navigator was telling him about the traffic situation and what course to steer to avoid collision. This, at times, put enormous responsibility on the 2nd officer and in effect, in poor visibility, put him virtually in command. The captain could make decisions but nevertheless had to rely heavily on the information he was being given. Like Robin, early captains had to rely on newly-recruited navigators without the luxury of long acquaintance and therefore trust, but in the rather nerve-wracking process learnt that speed was a highly useful tool for keeping out of trouble:

> *As one gained experience as a captain listening to the navigator calling out bearings and distances of targets, one was able to draw a picture in one's mind of the targets on the radar, provided there were not too many. Navigators were trained to focus on the relevant targets within the six-mile range. You only needed one target on a collision course to upset the applecart, because, having altered course for that one, there were many occasions when you brought one or more others into play. However, speed was our great asset and, if it became all too complicated, one was able to make a speedy diversion to clear the lot.*

In the early days John Syring, a Seaspeed captain was untypically cautious:

> *I would never go over 40 knots in fog. Most people would say, "Fog, jolly good, flat calm, we can go jolly fast because we've got excellent navigators." I agree we have excellent navigators – second to none – and they were navigating every day of their lives. They could see things on the radar that nobody else had seen, but at 40 knots you could close the throttles in an*

emergency and the craft would settle down like a swan. Once you started going over 40 knots you got severe deceleration, with the result that you spoiled the skirt, spoiled the people or spoiled the cars. So I was in the minority over that, but I felt happier at that speed.

One could build up a picture from the navigator singing out the bearings and distances of the targets. They were very, very good. On one occasion a navigator was muttering about some echoes he'd seen, or maybe he hadn't seen it – it was like a swarm. There it is – now it's gone. It turned out to be about 20 outboard motorboats or ribs, as we would call them nowadays, coming down from Ostend or France somewhere on a day trip across the Channel. All he was picking up were the outboard motors. Luckily the visibility was such – half to three-quarters of a mile – that we actually saw them and knew what he'd picked up on the radar.

This was, and remains, a totally unique operating environment at sea. No other vessel operates a system where one person has sole use of the radar and therefore sole appreciation of the traffic situation. It is a credit to the first flight crew personnel from both Hoverlloyd and Seaspeed who, from scratch, without the slightest prior experience or knowledge, produced a totally effective set of procedures which eventually resulted in a zero accident record over the 30-year period hovercraft were on the English Channel. Considering the nature of the difficulties involved such a clean record is something to be proud of.

Navigator Training

The natural consequence of such a set-up was the selection and training of flight crew in general and 2nd officers in particular – the first rung on the ladder – which was careful to the point of being meticulous. A number didn't make it through the training, with some complaining of the severe, if not bullying manner of their mentors. The failure rate over the years was remarkably consistent at about 1 in 8.

It was difficult to get used to the speed involved as we used the six-mile range on the radar, which meant, at 60 knots, another vessel or fixed object would cover that distance to a potential collision in six minutes. If it was another hovercraft, at the same speed, a collision was possible in three minutes – a very real cause for concern.

Strangely, what was noticeable with many first-timers was the tendency to 'over-speed', the feeling that things were happening so fast, there was a need to think and move fast to keep up. After a few hours watching the movement on the radar most trainees 'throttled back', were less hassled and anxious in their demeanour and handled things more deliberately. It was obvious they were matching their speed to actual progress of events.

Most failures were due to not coping with the speed and the aircraft environment in general, but there were one or two odd incidences of failures of spatial awareness.

When standing on your head, do you still know which is left or right?

Our radars, being north-up presentations, did in fact create that sort of left/right problem. When going on a south-easterly course to Calais the heading marker line is pointing towards you, down the page as it were. Like most maps, compass north remains at the top of the screen.

It was always noticeable when anyone was having this sort of spatial difficulty, because their performance going north was significantly better than proceeding south. This lack of spatial awareness is probably less obvious on a slow-moving ship as there is plenty of time to rethink the orientation. Approaching the Calais hoverport, at still a good speed on the SR.N4, there simply wasn't the time.

No doubt our future colleagues were put through the wringer. Over a period of two to three weeks new recruits were sat at the radars with black-out curtains around them and expected, rather in the manner of a horserace commentator, to keep a continuous patter of everything they saw on the radar, identify a host of objects correctly as either moving traffic or fixed navigation marks and give course alterations to avoid collision as necessary. All the time the rest of the crew were firing off a stream of questions at the slightest hesitation or suggestion that the navigator was lost.

It wasn't always pleasant, but considering the safety issues at stake, absolutely necessary. We had to know whether the recruit could do the job but, just as importantly, also know whether he could operate under pressure.

Apart from the radar expertise, which at least had been partially learned previously, the race commentary aspect would have been a totally new skill to acquire. George Lang says he practised even when he was off duty:

> *I used to practise the spiel by 'navigating' out loud whilst walking around Margate, treating pedestrians as other vessels!*
>
> *Just a variation on the safe driving technique recommended by the Institute of Advanced Motorists.*

As might be expected, our approach to training improved over the years. At the start, our somewhat *ad hoc* method of crewing meant the early navigators rarely stayed with the same crew throughout their training. Apart from the inevitable inconsistency of a training format, which would have been unsettling enough for the new recruits, it also resulted in some marginal candidates not being noticed until it was almost too late. Trainers were reluctant to come down too hard on what may just have been one bad run in a series of good performances, but with little previous experience of the trainee, they had no means of telling.

The converse of this scenario was potentially the most dangerous where a thoroughly incompetent navigator could put on a fluke display, maybe on a day with little traffic and few problems, and find himself checked out as fully accepted and qualified. Luckily for us all, these were rare occurrences, but when such a situation did arise it was difficult and embarrassing for all to have to admit to the mistake.

Later crews were teamed up as one entity and stayed together for the whole summer season. This was initially for purely logistical reasons, but the change had the added bonus of making sure trainees stayed with the same consistent training regime, mentored by the same people, who could gauge clearly whether progress was being made or not.

Every effort was made to help recruits get through the course. Nobody likes to fail, but it was interesting to note that in most cases, when the time came to call it a day, there was almost a sigh of relief on the part of the trainee. A very small number argued that they had been unfairly treated, but this was unusual and, once the decision was made, regardless of the argument, for our own peace of mind it wasn't going to be reversed.

Seaspeed had similar issues, but had a slightly different approach. Peter Barr:

> *The trainee doubled up with an existing navigator for a while for whatever number of hours was considered appropriate at the time, depending on the individual. He'd already completed the safety courses and when you thought he was ready to go you checked him out. We tried to keep the trainee on with the same qualified navigator for the first ten hours and then we shunted him around as it seemed a good idea that he should meet everybody and see the different ways of doing things.*
>
> *The failure rate depended on how desperate we were. There were one or two over the years who even on a calm day, when you could see forever, who were a danger to themselves and*

everybody else. I can only recall personally sacking two, but I didn't do all the checkouts in those days. We had more permanent staff because of the Cowes services, who were experienced hoverers before they came on the N4. That made a huge difference, but unfortunately we always had one or two captains who couldn't be bothered with the training aspect. They'd go hurtling across to France taking no notice of the navigator and when the fog came down they were the first to say the navigator didn't know what he was doing.

Checking Out

One other airline procedure that was totally new to us was the process of 'checking out'. No one could be accepted as fully qualified in any function, captain, flight engineer or navigator, before they had clocked up a certain amount of 'hands on' time – usually around 30 hours[4] – and then successfully completed a round trip under the scrutiny of someone experienced.

The check-out was carried out with as much rigour as to that of an airline pilot's and, considering the safety issues, was taken very seriously indeed.

In reality it was rare a recruit arrived at the check-out stage unless it had already been concluded by the staff involved that the particular trainee was clearly up to the job. If he hadn't been already judged capable some time before, he would have already been weeded out.

Here it's essential to emphasise that failure, although upsetting for the trainee and unpleasant for the staff, was not a lifetime judgement on someone as being wholly incompetent.

Probably the major failure was most likely one of maturity. What was required in navigators particularly was spatial awareness, good appreciation of radar images, ability to think quickly, but above all, strength of character, which can only come with experience and maturity. After all, we were asking the most junior member of the team to tell the captain what was wanted and, if necessary, stick to his guns. Most of those we were recruiting were in their late twenties at most, some of whom had that sort of maturity at that young age and some didn't. Those who didn't, no doubt were fine in a few years time, and most likely went on to be excellent senior officers and captains at sea. Their only misfortune was turning up to us at the wrong time.

---

[4] Later captains, although checked out in much the same time period, didn't usually go solo with passengers with less than 100 hours experience. In the first years, with the need for qualified staff a priority, we couldn't afford such luxury.

The Captain's View

From the captain's perspective, well trained and experienced navigators, who had the patter down to a fine art, were a pleasure to listen to, except at times perhaps they were too good! One day Roger, totally mesmerised by the navigator's commentary, ended up in real trouble:

Approaching Ramsgate on a fine summer's day, absolute dead calm and of course, typical of those conditions, thick fog. I mean really thick, the sort where you can barely see the water. Also means we're rattling along around at 65 knots plus. It was difficult to stop the acceleration in those conditions.

My navigator, new that year, is really doing a great job. He's got the patter off to a tee.

"Cleared the Brake Buoy, First Buoy to port bearing three one five, tracking six cables[5] to port."

"Come on to three one zero."

"First Buoy clearing three cables to port."

"No traffic, looking good."

"Come around to three zero five."

"Still looking good, nothing on the approach."

I'm fascinated, thinking to myself, we've really trained this guy well, this is what I like to hear; he's really crash hot.

It's probably low water springs, we come on to the sand just past First Buoy, I glance down at the panel to see the ASI leap up to 70!

HELL!

In all this time being mesmerised by the navigator's patter, I've forgotten to pull any speed off!

About a little over a mile to the ramp and we're doing 70!

I asked for full power and then some, and slapped every pitch

---

[5] For the non-nautical, a cable – roughly 600 feet – is one tenth of a nautical mile.

lever in reverse. My biggest worry at that point was that the props wouldn't reverse evenly and we'd start to slew out of control – the difference between smashing into the building head first or stern first. It wasn't going to matter a lot really, except that if it were the latter, I'd probably be still around to explain myself to a tribunal.

As it turned out we stayed straight, with all props roaring like crazy in reverse, with me watching the speedo like a stricken rabbit with a ferret.

My man on the radar is still doing a great job.

"Sandwich ramp bearing two nine zero, one mile."

"Now bearing three one zero, three cables."

"Sandwich ramp bearing three five zero, two cables, clearing to starboard."

(We're passing it and heading for the Sandwich Road! Oh heck!)

"Sandwich ramp astern to starboard – I think we have the road ahead!"

The luck of the Syms prevailed; by this time the speed was down to 20 odd knots and coming off.

We ground to a halt with a branch of one of the roadside trees, lining the Sandwich Road, waving at me out of the fog, no more than ten feet in front of the control cabin.

Phew! I feel shaky now just writing about it.

The best of it was, once I'd turned her around and waffled our way back up onto the pad, I pitched up to the Radio Room to be asked by a number of the guys, what happened? "We heard you getting louder on the approach and then you faded away again."

I can't remember what I said now but I'm not sure I told the whole truth.

View from the captain's seat in good visibility.

Flight Engineers

Part and parcel of teaching mariners to drive aircraft was teaching them about aircraft structures, systems and engines. Not only was this totally beyond our previous experience, but also presented a completely different approach to training that none of us seafarers had experienced before.

Although part of our marine training included the subject 'Engineering Knowledge', this was very generic, dealing with calculations of shaft and brake horsepower, fuel consumption and the like.

The other subject dealing with the structure of the vessel was 'Ship Construction', and this again was not only generic, but in the '50s and '60s the instruction given and examined for our qualifications was hopelessly out-of-date. In an era that had already moved to welded hulls and the advent of bulk carriers and containerisation, we were still expected to draw diagrams of riveted shell-plating and 'tween decks.

The very idea that ship's officers should have an intimate knowledge of the structure and operating systems of the particular vessel they are expected to drive has never occurred to the marine world and, with very rare exceptions, remains the norm to this day.

Our hovercraft training proved to be a revelation. It commenced with two weeks systems and structures training and final examination at the BHC assembly plant in East Cowes under the excellent tutelage of Peter Habens. Under his careful and thorough instruction we crawled through the hovercraft under construction absorbing detail of every systems linkage, pop-rivet and skirt segment.

This was followed up with a course on the Proteus gas turbine at the Rolls Royce establishment at Ansty. Emerging from this follow-up with a comprehensive knowledge of the engine and its capabilities, going through the 30 hours 'in the seat' and being checked out, we ended up as qualified SR.N4 flight engineers, fully versed in every technical detail of the craft.

This was no nominal or paper qualification merely there to satisfy some regulation or other, but rather a learning process that brought two enormous benefits to the operation of the company. Firstly, it gave us as flight crew a greater sense of the limitations and vagaries of the machine we were driving and thus made us more sensitive and less gung-ho in our approach. If nothing else, such knowledge must have helped to prolong the life of the craft and its systems.

Secondly, and probably more important, it gave us a better rapport with our engineering department. When snags occurred we could report back to base with some technical knowledge of the problems we were encountering – we were talking the same language. In Hoverlloyd we managed to maintain excellent relationships between flight crew and engineers with none of the 'oil and water' antagonisms which remain very much a feature of life at sea to this day.

Learning to Drive

At first glance the controls of the SR.N4 were no different to any other aircraft flight deck; rudder pedals, joystick and yoke, together with four propeller pitch levers and engine speed controls. It may have looked the same but the resemblance stopped there. What each control did was an entirely different matter.

The rudder pedal and the yoke, which in an aircraft control the rudder and the ailerons and elevators respectively, moved the propeller pylons. The rudder turned the front and rear pylons in opposite directions imparting a turning moment. The yoke when moved turned all the pylons the same way adding a sideslip element. It also had another function; when pulled it reduced pitch on all four propellers – a very useful means of quickly reducing speed.

Both operations were summed in the control system to provide an almost infinite combination of pylon configuration.

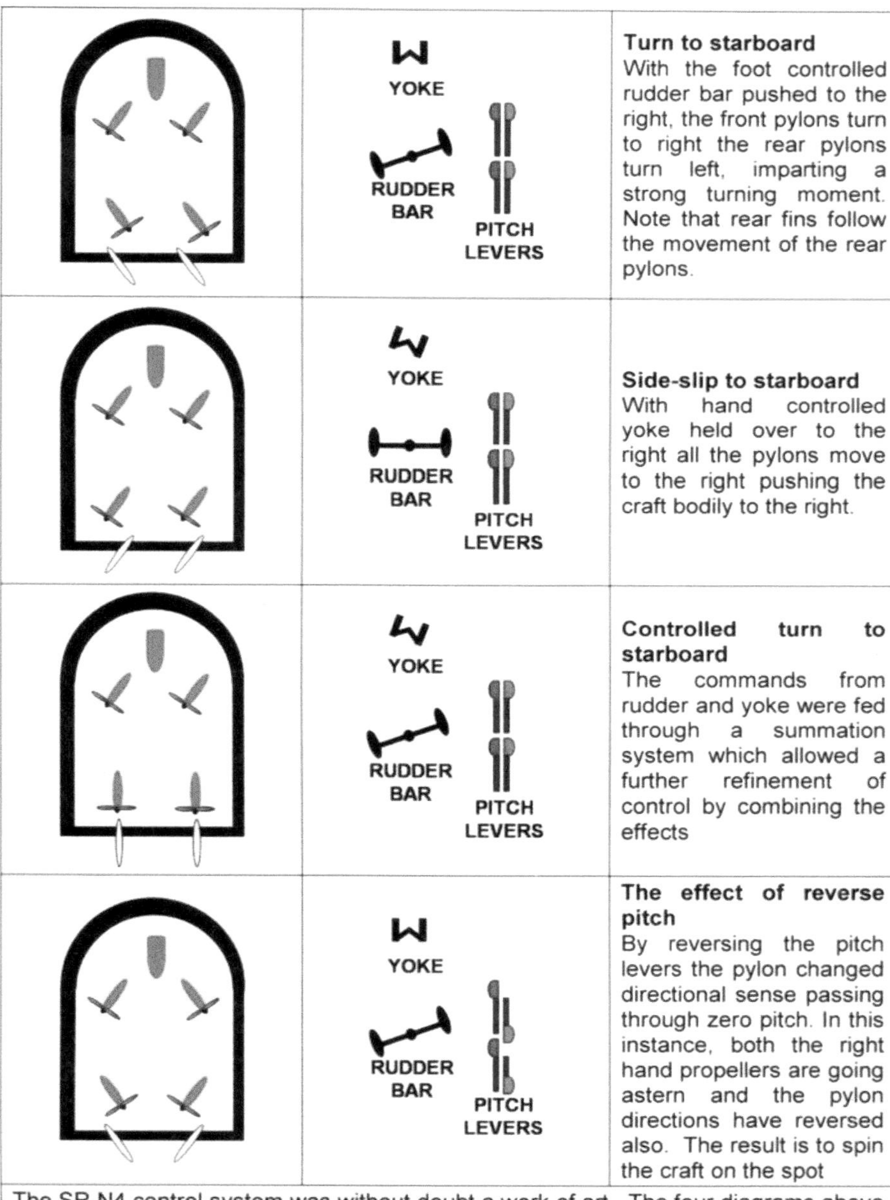

Turn to starboard
With the foot controlled rudder bar pushed to the right, the front pylons turn to right the rear pylons turn left, imparting a strong turning moment. Note that rear fins follow the movement of the rear pylons.

Side-slip to starboard
With hand controlled yoke held over to the right all the pylons move to the right pushing the craft bodily to the right.

Controlled turn to starboard
The commands from rudder and yoke were fed through a summation system which allowed a further refinement of control by combining the effects

The effect of reverse pitch
By reversing the pitch levers the pylon changed directional sense passing through zero pitch. In this instance, both the right hand propellers are going astern and the pylon directions have reversed also. The result is to spin the craft on the spot

The SR.N4 control system was without doubt a work of art. The four diagrams above show the responses that could be achieved by differing combinations of yoke, rudder and pitch on each propeller and pylon.

The computer control system was attached to the rear of the control cabin and was so 100% reliable that the engineering department almost forgot its existence. Emrys Jones, Head of Engineering, once asked Ray Wheeler, the N4's designer, why there were no spare parts available, to which Ray's response was, "Why? Have you ever needed any?" Emrys had to admit he hadn't.

The computer controlling the pylon, fin and variable pitch propeller movements, based on the *Black Arrow* and *Black Knight* rocket systems, located at the back of the control cabin. 1969. Of course this whole system would be matchbox size today.

The propellers were variable pitch and the pylons turned 30° either side of the centre line. Note the bait diggers over the stern of the craft digging for lugworms in the Pegwell Bay mud.

Ray Wheeler explains why:

> *You know the computers that were in the control systems – they were based on the* Black Knight *and the* Black Arrow

rocket systems. They had been extensively tested, so they were extremely reliable and I was responsible for it. It was a 'fly by wire' system. You see, it had to be tested for the environment of the kinetic heating as it was launched up to 1,600 miles an hour, or whatever, and out into space at very low temperatures. It had to withstand the vibration of these huge rockets so it was tested to death. It was totally reliable and never gave us any problems on any of our hovercraft.

The fact that when a pitch lever was moved into reverse it also reversed the pylon sense, added another multiplication factor to the host of possibilities, all of which made the SR.N4 not the easiest vehicle to master.

Perhaps the best analogy is a car with independent steering, gear shift and speed on all four wheels – imagine the possibilities of that!

It has to be said, however, that once mastered the magic machine was amazingly controllable and it could be turned literally on a sixpence. The BHC designers may have got a few things wrong, but, from an operational point of view, the control system was, without doubt, a work of genius.

Mastering it was the problem. For the fledgling captain, feet on the rudder pedals, left hand on the yoke and right hand on the pitch levers, the first hours of learning to drive were a testing, if not terrifying experience.

It's generally acknowledged by those who suffered the trials and tribulations of learning to drive, that once you had been checked out as a competent SR.N4 captain and received your licence, this was not the end of learning at all.

Out on your own, with no one else to guide you, was when the learning really started. Trying to control a totally free-moving vehicle – a bit like managing a golf ball in the middle of an unsteady tea tray – was not a question of knowing how to control the beast, but how to apply it.

There was one added complexity, which only added to the new pilot's anxiety. It was due to the pylon control system, cleverly designed as it was, with the pitch change linked to the propeller's angle of direction – the change didn't occur half as fast as he would have wished. When the propeller pitch travelled through zero on the way from ahead to full reverse, most likely due to a panic reaction in an already sticky situation, the slower travel of the pylon change would result in a ten-second period when the propeller was still in the old direction.

When one was desperately trying to get out of trouble this small time delay of a few seconds, in which the control was not giving the desired effect, seemed more like hours and, in the early days, did in fact land a few of the inexperienced captains in difficulties.

It was all very well knowing how to move the machine onto the pad at the hoverport, but did you have the skill to place it exactly in the right position? Roger still has the occasional nightmares remembering:

> *The first months of driving were a series of incidents on the learning curve, some of which bordered on the terrifying. There were a few times I walked off the machine shaking like a leaf, only too glad not to have damaged anything.*

A right side hydraulic ram for turning the pylon can be seen at the top of the picture and the centrifugal lift fan is just visible at the bottom of the fan air intake.

Getting to Know the Foibles

The fact is the hovercraft is unique as a vehicle; no other presents the problems of control in quite the same way. It is for a start the only truly amphibious vehicle ever invented. Others that have been produced to operate on land and sea change the mode of operation when transitioning from sea to land. For instance, the amphibious landing craft comes off the water and immediately has to shut down the propeller propulsion and engage the wheels. The hovercraft is not inhibited by whatever lies beneath; water, sand or tarmac, it's all the same.

Hovering on a cushion of air makes it, to all intents and purposes, frictionless, and it is this ice skating potential which produces most of the problems.

The first trick any potential new captain had to learn was to anticipate which way the thing was likely to move once the engines have been revved up for departure, sufficiently for the SR.N4 to come up 'on the cushion' and 'unstick' itself from the ground. Two things were likely to happen; either the craft would start to fall down the slope – the hoverport pads on average were built with a 1:100 slope – or, if there was sufficient wind off the sea, the craft could just as easily be blown against the gradient up the slope towards the terminal building.

The wind also tended to dictate the way the craft turned – for ease of car and passenger movement on and off, they were always parked facing the building, which necessitated on departure backing off and turning 180° to face seaward – on occasions the large tail fins, acting like sails, would dictate the way the machine commenced the turn. The wise let it go the way it wanted; trying to fight it was never a good idea.

The sum total of all this was that it didn't matter how experienced the captain; starting to come up to hover was always a period of uncertainty.

Surfing

Another aspect of the propensity to fall down any sort of slope was the disconcerting tendency to surf sideways in bad weather.

The south-west/north-east orientation of the Dover Strait, in conjunction with the prevailing south-westerly gales, meant that during times of heavy weather large swells were right on the beam on our south-easterly course to Calais.

The beam-on swell was not a problem in itself. In fact it was more comfortable and we liked to keep the weather on or abaft the beam if we could, because there was less risk of butting into the waves and causing damage to the skirt or structure.

The surfing incidents invariably occurred when the tide was also running with the wind, again our preferred circumstances as the wave pattern lengthened out and made for a more comfortable ride. However, that was when to expect surfing as well. The craft would sit up on the side of one of these moving slopes of water, cant over what seemed to be 20° or 30°, but in reality was only a few degrees, and, like any good pro surfer, rocket sideways down the slope. This was no real problem, apart from that vague feeling that as captain you weren't quite in control. It didn't feel at all right.

In addition, the 2nd officer, sitting at the rear of the flight deck with the radars, had the rather comical view of his colleagues up front leaning into the wave, presumably in some vain attempt to counteract the heel. A combined body weight of maybe 200kg – the 2nd officer would have also given in to instinct – was hardly going to have much effect on 200 tons of loaded hovercraft, but the crew did it automatically, every time.

Like the sport of surfing there could also be a dangerous element to this phenomenon. The situation was perfectly safe provided the captain never gave in to the quite natural response to turn up the slope. That could have the catastrophic effect of accelerating the back end down into the trough and doing damage to the hard structure, or even worse, damaging the engines, which were positioned at the rear of the hovercraft.

Dealing with Drift

One of the first things we learned when driving the hovercraft was the simple fact that it was rarely ever going where it was pointing. On the tarmac with the SR.N4's amazing control system the craft could be steadily pointed in one direction and moved in any other way you liked. One of the great sights was to see a machine being moved astern back out of the jacking area and skidding sideways along the front of the hoverport building to its loading position, all the time maintaining its north-westerly orientation.

Note how *Swift* appears to have altered course, but is still travelling in the direction of its momentum – the navigator's nightmare in fog.

Moving slowly around the hard standing was comparatively easy to master but moving at speed at sea was a different matter. Without any instrumentation to measure drift and actual direction the machine was taking – it would be decades before GPS[6] was available – captains had to develop their own sense of direction, which meant also developing an acute sense of what was safe and sensible. If you were to pass any fixed object close to, it was common sense to pass downwind and certainly never with it on the outside of a turn.

Where turns close to obstacles could not be avoided and, what is more, had to be negotiated accurately, such as the severe turns needed to approach the Sandwich ramp at Pegwell and the Eastern ramp at Calais, they presented a particularly anxious part of the learning process to anyone starting out.

Crash Stops

Although driving the N4 required a certain amount of nerve, what we certainly didn't need were dare-devil heroes. Our personnel were encouraged to be quite the opposite, particularly so as far as our navigators were concerned. If a navigator found himself in a difficult position, or was in the slightest doubt at the way a situation was developing, he was expected without any hesitation to call out, "put her down". In other words, to crash stop immediately by dropping the lift power and slamming the craft into the water, which in effect became a voluntary plough-in. If a navigator at any time shouted "Down!" there was no hesitation on the captain's part – down we went. The 1st officer was instructed to slam the throttles shut and, with the lift power dropping almost instantaneously, the machine hit the water like a ton of bricks, the stopping distance from flat out 60 knots was around a length and a half – 200 feet.

After such events there may have been post mortems, but never recriminations. No navigator was ever criticised for doing the prudent thing, even when subsequent events may have suggested the drastic manoeuvre may have been unnecessary. Safe, not sorry, was the byword.

The truth is, especially in the first year of operation, drastic manoeuvres such as this were frequently all too necessary. It was an unfortunate outcome of the learning-from-scratch process – navigators 'going solo' after a minimal training period trained by people who had very little experience themselves. There were a number of near misses, not just with the ships in the Strait but between ourselves. Two hovercraft at 120 knots closing speed is the stuff of nightmares.

---

[6] Global Position Systems (GPS) measure geographic movement regardless of which way a vehicle is facing.

The term 'unfortunate' above is probably the wrong adjective. They were 'misses' however near, something we survived from with lessons learnt to ensure it never happened again. We were in fact fortunate to be given the chance. The alternative scenario would have ended in this book never being written.

We all had our white-knuckle episodes, this was Roger's:

> *Coming out of Pegwell as navigator one winter's evening during the first year of operation, just as it was getting dark in poor visibility, for reasons I don't remember, the captain was initially having his own problems with control. Any course I asked for I was lucky if he was within 10° of it. Worse than that, he was sheering from one side to the other, so even if it was the wrong course it wasn't steady either.*
>
> *As we approached the Gull Stream, the small vessel channel inside the Goodwins, there was a coaster near the Brake Buoy going north. I was trying to clear astern of it, continuously asking the captain to maintain a steady course. As we got within a mile or so he managed to steady up for a short while, long enough for me to realise with horror that the bloody thing was coming south not north, by which time we had closed to half a mile – 30 seconds to collision – and pointing across its bow. I did the only thing I could do at that point and told him to drop it in. I don't think that he or the 1st officer actually saw it pass as no one declared it visual.*

Hitting the water in such a manner was spectacular, a great fan of white water pushed ahead, sometimes as high as the control cabin. The deceleration was teeth jarring, but at least we on the flight deck wore seat belts. Not so the rest of the crew and the passengers. Right from the early days when seat types for the SR.N4 were being considered by Seaspeed for the SR.N4, and agreed by Hoverlloyd, seat belts had been ruled out as being a nuisance to passengers and difficult to control by the cabin crew. We certainly sustained occasional injuries, but, surprisingly, nothing really serious.

After any crash stop, the first priority was always to check if the car deck crew and our passengers and cabin staff were okay. There was always a real possibility of the car deck personnel being injured by vehicles breaking adrift from their retaining straps. There were one or two broken limbs in the early days but as time went on this fortunately became a rarity. The car deck crew quickly learnt where not to stand and, with a combination of the growth in the flight crew's competence, together with gradual improvements in skirt design, plough-in itself became more and more infrequent.

Despite the lack of passenger seatbelts, passenger injury was mercifully uncommon. The biggest source of danger, until the Mk.2 modifications redesigned them, were the forward facing bond lockers of the Mk.1[7]. The lockers were always open during the crossing while the stewardesses sold duty-free goods. This meant in the event of any crash landing there was an almost guaranteed shower of missiles propelled forward down the centre walkway. Packets of 200 cigarettes would have been little problem, but flying Scotch bottles could have been lethal.

Plough-in

Slamming into the water in such a fashion, deliberately inducing plough-in was also something which, in the early days, could happen quite naturally without the captain's help.

At least when we dropped it in ourselves we were prepared for the consequences. When the machine ploughed-in itself, it often came as a nasty surprise. All hovercraft are prone to it, as it's the natural result of the propeller thrust being some distance above the major source of drag – the fingers at the bottom of the skirt. This push-pull coupling was a 'trip up' waiting to happen.

A lot of development went into countering the tendency, in addition to the initial introduction of the 'anti-plough bag', which allowed air flow to be contained as the bow dropped, to act as a giant flotation fender. The skirt line was also tapered towards the stern to enable the propeller thrust to be pointing slightly upward.

As skirt geometry on the SR.N4 was gradually improved over the years, eventually air-flow under the machine was balanced so that there was always a greater pressure forward than aft and unexpected plough-in gradually became a thing of the past.

In the early years the worst times for plough-ins to occur were on days when the sea was particularly calm. This meant the machine was travelling at its fastest, which also meant the effect of drag would be at its greatest too.

In such circumstance the worst problems were the wakes of ferries. These vessels, twin-screwed and comparatively fast for conventional ships, threw up a large wake astern which could be as much as two metres in height. Hitting what was, in effect, rough water from our previous billiard table environment caused the skirt drag to increase

---

[7] The Mk.2 lockers were placed sideways, fore and aft.

quickly and the consequent slowing of the bottom of the craft while the top kept tearing along at its previous speed, resulted in inevitable plough-in.

Over the years, as our experience gradually moved up the learning curve, captains knew to slow down long before the wake hit. It was always big enough to be seen clearly when visibility was good and even stood out on the radar when it was bad. Nevertheless there were still times when we were caught napping. This is Robin's sorry tale:

> *On one occasion we were running behind schedule and I was keen to catch up. Visibility was down to about 300 yards in a slightly choppy sea and we were moving along at 55 knots about four miles off Calais.*
>
> *In my impatience to get us back on schedule I had already ordered the car deck supervisor to start unlashing the cars.*
>
> *The navigator had picked up what was obviously a ferry on its way from Dover to Dunquerque, but for some reason he hadn't picked up the dreaded wake on the radar. When I did see it visually, as we were about to pass astern of the ferry, it was all too late.*
>
> *There was just not time to warn anybody of the impending 'crash', but fortunately I had a very experienced car deck supervisor, who recognised the on-coming plough-in, and yelled to his crew to jump clear of the cars, which thankfully they all did. The most awful bit was being powerless to stop the bow sinking and hearing all the prematurely unlashed cars crashing into each other and wondering if there were any injuries amongst the crew. The pile up was so bad that cars had been shunted onto the bow ramp, changing the trim of the hovercraft.*
>
> *Having established that no one had been hurt we made our way slowly to Calais Hoverport. On arrival I asked the car passengers to remain in the cabins and, rather shame-faced, descended to the car deck to survey the wreckage. It took two hours to unload the cars through the stern door and complete damage reports. I can't recall if there were any write-offs, but sadly there were several very unhappy passengers. We never did get back on schedule.*

Of course Robin wasn't the only one to get into this sort of trouble. In our eagerness to keep on schedule, we all tended to call for cars to be unlashed at times when it might have been wiser to hang on a little longer. Luckily, most of the time, we got away with it.

Sheer Exhilaration

Undeniably the N4 was a difficult machine to learn to drive, but once mastered, the skill became automatic and one hardly thought about it. It was then that the feeling of flying this wonderful machine produced a sense of sheer exhilaration.

Many of our old colleagues have described the sense of euphoria, very similar to the sensations that professional sports people must experience when involved in what are termed today 'extreme' sports. Activities that combine speed, danger and difficult manoeuvres, such as snowboarding, skydiving and base jumping for example.

Hovercraft, and especially 'the big one', combined all those adrenalin-pumping ingredients. At Pegwell there was another factor involved – performing to an audience, where the pleasure was so obviously mutual. Once the viewing platform was placed on the roof of the hoverport, we on the flight deck, as we rose eight feet into the air before turning towards Calais, came face to face with the public. Hardly without exception we saw a crowd of smiling faces. We smiled and waved back – performers in a magic circus – everyone delighting in the show.

The hovercraft had that effect on people. For many of us, driving the N4 wasn't simply a job; it was a fantastically enjoyable piece of theatre.

Chapter 16

Avoiding Collision

The Risks Involved

It cannot be emphasised too strongly how vital the navigation of the SR.N4 was to the overall success of the cross-Channel operations. It has already been mentioned, more than once, how important it was to recruit navigators who had a strong background in maritime radar. The whole efficiency and safety of the undertaking fully depended on the design and, what is more, the speedy evolution of a totally new approach to radar collision avoidance.

Although the 2nd officer was always referred to as 'the navigator', in reality his role was essentially 'collision avoidance officer'.

Considering the previously described constant danger of collision in the Dover Strait at the time, safety was the keyword. For such comparatively fragile craft as the SR.N4 to be involved in a high-speed collision with anything at sea would have spelt the end of not only many lives but the end of the industry.

Just imagine the flimsily constructed SR.N4, with 200 plus passengers onboard, slamming at its 60 knot cruising speed into the side of a mild-steel constructed 100,000-ton tanker. Not a lot different from an average-sized aircraft nose-diving onto the tarmac. The consequences are unthinkable. Perched right forward, in their small fibreglass shell, nearest the point of impact, it's doubtful the flight crew would have survived at all. Cars would have broken adrift on the car deck, with the added risk of tanks full of petrol igniting. As for the cabin crew and passengers, all would have been thrown violently forward, a tangled mass of dead and injured.

If one might think that this is somewhat alarmist and melodramatic it has to be said that something similar actually happened in 1991. The Norwegian high-speed catamaran SeaCat, on route from Selje to Bergen, collided with the vertical face of a rock at full speed. The impact tore seats from their moorings and large pieces of equipment, such as microwave ovens and other cooking facilities, were flung around the passenger cabin. Out of a passenger complement of 146, two died and 74 were injured.

It is sobering to reflect that this vessel was travelling at 36 knots when it hit, not much more than half the cruise speed of the SR.N4.

On the Edge of the Law

If this was not worry enough, the hovercraft was, in some respects, operating somewhat outside the law.

Although much effort had been put to ensuring this new invention was amply covered by legislation in the most important area, that of safety at sea, it was not covered in any satisfactory manner at all[1].

The international laws laid down by the International Maritime Organisation to ensure the safe conduct of all vessels at sea, regardless of size, are *The International Regulations for Preventing Collisions at Sea* (Colregs) and have been in existence since the mid-nineteenth century.

Over this long period the world had seen the transition from sailing vessels to steamships, from small 5,000 ton vessels to 500,000 ton leviathans and from sextants and magnetic compasses to electronic navigation in all its modern forms. At each evolutionary point in seafaring history the occasional international conferences adjusted the regulations to match. However, when it came to the sudden appearance of this unique amphibious vehicle capable of previously unheard of speeds, the best that could be done was to include this 'non-displacement craft' as just another 'power driven vessel'.

The Colregs apply to everything that floats from the smallest rowing boat to the largest bulk carrier, and the hovercraft, despite its unique construction, speed and amphibious capabilities, is no exception. The problem is that rules designed for 15 knot vessels don't adapt well to something going four times as fast.

A large part of the rules, 38 in all, are taken up with the type of lights and signal shapes various types of vessels will show in differing circumstances. For instance, any air-cushion craft is required to show a flashing yellow light as well as sidelights and masthead lights for a vessel of its dimensions[2]. There was no problem complying with this part, although most of the time the cross-Channel hovercraft were in total contravention of the 'Steering and Sailing Rules', the part containing the rules describing the conduct required of vessels in order to avoid collision.

---

[1] Although covered in Appendix 2, it is worth mentioning here that the Hovercraft Act of 1968 was an enabling act, which gave a legal definition of a hovercraft, and from July 1972 the hovercraft stood as a vehicle in its own right.

[2] On the Mk.1 and Mk.2 the length of the craft only required a single white light which can be seen on the mast attached to the rear end of the control cabin. Extending the length of the Mk.3 by another 55 feet meant that the rules now required another white light in line with the other, hence the extra mast towards the rear of the craft.

The principal difficulty for the hovercraft was trying to comply with Rule 7, which makes it compulsory for all vessels with operational radars to use some means of plotting other vessels on the radar screen, in order to confirm their closest approach.

The radar installations in the first craft were in fact fitted with what was called a 'reflection plotter'. This comprised a clear Perspex screen fixed over the radar picture on which the navigator could mark with wax pencil the positions of the various targets[3], which in turn the clever optics reflected down onto the screen. This was a very simple and useful aid and many of our recruits would have been familiar with their use on conventional vessels.

Unfortunately it was of no use on the N4. It was realised almost immediately that at the speeds the hovercraft navigators were dealing with, the whole process of marking the targets was a laborious process and simply not fast enough to produce the instantaneous information the captain required. The whole idea of plotting targets was dispensed with and Rule 7 was ignored.

Naturally the biggest risks of collision occur in fog, and poor visibility was a fact of life in the Dover Strait. It is not unusual for thick fog to remain in the area for several days at a time. Particularly in summer, fog tended to coincide with periods of high pressure, which invariably meant very calm seas. The result for the hovercraft, which could be said to have been designed to run on plate glass, meant the machine was capable of its highest speed when it was difficult to see the water, let alone any distance ahead.

This state of affairs brought the N4 once more in conflict with the collision rules, which required all vessels in poor visibility to go at a 'moderate speed', a requirement that expected the average vessel to reduce to half speed. Such a reduction would expect most ships to be running at around six or seven knots, not blasting along at 60 knots plus.

That this practice was known about and accepted by the authorities can be put down to a general acceptance by the marine courts that an acceptable definition of moderate speed was one which allowed the vessel to stop within the limit of the visibility. So taking into account the fact that the N4 could stop somewhere within a length and a half from going flat out was considered acceptable. The reality that we were tearing along in the thickest fog at the fastest speed, when one could hardly see a hand in front of the face, was quietly ignored.

---

[3] Reflecting the original military nature of radar operation, the term 'target', meaning anything appearing on the radar screen, remains in use in non-military vessels to this day.

Like all government bodies, if anything serious had happened there would have been no hesitation in proclaiming that we were never authorised or sanctioned to do so. 'Twas ever thus.

Fundamental to the steering and sailing rules is the conduct required of vessels in the three basic confrontations, namely: meeting head-to-head, vessels in crossing situations and overtaking. Hoverlloyd's route meant that most of the scenarios involved crossing the transiting traffic in the Channel, although Seaspeed had the added complication of meeting the end-on ferry traffic in and out of Dover.

Peter Barr describes the sometimes chaotic scene off Dover and the pressure it put on the navigator:

> *As you were closing in on Dover it got to be too busy at times because you had ferries going around in circles off the Eastern Entrance and similarly off the Western Entrance traffic shooting in and out, so you left it totally to the navigator.*
>
> *You had to call Port Control ten minutes off and he would tell you if there was anything significant other than the norm around the entrances or if you were likely to be delayed, and what ferries you should look out for. We were often delayed getting in and out of the port in fog.*
>
> *The trick for navigators was understanding the whole picture on the radar. I mean everything would be going nicely for them and we were, say, ten or five minutes off. Then things would start happening, such as you'd be told there was a ferry not ready to leave and it would be another five minutes, so you'd slow right down, keep spot on the approach line the navigator wanted, and then you'd realise you were getting too close and the ferry still hadn't left, so you'd have to peel off.*
>
> *The poor devil at the back, working in relative motion, saw his picture going round in circles and had to start to make sense of the picture all over again – no fun at all especially if the captain starts getting edgy and starts questioning him about this, that and the other.*

In cases of crossing conflicts the regulations require the vessel with another on its starboard side to make an alteration to starboard to pass astern of it. Considering the density of movements through the Straits, where the hovercraft had to thread its way through long lines of comparatively slow-moving ships, any set of hard and fast rules of this nature were always going to be impractical. The general rule of thumb adopted on the N4 was to pass one mile ahead and half a mile astern and alter to port or starboard towards whatever was the clearest gap in the line.

Again we followed our own rules, in many instances preferring to pass ahead, for the good reason that it was the safer and more prudent course of action. If the slow mover started to panic and made its own drastic manoeuvre it would be impossible for it to aggravate the situation and bring about a collision, whereas with the hovercraft passing astern, such a panic manoeuvre could close the gap very quickly.

The Bentley Club Fiasco

With such an *ad hoc* regulatory environment it was inevitable that sooner or later we would come unstuck.

Robin, who was the captain in this instance, has the distressing details:

> *Approaching Pegwell Bay the navigator had spotted a small boat coming out of Ramsgate Harbour at high speed and, because he was going to pass clear on our port side by the time we got there, the navigator asked me to alter course to port to line up with our fairway marker. This manoeuvre still allowed us to clear close astern of the crossing boat still heading towards the Stour.*
>
> *As we got closer to this pleasure craft, it suddenly, without warning, turned through 180° and came charging back across our bow. Still travelling at 55 knots (63mph), I had no chance of avoiding the craft, so my only option was to give the dreaded order, "put her down". The 1st officer slammed the throttles shut, with the inevitable result of a spectacular plough-in followed by the dread sound of cars cascading into one another.*
>
> *They were not just any old cars either. The unfortunate fact was we were carrying on the car deck The Bentley Club on an outing to Europe. In addition to the Bentleys there were vintage Rolls, Bugattis, Maserattis, and a whole catalogue of other expensive rolling stock you could bring to mind. All shunted nose to tail down the car deck in fine style.*
>
> *My immediate reaction was one of profound relief that I hadn't hit the boat, followed by seething anger as to how anyone could have been so stupid and irresponsible. On arrival at the hoverport I stormed into David Wise's office and related my tale*

of woe, demanding something be done. David, as always, was as cool as a cucumber – that was probably why he was Operations Director. He calmed me down and said he would look into it.

Having just passed Ramsgate Harbour, the hovercraft is on the Pegwell Bay mud flats on the final approach to the hoverport.

Roger, who in his subsequent academic career, specialised in the collision regulations and acted as expert witness in a number of court cases, has a particular interest in this story:

Small vessels driven at speed by small-brained idiots were always going to be a problem. Robin's unfortunate confrontation was not the first but certainly the most serious. This particular twit had already accumulated a track record. Usually after a few gins on a Sunday lunchtime, he would come tearing out of Ramsgate Harbour in his souped-up speedboat, at 30 knots plus, with the express purpose of racing the hovercraft.

This happened around lunchtime and by the time I went home for dinner Dave Wise's phone had rung red hot for most of the afternoon while the insurance claims rolled in. I don't recall now the final figure but I do know it staggered me at the time.

We passed Robin's report of the incident along to some maritime lawyers in London – aptly named Crump and Partners as I recollect.

We knew who the speedboat idiot was and we got hold of Crump and Co with the idea of suing him. Sadly, after interviewing the crew they told us we didn't have a leg to stand on. The clincher was the navigator saying, that some five miles off, he noticed the idiot tearing out of Ramsgate and altered five degrees to port.

The Crump people must have come to the conclusion that the navigator had altered to port to avoid collision, which according to Robin was definitely not the case. It's the only explanation I have to account for why they told us we had no case because we had gone to port in a crossing situation and it was against the rule. Either the crew didn't explain the situation very well or unfortunately the investigators totally misconstrued it.

This is the prime cause they said, altering to port in contravention of Rule 15 – the crossing rule. In other words, the first action was the prime cause of the accident and, after that first wrong move, it doesn't matter how many right moves you make it still turns out to be your fault.

It's worth noting that until I talked to Robin recently I was under the same impression as the Crump people that his navigator had altered to port ahead of the boat. The fact that this was not the case leads me to suspect that the major problem may have been a total misunderstanding between cultures.

The lawyers would have been used to straightforward disputes between ships and other conventional craft. The fact we lived in a totally different world and by that stage, without our realising it, talking a different language, quite possibly led to them misconstruing the crew's evidence.

Knowing Your Place

From a purely navigational sense, finding one's way was hardly a problem in a short 40-minute journey from Ramsgate to Calais when both ports could be seen on the 24-mile range of the radar. Although the new recruit had to learn the position and patterns of the multitude of navigation marks on the route, this was not primarily for position fixing but critical in the process of identifying collision risk.

Our new recruits were skilled in the use of radar, but the real problem was the equipment itself, which although the best available in 1969, was by today's standards primitive in the extreme. We were stuck with what we had – radar using pre-transistor 'bottle valve' electronics with an analogue presentation, which showed target movement trails by what was called 'afterglow'[4] – a far cry from the computer-calculated vector presentations of today.

What was really needed was radar equipment that was designed specifically for high-speed operation, but with such a small and uncertain market there was little incentive for manufacturers to incur the expense of looking at the specific problems.

Nevertheless, despite the shortcomings of the equipment, paradoxically it compelled the evolution of a navigation technique whose keynote was simplicity. The information provided was basic and the interpretation of that information was also simple and direct.

Bob Adams in the early days as 1st officer making a trip as navigator. The curtains which normally surround the navigator's station are drawn back and the radar visor, which enables the screen to be seen in daylight, is in place. If removed the curtains would need to be drawn.

---

[4] The early radars used cathode ray tubes, the display screens of which were coated with a compound that when hit by the stream of electrons produced by the returning echo of the target glowed brightly. It was designed to remain glowing for a period of time, hence as the target moved on the screen the 'trail' showed up as a gradually fading tail.

The navigator's position – Left to rght: Decca 629 with daylight visor removed; Sperry GB4 compass (note stopwatch), Decca 202 with visor in place; VHF radios; HF radio.

The key to the whole process was the accurate assessment of drift, the angle between where the hovercraft was pointing and the actual direction of movement over the sea surface. The average drift angle was around 10 or 15 degrees but in some circumstances it could be as much as 30 degrees. In the days before Global Positioning (GPS) the only way to ascertain the true track of the hovercraft was to line up on a known fixed object and watch how we tracked past it. Hence the need for the new navigator to learn quickly the patterns of buoys of which luckily there were dozens in that part of the Channel.

Once he had learned the patterns of the fixed marks a trainee was well on the way to an acceptable level of competency, but it certainly wasn't easy. The paradox was that although the radar range could be increased to identify the breakwaters of Calais on the French coast, the immediate problem for the navigator was where was he now in the near vicinity?

Until full familiarity with the area was achieved it didn't take much for the navigator to be totally lost. Once some experience had been accumulated it was surprising how different the various patterns looked if the craft was approaching from an unusual direction.

Not dissimilar to the captain's experience, navigators, once checked out and on their own, soon found that they had more to learn. This is to

George Lang's embarrassment but Roger remembers, after a couple of hundred hours or so, getting into much the same predicament:

> *The incident I do remember was when I got lost mid-Channel. We had been set to starboard by numerous encounters and then I could not make sense of where I now was. I went very quiet whilst trying to sort it out and the 1st officer had to get out of his seat to come back and see if I was still alive! Most embarrassing.*

The Technique

Once comfortable with knowing where he was, the navigator's job was relatively straightforward. Using the mechanical graticule, which was nothing more than a moving circular ruler, he lined them up against the echoes of the known fixed targets, which showed the actual direction that the machine was taking, and then compared that line with the other targets on the screen. He could then identify which way the moving targets were tracking, by comparing the slight angular differences.

The diagrams below hopefully provide a better understanding of what was involved.

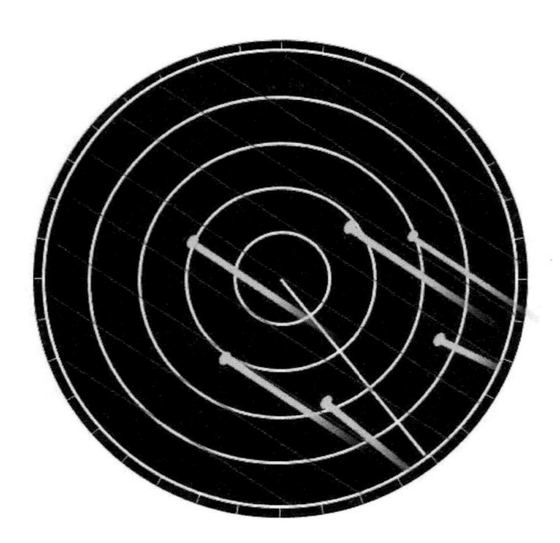

Figure 1. The Decca 629 screen at the edge of the Goodwin Sands, heading towards Calais. The usual operating range of the radar was kept at 6 miles; hence the rings are 1 mile apart.

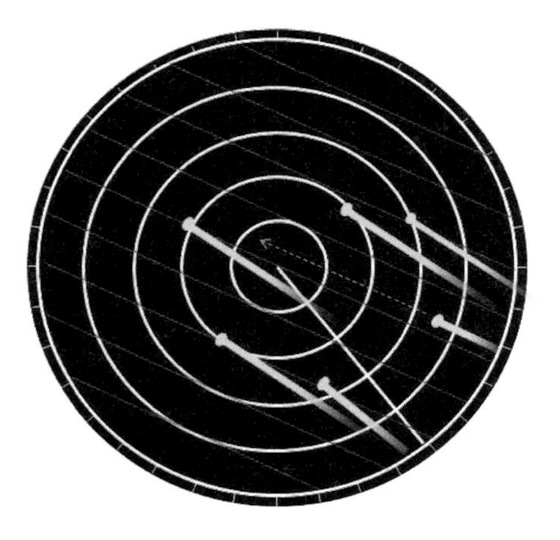

Figure 2. Checking the Closest Point of Approach (CPA) of the target to port.

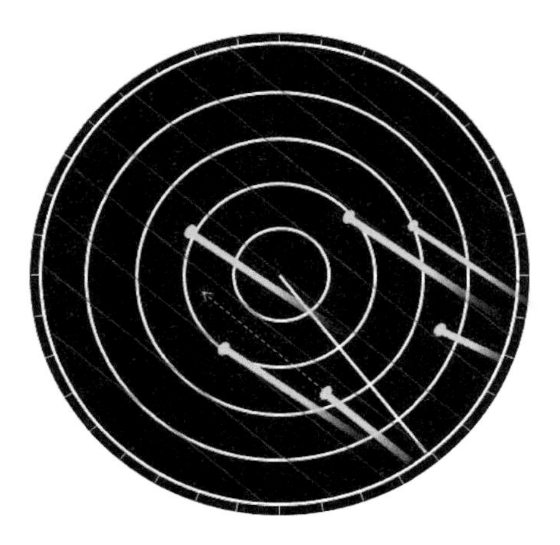

Figure 3. Checking the CPA of the target to starboard.

Strangely enough it was the speed involved which helped the navigator, as the length of the tails of afterglow fading slowly behind the target provided a very clear indication of target movement on the screen. In this instance, Figure 1, the four nearest targets were the wrecks and navigation marks at the edge of the Goodwin Sands.

The white line is the heading marker showing that the vessel is at present on a course of 141°. The navigator has lined up the red graticule lines on the fixed objects, which give themselves away by their pattern and movement parallel to each other.

The prevailing south-westerly wind is pushing the machine roughly 18° off to port.

The two non-parallel targets are obviously not stationary but vessels on the move. In Figure 2, the navigator's next move is to swing the graticule on to each in turn and judge, by running his eye down the nearest line, what the closest point of approach (CPA) is likely to be. In Figure 2 the graticule is lined up with the target to port which shows the target is moving away from the static line in a left-to-right direction, and when extrapolated back to the centre suggests this vessel will pass a little less than half a mile down the port side.

Figure 3. Using the same technique, the target to starboard has a CPA of around a mile and a half, and from its angle to the track line is moving right to left. In both cases the hovercraft is passing ahead.

The question now for the navigator is what to do about it? The target to port is passing too close and the easiest manoeuvre would be to alter 5° or 10° to port and slip astern. In an operation when at any time there are at least two craft at sea, there is always another fast craft somewhere ahead. We always kept a half a mile to starboard of our centre track line. These were two lines drawn on a Decca flight log so that a one mile separation was maintained at all times. Although it means crossing that track, the only safe manoeuvre is to come to 10° to port and go astern of the slow moving target.

Of course this was not the end of the navigator's working day. Over the edge of the six-mile range there would have been dozens of other vessels, all posing different problems.

The Biggest Problem

The system of identifying and avoiding collision worked well except for one situation, which was concerning to say the least. Perhaps terrifying would be more apt. The one object we were least able to identify was another hovercraft on the opposing route. With an average closing speed of 120 knots, and with the radar set on the six-mile range, the potential for collision was three minutes. The speed of the opposing machine was not immediately apparent because the tail left behind was no longer than the average target tail of a 12-knot cargo ship proceeding up the Channel.

At our first forays at the bottom of the learning curve, this inability to recognise the speed differential was totally unexpected, and led to one or two very near misses between hovercraft, which don't bear thinking about now.

Once the phenomenon was recognised, it then took some while to understand the reason. It has already been explained that all original radars used cathode ray tube (CRT) displays. Also, already explained, incoming echoes were marked on the tube surface by the echo, in the form of an electron stream exciting a chemical compound which produced a glow. The compound had a level of persistence that allowed the glow to die out slowly, usually over a period of at least four or five minutes. In theory, that should have produced a very fast approaching target clearly identified by its very long tail.

Why didn't it?

Although never discussed with the manufacturers, our conclusion was that the fast-moving target was 'overprinting' on the screen to a lesser extent than a slower target and therefore fading faster. This was borne out by the print left by a much faster low-flying aircraft, the imprint of which left a series of unconnected, quickly-fading dots. The speed of an approaching hovercraft image was not quite fast enough to show as dots but instead produced a string of dots that slightly overlapped. The result was a rather weak print on the screen, which faded faster than something a lot slower.

The Hi-tech Solution

The solution was the usual Hoverlloyd approach to things, namely a common or garden stopwatch bolted below the navigator's compass repeater. As soon as a suspicious target appeared over the edge of the screen, it's time of crossing over two range rings[5] was noted and the approach speed, dangerous or not, was ascertained.

---

[5] The range rings on the six-mile range were spaced at one mile apart.

Another Problem with Drift

It would be natural to assume that considering the extreme manoeuvrability of the SR.N4, an alteration of course would be immediate, but unfortunately not so. One thing that really frightened rooky navigators was that, yes, when the captain was asked for the required alteration of course, the craft moved to the new heading immediately. Again, when the navigator spoke the captain didn't argue, but the picture on the screen didn't change. To all intents the movement of the targets stayed much the same – the potential collision direction remained steady. After a few seconds – they were long seconds nevertheless – the relative movement of the target started to veer away and the navigator's heart rate could return to normal.

The reason for angst was once more the ice skating syndrome. The complete lack of friction made turning through the compass quick and easy, but the same lack of friction meant that it took some time for the old course momentum to change. This will be familiar to any car driver who has been unfortunate enough to be caught on a corner in black ice. The wheel can be turned as much as you like, but the car will refuse to turn and carry on forward, probably ending up off the road. It was a slightly worse sensation on the SR.N4 because it looked as if it was turning but it wasn't.

The sum total of all this was the turning circle, which, considering its size in small boat terms, should have had a tactical diameter within a hundred metres or so, but in fact achieved an average turning diameter of half a mile, something these days one would expect to be characteristic of a 200,000 ton VLCC (Very Large Crude Carrier).

Measurement Difficulties

One other problem, which was not so easy to deal with, was the developmental state of the radars we were using. Although by the end of the '60s transistor technology and 'solid state electronics' were well advanced, marine radars were, by and large, still using vacuum valves and klystron cavity resonators; technology of the '30s and '40s.

Measured against what was generally available at that time, the Decca radars we used were first class, extremely reliable and well thought out ergonomically. The difficulty that occurred was not so much a matter of reliability as one of accuracy.

Every so often it would be noticed that the range rings on the screen were no longer equally spaced. The one mile ring would be smaller than usual and the spacing out to the six mile ring would gradually increase to somewhat larger than normal.

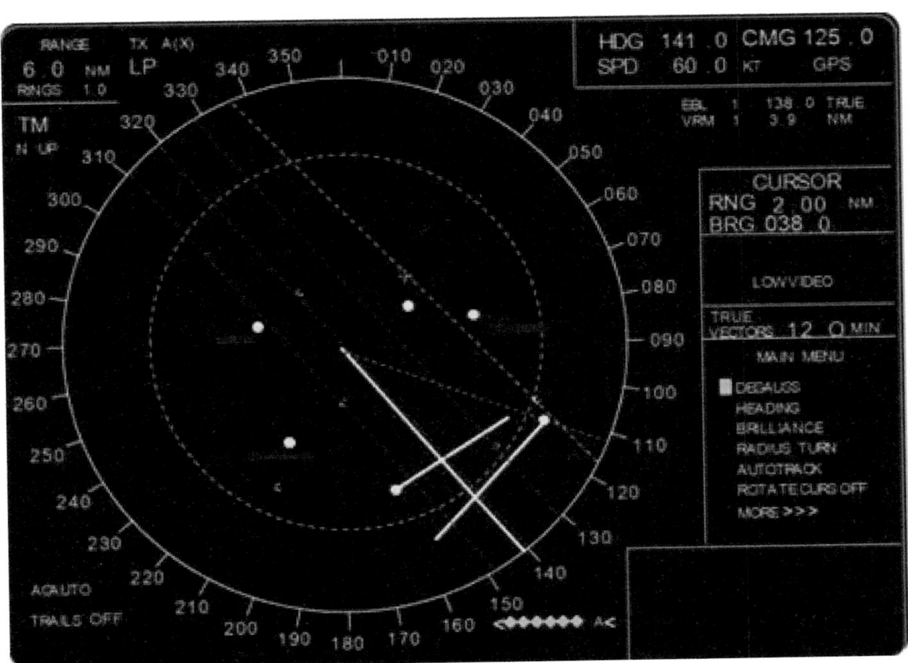

The Decca Bridgemaster – 40 years on from the 629 this is typical of the 'do it all for you' age. This is the same scenario as before but the fixed objects are shown as stationary and the two vessels show their actual course and speed. Maps are overlaid so the navigator no longer needs to remember the patterns to identify the navigation marks. (A hand-drawn facsimile – not a photograph.)

Although looking odd, it has to be pointed out that this had no effect on measurement accuracy, as a target on the three-mile range ring was definitely three miles away, regardless of where it physically appeared on the screen. On the average slow-moving vessel this could be tolerated, but with the high speed N4 it was a worrying problem because of the way we were using the set to differentiate target movement.

As explained above, the navigator judged closest point of approach, in other words potential collision, by lining up the parallel cursor against the target trail and extrapolating that movement towards the centre. The distortion of the range rings had the effect of curving any movement in towards the centre. Because of this curvature in target movement, a vessel appearing at the limit of the range, possibly clearing a mile down the starboard side, would initially appear to be on a collision course, prompting the navigator to consider altering to starboard to avoid it; a disaster waiting to happen.

Why this seemed to occur on the N4 so often is open to conjecture. Of course, it may simply have been we noticed it more because of the seriousness of the problem, but it could also have been due to excessive vibration having an effect on the delicate electronics. One of the misnomers of all time was to refer to the hovercraft as an 'air cushion vehicle', a cushion it certainly wasn't. A constant surprise to newcomers was the bone-jarring ride. This was especially bad on calm days, with perhaps just a small 'lop' of a few inches, when the craft would be at its fastest and the vibration would be at its very worst. It felt like driving a car with solid tyres over cobble stones; in fact we called it the 'cobble stone effect'.

Whatever the cause it was something we had to live with, constantly asking our engineers to readjust the rings. The later craft, the Hoverlloyd *The Prince of Wales* and the Seaspeed Mk.3 upgrades of the two *Princesses* were all fitted with new radars which, although still without daylight viewing screens, were solid-state throughout. However, right to the end of their life in Hoverspeed, *Swift*, *Sure* and *Sir Christopher* stayed with original 1960s' radar fit and it is assumed the range-ring spacing remained a problem.

An Art not a Science

It could be said, once one got used to it, collision avoidance at high speed was easy. A highly experienced N4 navigator didn't avoid collision by measurement and calculation – whether his own or a computer's – but by simple observation and perception.

Today mariners on any ship are taught, and indeed are required by law, to plot all approaching vessels, either manually or electronically, and calculate collision risks and the resulting manoeuvres. By today's standards the cross-Channel flight crews were breaking every rule in the book because they merely looked at the radar situation, judged from experience what was required and altered course accordingly.

Under expert tutelage a trained navigator with a few hundred hours under his belt could look at a situation, alter 5° or 10° and know precisely how the track of every target on the screen would change. It should be noted here that the illustrations referred to earlier are a gross simplification. In reality, the radar, at times, would have been cluttered with scores of other blips, so it wasn't just a question of sorting out the immediate situation but how the alteration would affect other conflicts further ahead.

If the above sounds like an excuse for pushing blindly on and hoping things would be okay, it has to be emphasised that this was a true methodology. It was to teach an art rather than a science, no different than, say, learning the violin. A pupil gradually develops through the learning process; from incompetence, through conscious competence, to finally, unconscious competence. The final stage is where the player no longer has to really think about what is needed to produce the music; it is produced automatically with the minimum of thought.

It is interesting to note in this photograph that, although the new radar is an Automatic Radar Plotting Aid (ARPA), replete with daylight viewing, vector displays, GPS input etc., the navigator has chosen to stick with a facsimile of the original 629 format. The target trails are similar to the after-glow trails on the old installation, the difference is that this is a computer screen and therefore the trail is digitally produced.

The Decca Bridgemaster Radar as installed in the Hoverspeed Mk.3 craft. The hovercraft is approaching Calais. Although set to the usual six-mile range, the radar allows the display to 'off-centre', which provides an extended view ahead.

The operator above is Roger Warren and, as an 'old hand', the display mode suggests that he is used to the KISS principle[6] of the old set-up. One suspects he is working with what he is used to.

On the other hand, it is not necessarily a question of old habits being hard to discard. There were genuine problems with the new system, resulting in the calculated vectors, seen in the previous picture, being far from accurate.

---

[6] KISS – 'keep it simple stupid!' – An astute American expression suggesting that the safest and most efficient way of doing something is invariably the simplest.

As always, progress is a matter of two steps forward, one step back.

Perception was the essence. No one needs a radar and a computer to wind their way through a reasonably spread-out crowd of people. As humans we have an intrinsic ability to execute avoidance manoeuvres and manage to avoid bumping into people. On the N4 the navigator's approach was no different; he avoided trouble by a finely-honed instinct, nurtured by first-class training and hours of experience. If there is any doubt about the validity of such a proposal, one only has to look at the 32-year record of high-speed operations across the Dover Strait without one accident.

Chapter 17

Looking after People

Our Amazing Budgies[1]

The Pride of the Fleet – Madeleine Marsden (of the hijack incident, extreme right) with, from left to right: Unknown, sisters Lesley and Vivian Thatcher, Lesley Lewis, Cathy Whitnall and Betty Dowle.

It is to be expected that many 21st century eyebrows may well be raised at the idea of Hoverlloyd's workforce being clearly segregated into 'boys' jobs' and 'girls' jobs'.

Although at this time feminism was a gathering force in the western world, it had yet to reach such social backwaters as East Kent.

Not that there was any conscious move to segregate the sexes into 'suitably appropriate' areas of employment, but it happened nonetheless. While the engineering department, the flight crew and car deck teams were exclusively male, the cabin staff was exclusively female; all young and recruited for their good looks, charm and deportment.

[1] The origin of the term is attributed to Captain Tom Wilson, one of the 'originals' in more ways than one. Common parlance in Britain in the '60s was to call young women 'birds', so this was Tom's version and it stuck.

Like it or not, our cabin department was a totally female preserve. Considering the sometimes horrendous working conditions, it is debatable whether the average macho male could have put up with it. Working either two or three consecutive return crossings per day, which nominally, if three, should have been a six-hour shift, could in severe weather, extend to eight or nine hours. During this time the girls had to maintain their feet on what was essentially a wildly bucking speedboat, deal with sick and sometimes very frightened passengers, do their best to remain calm, *and* sell as much duty free goods as possible to advance the company's profitability.

This, from Katrina Patchett, is but one small illustration of a fairly average bad day – and there were many of them:

> *I was on the craft that hit a 'hole' in rough seas and all the new 'safety glass' front windows smashed in – luckily no passengers were in the forward cabin and no one was hurt. I was in charge and had, before all this happened, told the other girls to remain seated since it was so rough.*
>
> *A French teenage boy halfway down the main cabin had started to panic and screamed, "Nous allons mourir! Nous allons mourir!"* (We are going to die!). *I went over and calmed him down, telling him the craft was so safe, etc. Seconds later, there was a smash, I turned around and saw a wall of grey water, floor to ceiling, at the front of the main cabin. I thought, "My goodness, he's right – we are going to die!" It was like something out of the* Poseidon Adventure. *The craft pulled back up so the water went down, and we returned to Ramsgate. Of course, I thought it all exciting – the joy of being young and thinking life goes on forever!*

Of course there were many days when all was calm and peaceful, but human nature being what it is, mainly it is the drama that is remembered, and that was more likely to occur in bad weather.

The Right Type of Girl

As illustrated above, the job of stewardess on a cross-Channel hovercraft may have appeared to the average young woman to be on the same level of glamour as being an air hostess, but in fact it required a far greater level of physical and mental fortitude.

June Cooper, who through most of her career at Hoverlloyd, was deputy to Joan Stroud the Chief Stewardess, quotes the requirements at the time:

It was essential to employ the right type of girl – she was required to be not only attractive and well groomed but have a lively personality and to be caring and level headed. With this in mind a minimum starting age of 21 was decided upon.

Clockwise from bottom left: Val Hughes, unknown, Inez Bremmer, Gill Spencer, unknown, Janet Goodbourne.

BETTY DOWLE

SO YOU WANT TO BE A STEWARDESS

AS pretty blonde Betty Dowle welcomes passengers aboard the Hoverlloyd hovercraft for their cross-channel flights this year, she'll be clocking up the first sea-miles towards her next target: for last November Betty made her 10,000th crossing with Hoverlloyd and received a gold watch to celebrate the event.

One of a team whose job it is to make you, the passenger, as comfortable as possible, with an eye on safety too, Betty has been with Hoverlloyd for seven years. Her sister also works with the Company. Married to an Able Seaman, Brian, who works on a ship on another cross-channel route, Betty lives at Sandwich, not far from the Hoverport at Ramsgate, and in her spare time helps her husband modernise the bungalow they have bought there. All the nice girls love a sailor, it seems, for when Betty married Brian, five other Hoverlloyd girls married sailors in the same year.

What's it like to work aboard the Hovercraft? Hectic, at times. In good weather the crossing can take as little as 30 minutes, and in this time the cabin staff have to take round drinks, sell cigarettes, perfumes, other duty-frees. And with a full load of 278 passengers as they so often have in summer, Betty describes it as "A bit of a rush."

The first job the cabin staff do (there are between 6-8 of them on each flight, according to the number of passengers) is to see the passengers on board. It takes twenty minutes to board the cars on a full flight. Then they count the stock in the bar, as they have probably taken over from another crew. This has to be done out at sea because of Customs regulations. After that it's time to serve the duty free goods and drinks. The cabin crew work in teams, on a shift system, two days on, two days off and, like air crew they are not allowed to drink alcohol for eight hours before they start their duties.

Hoverlloyds cabin staff have two weeks' training on the ground, once they have been accepted, before they are allowed to go on flights. This includes a first aid course and special training in safety routines and regulations and emergency procedure, so they know exactly what to do, instinctively, should the occasion arise. The girl members of the team are also given a course in beauty and poise, too, learning how to put on make-up professionally, and keep it that way ("after two flights, it needs renewing" says Betty). They also learn how to walk properly, and how to wear their uniforms correctly. Perks of the job include almost free cross channel travel and in the case of female cabin staff a special allowance for tights. Uniforms are cleaned free of charge.

What makes you good at your job? "You've got to be flexible to cope with the shift system," says Betty. "Apart from that you need stamina, and you must be cheerful and helpful." Does she ever get seasick? "I've only felt ill once," says Betty. "And that was during my training. You get used to it. But I remember going out in a towing boat once, and feeling sick, and I thought 'This is ridiculous . . .'"

Do they ever get strange requests from passengers? "Well, Americans always ask if they can have eggs and bacon, or champagne, or special cocktails served to them," says Betty. "But we'd scarcely have time to serve that. Some people who have booked through to Paris think that the Hovercraft continues there, overland, when we arrive at Calais. And one old lady stayed sitting up front at Ramsgate, the other day, waiting to hover on to London."

Betty has never flown in an aircraft, but says she would far rather be a stewardess on a Hovercraft than in the air. Does she ever get bored with the constant to-ing and fro-ing? "It's different every day," says Betty. "Besides, I like dealing with people, and the other team members are very nice – it's a very social life."

ROBERT WRIGHT

MADELINE

No mention of the need for the sea legs of an old China clipper sailor, the strong stomach of a paramedic and the calm assurance of a lion tamer. It is hardly surprising therefore that many aspirants taken in by the glamour were sadly disillusioned with the reality.

It was very soon apparent that the 'right type' was not so easy to find, and, with the expansion of the service over the decade, recruiting and training around 100 extras to add to the 36 permanent staff for the peak season was becoming a bigger and bigger headache. As time went on inevitably the criteria became more and more flexible. If a girl was presentable and showed obvious fitness and strength of character she was in.

Jacquie Case was clearly of the necessary stalwart material, still undeterred after this experience:

> *My worst day was a gentleman making a dash for the loo, to be sick, who caught me in the aisle and wrecked my uniform.*

However, two of her aspiring contemporaries were obviously not:

I started in March when the weather was still fairly wintery. The first training trip on the craft was a rollercoaster ride, where we immediately lost two of the group who swore they would never get back onto a hovercraft in their lives again!

It could be said there was really no need for searching interviews, one trip in bad weather was enough to sort the wheat from the chaff.

AS pretty blonde Betty Dowle welcomes passengers aboard the Hoverlloyd hovercraft for their cross-channel flights this year, she'll be clocking up the first sea-miles towards her next target: for last November Betty made her 10,000th crossing with Hoverlloyd and received a gold watch to celebrate the event.

One of a team whose job it is to make you, the passenger, as comfortable as possible, with an eye on safety too, Betty has been with Hoverlloyd for seven years. Her sister also works with the Company. Married to an Able Seaman, Brian, who works on a ship on another cross-channel route, Betty lives at Sandwich, not far from the Hoverport at Ramsgate, and in her spare time helps her husband modernise the bungalow they have bought there. All the nice girls love a sailor, it seems, for when Betty married Brian, five other Hoverlloyd girls married sailors in the same year.

What's it like to work aboard the Hovercraft? Hectic, at times. In good weather the crossing can take as little as 30 minutes, and in this time the cabin staff have to take round drinks, sell cigarettes, perfumes, other duty-frees. And with a full load of 278 passengers as they so often have in summer, Betty describes it as "A bit of a rush."

The first job the cabin staff do (there are between 6-8 of them on each flight, according to the number of passengers) is to see the passengers on board. It takes twenty minutes to board the cars on a full flight. Then they count the stock in the bar, as they have probably taken over from another crew. This has to be done out at sea because of Customs regulations. After that it's time to serve the duty-free goods and drinks. The cabin crew work in teams, on a shift system, two days on, two days off and, like air crew they are not allowed to drink alcohol for eight hours before they start their duties.

Hoverlloyds cabin staff have two weeks' training on the ground, once they have been accepted, before they are allowed to go on flights. This includes a first aid course and special training in safety routines and regulations and emergency procedure, so they know exactly what to do, instinctively, should the occasion arise. The girl members of the team are also given a course in beauty and poise, too, learning how to put on make-up professionally, and keep it that way ("after two flights, it needs renewing" says Betty). They also learn how to walk properly, and how to wear their uniforms correctly. Perks of the job include almost free cross channel travel and in the case of female cabin staff a special allowance for tights. Uniforms are cleaned free of charge.

What makes you good at your job? "You've got to be flexible to cope with the shift system," says Betty. "Apart from that you need stamina, and you must be cheerful and helpful." Does she ever get seasick? "I've only felt ill once," says Betty "And that was during my training. You get used to it. But I remember going out in a rowing boat once, and feeling sick, and I thought 'This is ridiculous. . . .'"

Do they ever get strange requests from passengers? "Well, Americans always ask if they can have eggs and bacon, or champagne, or special cocktails served to them," says Betty. "But we'd scarcely have time to serve that. Some people who have booked through to Paris think that the Hovercraft continues there, overland, when we arrive at Calais. And one old lady stayed sitting up front at Ramsgate, the other day, waiting to hover on to London."

Betty has never flown in an aircraft, but says she would far rather be a stewardess on a Hovercraft than in the air. Does she ever get bored with the constant to-ing and fro-ing? "It's different every day," says Betty, "Besides, I like dealing with people, and the other team members are very nice – it's a very social life." □

Cabin Staff Organisation

The dual nature of the passenger accommodation split by the centre car deck meant the cabin staff operated as two separate entities. Each crew of six stewardesses was labelled 'A' to 'F', and split three each side. The 'A Girl' was in overall control, similar in position and responsibility to an in-flight supervisor in many airlines today.

Her principal duties were to make sure her crew were well turned out and were following the rules and protocols laid down in the *Operations Manual*.

The very first thing was to ensure that the cabin staff in her charge checked in on time – an hour before their first scheduled flight – and, if necessary, arrange replacements from the standby crew pool should there be any 'no shows'. Once satisfied that all were present and correct she reported her crew to the captain, and received a briefing from him as to what could be expected for the day.

On the craft she ran the starboard side cabin while her second in command, the 'B Girl', was in charge of the port side. Both operated autonomously, the only extra responsibility for the 'A Girl' was to liaise with the flight deck should anything untoward occur and to keep 'B' informed.

The Pre-Computer Age

If dealing with occasionally frightened and seasick passengers wasn't enough, there was also the difficult task of selling duty free. The stewardesses were expected to memorize the exchange rates of up to ten different currencies; no such thing as pocket calculators back then.

These days, with our complete reliance on computers of all shapes and sizes, it is very likely that a job our amazing girls saw as merely routine would have been a difficult task for mathematics graduates.

At least there was an incentive in the form of a commission on sales, which was divided up and distributed amongst the girls monthly.

In addition to the requisite attributes listed above, ability in a European language was favourably looked upon in any new recruit, as was nursing experience. The very sensible requirement laid down by the company from the beginning was that there should be least one French speaker and someone medically trained in each rostered crew of cabin staff.

Both senior staff, 'A' and 'B', were required to give a safety briefing in English and French prior to each departure.

Going back to the captain's briefing, at the commencement of the day, especially important for the cabin crew was the likely state of the weather. Very good weather, calm flat sea and very little wind meant a fast crossing, sometimes as little as 30 minutes, and less in a few isolated cases.

Betty Dowle recalls:

> *On a calm day, if it was a full craft, you had to get people served quickly, as the crossing time was likely to be shorter than normal. You didn't know whether to do drinks or the duty free first. If they didn't get one and they wanted both we'd be in trouble because we just didn't have time to serve both. We used to phone up the flight deck and ask them to slow down because we hadn't served everybody, but we didn't always get the cooperation. Robin was the worst offender. Ted Ruckert was bad too – he was 'Rocket Man'.*

In contrast, if bad weather was expected where the girls would have had difficulty keeping their feet, it might be advisable, as Katrina's earlier story suggests, not to attempt to sell drinks and duty free at all, but remain seated themselves and hang on.

A newspaper cutting (paper unknown) of film star Michael Caine sitting beside Betty Dowle. The hovercraft had been turned into a film set for a scene from *The Black Windmill.*

Changes of Uniform

For the first two or three years there were a number of changes to stewardess uniform, none of which seemed to completely fit the bill, until in 1974, when the very stylish red ensemble, shown at the beginning of this chapter, was introduced.

In 1969 the first uniform was not a question of choice of style but driven more by necessity to keep some restraint on the budget. Consequently, a job lot of second-hand uniforms were obtained from the recently failed airline Eagle Airways – what June Cooper refers to as 'hand-me-downs':

> *Obviously, to look the part, a stewardess needed a stylish but practical uniform. In the first months of operation, the girls were issued with a not very flattering two-piece maroon and black check suit, a hand me down from Eagle Airways.*

However, by the time the summer season began, a very attractive burgundy outfit designed by Peta, one of the daughters of Managing Director Les Colquhoun, had been introduced. This was a short-sleeved dress with a high neckline worn at least 3 inches above the knee and a side buttoning matching jacket. It was worn with a blue pillbox style hat sporting a red pompom. The girls loved the dress and jacket, but not the hat, though looking back at the pictures, it did add a certain flair.

Not sure why the girls were not keen on the pompom – in retrospect it looks rather fetching.

1969 Summer Uniform – Val Hughes, Sandra Sparrowhawk and Ann Deverson *(Hoverlloyd publicity brochure).*

The next uniform introduced in 1971 was the first to be professionally designed. Despite the 'professional' label, it didn't appear to be all that stylish or flattering, and apart from anything else, the turquoise colour tended to clash with the predominate red of most of Hoverlloyd's colour schemes.

It was also meant as a design for all seasons but, as June points out, it was hardly sufficient to deal with winters on the east coast of England:

The turquoise suit served both as a summer and winter uniform though still worn well above the knee. It consisted of a jacket and skirt with a matching detachable collar, replaced daily, and white gloves and scarves.

In winter the same suit was worn with long-sleeved, white round-necked jumpers though polo necks were strictly forbidden – heaven knows why. We would all have been much warmer – it was incredibly cold out on the pad in winter. A later addition of a thick Henry Lloyd yachting jacket was a godsend and I know that after 30 plus years, some girls still have one.

Jacquie Case wearing the turquoise uniform.

Finally, in 1974 the eye-catching red uniform was introduced and remained the preferred stewardess dress for the next few years. The hat June refers to as a 'trilby', was in fact modelled to be reminiscent of the Beefeaters hats at the Tower of London, presumably to give the whole ensemble a decidedly British feel.

Again there was the problem of cold winters:

> *Whilst we enjoyed the uniforms, the cold affected us badly and many requests were made for trousers in the winter. The male hierarchy seemed reluctant to agree this change, but eventually a striking winter uniform was introduced by the same designer, consisting of a red blazer and well-fitting black flared trousers worn with black polo neck sweaters (at last!) black gloves and a black trilby style hat again.*

1974 uniform – winter version. Chief Stewardess Joan Stroud and June Cooper with the designer Ray Stringer.

Mo Langford.

'The Duchess'

All this regimentation needed organising and, by the time the third craft was delivered in 1972, there was a need for a full-time person to ensure that standards of dress and protocol were kept up to scratch. Maureen 'Mo' Langford, always beautifully turned out herself, was a good choice:

> *The Chief Stewardess, Joan Stroud, asked me if I would consider taking on the position of training stewardess. After some soul searching I accepted. It was a wonderful experience working alongside Joan, and June Cooper. Whenever I decided to do a 'check flight' the flight crew and car deckies would see me marching across the pad and put out a warning, "On your*

toes girls, here comes the Duchess." But I think I was very fair in my reports.

I enjoyed every single minute of my 17 years with Hoverlloyd and Hoverspeed, and made some great and lasting friends.

Evacuation

Seaspeed's first evacuation demonstration. Life rafts 1, 3 and 5 should have been launched first, but 4 was inadvertently launched at the same time, resulting in it sitting on top of 3 and 5. July 1968.

Although the sales of duty free goods were an essential element of the company's profitability, from a regulatory point of view the stewardesses' primary role was to look after the safety and comfort of the passengers. In the event of a disaster their task was to help the passengers evacuate the hovercraft in a calm and efficient manner.

None of the craft could operate without an Air Registration Board 'Permit to Operate', a certificate that was only issued after a thorough investigation of the craft's structure and systems, with specific attention to the safety features and equipment, culminating in a real-life demonstration of an evacuation.

No. 4 life raft was punctured and rendered unseaworthy, so a replacement raft was launched from the port side.

Like everything else to do with this new mode of transport there was no previous model for what was required. The requisite safety arrangements for ships – especially since the *Titanic* disaster – were well established by the second half of the century, as were those for aircraft, but what was suitable for this new vehicle? Lifeboats, the usual provision for ships, were too heavy, and aircraft evacuation chutes would not have worked so close to the water. Inflatable life rafts were the only realistic solution, but even they had their difficulties.

The BHC design team equipped the SR.N4 with ten RFD Beaufort H.30 type inflatable life rafts. Each was capable of carrying 30 persons, but with an emergency overload capacity of 40. They were secured in cradles on the outside deck, five on each side. Once released and pushed over the side they automatically inflated.

There were two possible exits from the cabin, the main passenger entrance and a small emergency exit door forward at the turn of the bow. There was a similar emergency exit door right aft, but this had to be accessed by a rather tortuous route via two small doors through the electrical bay room and the rear fan compartment. It was listed as an extreme emergency option and not considered in the normal evacuation plan.

The first tests for this arrangement were undertaken in 1968 when Seaspeed took delivery of the prototype *The Princess Margaret.* As might have been expected with something so new, it was not particularly successful, to put it mildly.

John Lefeaux, Seaspeed's chief engineer[2] at the time, in his book *Whatever Happened to the Hovercraft?,* notes that the Board of Trade's criterion for a successful evacuation was that the craft should be emptied in three minutes. In retrospect this was either a typographical error or BoT's vision of what was possible totally misguided. If everything was designed to perfection and went according to plan it would be amazing if the machine could be evacuated in 15 minutes[3]. It is worth mentioning here that the 127 'passengers' used in the demonstration off Dover, in anything but ideal weather conditions, were fit young soldiers – hardly a realistic sample of the average population at all. Nevertheless, they managed to evacuate in 13 minutes 20 seconds. In these circumstances, with 'real people', even 15 minutes is probably too much to expect.

From a design point of view there was a major problem with how the life rafts were placed. It was supposed at the time that the whole bow area would be hazardous for both life rafts and passengers. Consequently life raft no. 1 was situated abaft of the forward emergency door resulting in a squashing together of the remaining four.

Their spacing was such that all the life rafts could not be launched or loaded simultaneously. The procedure was supposed to be, launch, load and cut clear 1, 3 and 5 first and then follow with 2 and 4.

Unfortunately, in the trial, no. 4 was launched at the same time as 3 and 5, an understandable error in a first attempt with an untrained crew. It resulted in no. 4 raft being squashed between 3 and 5 and forced vertically up against the folding stairs of the main passenger entrance. The eventual result was the life raft being severely punctured by the stair mechanism and rendered unseaworthy.

No. 3 was also punctured during launch. This was believed to have been caused by protruding pop rivets, but the damage was not severe enough to seriously affect the raft's buoyancy.

---

[2] John Lefeaux took over as managing director of Seaspeed in 1973.

[3] The only time in Hoverlloyd's history that passengers were actually evacuated into the life rafts, the whole procedure took an hour. The grounding of *Swift* on the beach at Sangatte west of Calais is recounted in Chapter 21: *Hoverlloyd's SR.N4 - The First Years.*

The teething troubles with the life rafts were not the only ones. Several of the life jackets inexplicably blew up when inflated, not the most reassuring thing to happen to an already frightened passenger in an emergency.

Seaspeed's *The Princess Margaret* on her first trial to Boulogne. The life rafts can be seen close together in their original configuration before modifications were made during the winter of 1968/69.

One last problem, conceivably minor but extremely serious in lifesaving terms, was the over-long bowsing lines. These are lines that remain connected to the raft after launch and hold it firmly alongside so that passengers can board. The lines as designed were too long, leaving a wide stretch of water, impossible to leap across, except for the young and athletic.

Despite everything going wrong that could go wrong, the evacuation only took 13 minutes 20 seconds, which, considering the difficulties encountered, was extremely well done.

Needless to say, when compared with the BoT's three-minute target, the trial was declared a failure and a repeat demonstration was arranged. In this second attempt, four days later, the ARB softened its stance considerably and this time only required 30 passengers of both genders and varying ages to evacuate into one raft only – in other words 'real' people. Even so, an illustration of the difficulties of evacuating a hovercraft with your average passenger at sea in virtually calm conditions was noted in the second report, which said:

> *All went well, except for one young female passenger, who was reluctant to walk along the narrow outside walkway on top of*

the plenum chamber and even more reluctant to board the life raft. Verbal persuasion and a firm hand eventually achieved the desired result, but 20 vital seconds were lost.

This evacuation was achieved in 5 minutes and 20 seconds, a great improvement on the previous exercise, allowing the ARB to tick the necessary boxes and give Seaspeed its SR.N4 'Permit to Operate'.

Hoverlloyd's Turn

Obviously, following the first debacle, a number of modifications to the safety configuration were made, such as the spacing of the life rafts to allow simultaneous launching, and by the time *Swift* turned up at Pegwell Bay in 1969 the evacuation process promised to be a routine one. That is not to say, however, that there still remained some design snags which would require major modification at Mk.2 level before they were sorted out.

Swift departing Calais during trials in 1969 with the much modified life raft configuration. The no.2 life raft is now where the no.1 was in the earlier format, thus allowing a greater spread of positions and no need for staggered launching.

The major remaining difficulty was that the side structure of the craft, over which evacuating passengers had to pass along to arrive at the life rafts, could hardly be described as a 'deck', as in the original Mk.1 design this was a mere one-foot wide ledge for much of the craft's length.

So, in an emergency the passengers were expected to edge their way along this narrow ledge and then negotiate a one-metre step down into the life raft – not the easiest thing to negotiate, even for the fittest. For the elderly and those carrying infants it was a big ask. To be fair there was a handrail, but it still wouldn't have been easy.

Risks of accidental falls in the water aside, all went more or less well with the 1969 demonstration to the ARB and BoT, and in good time, but there was one small hitch. No one had double-checked the inner cabins, and one held a number of passengers still sitting there[4].

In his capacity as operations manager, Bill Williamson observed the event:

> *To start with all went well and we, on the roof, saw the 'passengers' stream out of the main and forward cabin doors and into the life rafts. Then the flow of people stopped, but there was still an empty life raft alongside. A bit strange I thought. When we went down to the main deck the starboard inner cabin was still full of seated marines.*
>
> *The ARB rep, who was observing in that area, said that the 2nd officer had come to the door of the inner cabin and shouted, "Follow me!" at which point just two of them stood up and followed him out. The rest sat totally mesmerised in their seats. The ARB rep described it as unbelievable.*

One rumoured explanation for the debacle was that the ARB observer 'sabotaged' the event by telling the marines to stay put after the 2nd officer had told them to evacuate, to see if he would come back to check that the cabins were empty.

Whatever the explanation, before obtaining their 'Permit to Operate', Hoverlloyd was obliged to repeat the drill on the pad for the benefit of the authorities without launching the life rafts, but with just the passengers leaving the craft via the embarkation steps.

Car Deck Staff

The 40-minute average crossing time between England and France and the neat schedule departing from either side on the hour, meant a mere 20 minutes for the car deck and cabin staff to move 36 cars and 280[5] passengers on and off. The passenger total was obviously fixed by the seating but the vehicle numbers could change depending on the individual size.

---

4 Like the Seaspeed trials, Hoverlloyd used Royal Marines as demonstration passengers.

5 These are figures for the SR.N4 Mk.2.

With such a variable set of circumstances, filling the car deck to capacity to maximize the load factor[6], but minimizing the turn-round time, required a high level of orderliness and efficiency on behalf of the car deck supervisors in particular. Hoverlloyd was fortunate that the first generation of car deck supervisors, who were to set the standard for the succeeding years, were almost all ex-Royal Navy or RAF. This not only ensured that loading and discharging was accomplished on time with military-style efficiency but the presentation and proficiency of the crew as a whole was of a consistently high standard.

The SR.N4 Mk.2 car deck looking from the stern to the bow. Note the 13-foot ladder leading to the flight deck. One can just make out the recesses on either side where the inner cabins used to be in the Mk.1, but now make space for three additional cars in each recess.

Minis and 2CVs

No matter what the load, a Mini could always be squeezed in somewhere. Hoverlloyd's car deck superintendent, Ted Unsted, claims the record for a full load of Minis:

> *I believe the record for Minis was 52. Mind you, we couldn't cross from one side of the deck to the other without going*

[6] During the peak months it would not have been unusual for most flights to depart with load factors of 100%.

across a few bonnets[7].

As far as I remember it was for the Mini Club of Great Britain and in total I think there were about 100 of them to transport that day. As it was flat calm we settled for ten rows, five abreast and a couple on the incline. Drivers were very good and for the last row in, the drivers climbed out of the window as the doors were too tight against the bulkhead to open.

Initially in the early '70s there were few difficulties with abnormal sizes of vehicle but as time went on, coaches in particular got much larger as did the introduction of Winnebago-style campervans into Europe. All these innovations in holiday transport must have made fitting the car deck load more and more problematical.

The Citroen 2CV.

Funnily enough the vehicle that gave the most trouble in the early days was one of the smallest cars, the Citroen 2CV, commonly known as the 'dustbin'.

The normal method of securing the vehicles was a webbing strap with one end looped around each car wheel and the other hooked into a 'D' ring on the deck. A ratchet lever tightened it up. In the majority of cars the suspension coped with the movement at sea, allowing the car to move while the wheels were firmly secured to the deck.

---

[7] This was strictly against our safety policy, but in our eagerness to keep our load factors up and the company profitable, nobody 'noticed' at the time.

The difficulty with the 2CV was two-fold. The bodywork covering the rear wheels made securing a lashing impossible and the suspension – described by some as 'comically soft' – in the same circumstances, could allow the body to swing around violently and give every impression of parting company with the wheels. The only solution was to open both doors and pass a car lashing across in front of the seats and strap the whole vehicle down.

Alan Mitchell, Car Deck Supervisor (centre), with his crew of (back left clockwise) Barry Nicholls, Bill Cook, Jeff Igoe, and Dave.

Difficult Passengers

Compared with the cabin staff the car deck had minimal contact with passengers; only in the short time it took for people to move from their car to their seat in the cabin, and vice versa. In fact, if things were going well, the shorter the better.

More often than not the situations they had to deal with were not their own confrontations so much as having to come to the aid of the stewardesses up against obstreperous passengers – invariably men – in the cabins. They provided a reassuring male back-up and, in extreme cases, a bit of extra male muscle.

Bill Cook, Car Deck Supervisor, and his crew.

Car deck supervisor Dennis Holness recalls a story about Alan Mitchell, one of the original stalwarts, who probably did more than most to mould the character of Hoverlloyd's car deck over the years:

> *If I remember rightly it was to do with this guy wanting breakfast, which of course we didn't do. He was so much trouble and starting to threaten violence – one wonders what he had already had for breakfast – that the girls called for help. From what I understand Alan Mitchell and another car decky held him down until it was time to get off.*

Another version of this story, apocryphal or not, suggests that Alan strapped the chap down on the car deck with the car lashings. Sadly, the truth of this anecdote cannot be verified as Alan is no longer with us. Authenticated or not, it sounds typical of the man, and for that matter, typical of his time. In those days you could get away with initiatives like that without being sued for ill-treatment and trauma. In this century the man's lawyers would have been knocking on the door that afternoon.

Less risky in legal terms there was an even more effective – and sneaky – method of dealing with the not necessarily violent, but the above average rude and difficult. Dennis Ford calls it the 'don't mess with my crew' method:

> *One of the great aspects of working for Hoverlloyd was that we enjoyed a very good relationship with the customs and immigration officers.*
>
> *Calais to Ramsgate, mid-morning, pleasant weather, all going well, until some middle management type (accompanied by his wife) decided to make an issue of our duty free prices on board. His protracted tirade resulted in the senior stewardess (very experienced and one of the nicest that Hoverlloyd employed) being reduced to the verge of tears.*
>
> *I'm not sure how the flight deck heard about this, probably via the car deck, but the captain was far from impressed with this man's behaviour.*
>
> *Somehow, customs also got to hear about the passenger in question – possibly the captain 'mentioned' it in a radio call! However, two hours later, after one more return trip, as we passed the Customs Hall on the way home, we noticed this same gentleman with the entire contents of his car spread out all over the floor.*
>
> *When he saw the smile on our faces (despite our efforts not to be too obviously delighted) it might have dawned on him that upsetting one of our best stewardesses for no good reason was not such a bright idea!*

The efficiency of the car deck was such that captains were rarely called to intervene. When they were it was usually in cases where company liability might be an issue.

Roger found himself dealing with one potential issue with a surprising outcome:

As a result of a mechanical problem I had to abort one departure from Calais when only a few miles out. We returned to Calais and, because it was obviously a large enough snag for engineers to come across to fix, we parked it high up on the western end of the pad and the decision was made to transfer the load on to the next craft in.

We had hardly started to disembark the cars when the supervisor let me know there was a complaint about car damage and could I help sort it out? When I arrived on the car deck, it turned out to be an Italian family, really upset about a large soup plate size dent in the front passenger door (right-hand side as it was left-hand drive). The family consisted of a couple of kids; papa, small and wiry; and mama, large and formidable. He didn't seem too upset, but mama was looking daggers.

I was really puzzled about it all as the weather wasn't in any way rough and we had hardly been at sea for more than ten minutes. The only possible cause of the damage I could think of was someone in the adjacent car not being too careful when opening their door.

Luckily it hadn't yet been moved so I opened the next-door car's door to see if it matched up. I remember saying to papa, "If I open this door it doesn't match up." I then opened his door to demonstrate that swinging his door out wouldn't have matched either. I pulled on his door handle and had just started to say, "and look if I open your…", when with a loud 'dong' noise the dent came out, leaving the panel totally pristine.

Papa was all smiles. "Scusi, scusi!" he says. Mama was a different matter. I got even more daggers from her. I think that she was convinced I had got away with liability by some clever English trick.

On reflection the panel of the door was that flimsy, anyone leaning against it would have caused the dent. It might well have been mama herself – it was roughly the shape of her bum.

Sickness and Other Embarrassments

As the above would suggest, passengers came in various sizes, shapes and temperaments – both individual and cultural – varying abilities to cope with seasickness and/or alcohol and in many cases a combination of both. Our patient cabin and car deck staff dealt with them all.

Seasickness was probably the major problem. In rough weather the movement of the hovercraft was erratic, bouncing from one wave to another with a jarring, unsettling motion, totally unlike the rolling motion of a ship. Because of this unusual movement it was noticeable that many hardened sailors found it just as upsetting.

Seasickness tales abound, the most common occurrence it seems was the loss of false teeth in sick bags. Incidents of departing passengers returning to the cabin and saying, "I can' fime me teef", were innumerable.

The girls had various strategies for dealing with it. If at all possible, one ploy was to pretend not to notice in the hope that some other colleague would have to deal with it. Either that or pass the problem to someone else, more often than not the ever chivalrous and obliging car deck crew.

Lynn Gibbons

Jacquie Case opted for the feminine charm, but it couldn't have been pleasant searching in the first place:

Seas sickness was a problem. Several sets of false teeth were lost over the years into the sick bags. We took it in turns to squeeze the sick bags, and when we found the teeth we had to sweetheart one of the car deck to fish them out for us.

Even then, Lynn Gibbons says, it didn't always work as planned:

Lovely Andy Holland[8] *volunteered to search through the bags and, when finding 'the one', emptied it down the forward loo, fished out the teeth, wrapped them in a cloth and gave them to a stewardess who returned them to their owner.*

When Andy asked later if she had explained that they would need a proper clean, she replied, "Well I was just about to when he popped them back in his mouth!"

Ugh!

---

[8] A car deck supervisor and a Hoverlloyd legend. He and his three sons were the veritable backbone of the car deck. One of his sons, Alan Holland, was promoted to car deck superintendent when Hoverspeed was formed.

Some days it wasn't only the passengers, it could hit the staff too, particularly each season's 'new girls', who were still to find their sea-legs.

Jacquie Case:

> *On one occasion I was virtually the last stewardess standing. The summer temps all succumbed to sea sickness, along with most of the passengers.*

If it wasn't seasickness then it was just sheer fright. It has to be accepted that many things some people find exhilarating others find terrifying. The hovercraft rocketing through heavy seas was not everyone's cup of tea.

Fear and its consequences, panic and irrational behaviour, are not so easy to deal with, but invariably our amazing girls, calm and unflappable, rose to the occasion.

Katrina Patchett

Katrina Patchett again:

> *One incident that I do remember was on another very rough day with a full load – a big man confronted me by the main cabin door. He had gotten out of his seat and towered above me, and was hysterical and panicking. He demanded that I get the captain to stop as he wanted to get off! I thought he was going to hit me. I just looked at him and calmly said, "Well, there's the door", and pointed to it. Luckily, it brought him to his senses and he sat back down (and luckily, he didn't open the door!).*

Usually some calming words from the captain would help, sometimes not. Hugh Belasyse-Smith thought he'd done the right thing by keeping everyone informed, but Katrina says it didn't quite have the desired result:

> *Hugh Belasyse-Smith told passengers before leaving Ramsgate that we had only three engines and, since it was very calm, we would still go, taking an engineer with us to work on the defective engine during the flight.*

A party of about 50 American tourists panicked and demanded to get off – it was like a riot.

I called Hugh and asked if he could say a few calming words, which he did, they settled down and we set-off.

A new intake. Joan Stroud, chief stewardess, on the left with June Copper, her assistant, on the right.

In stressful situations it does seem that different nationalities and cultures react differently. In the example just mentioned, Americans tend to act as a group and make demands.

The French act in a similar manner but shout louder.

In total contrast the Japanese sit tight and 'freeze'; in some cases faint. In this instance the girls found that good old-fashioned smelling salts were the order of the day.

The British were much like the Japanese, silently holding tightly onto their drinks or sick-bags in an almost catatonic state – perhaps mentally composing a letter of protest to *The Times.*

The more religiously-orientated cultures simply prayed. One more thing for the girls to remember; which way was Mecca? Betty Dowle recalls an incident involving a particular South American tour operator and, by total contrast, the British senior citizens:

We had the Abros from South America – I think it was the name of the coach tour that regularly took these people across – and some of them would go down on their knees and pray, they were so scared.

On the other hand:

The little old ladies on the day trips used to amuse me, because most of them were fine and even enjoyed rough weather and still wanted their drink.

In later years the company's requirement for at least one medically trained person per crew was getting more difficult to achieve as crew numbers expanded. Nevertheless, all cabin staff were expected to have first-aid training. In this instance Jacquie Case was called for:

A French man drove onto the car deck with a smashed windscreen. On getting out of the car he managed to gash his arm very badly on some loose glass inside the car. Being French, he wanted to get to Calais to repair the car.

He badly needed stitches, probably about 20, but refused to disembark, and went into the cabin leaving a trail of blood behind him. I was on the car deck, so followed him into the cabin to inform the senior stewardess. Unfortunately she was squeamish, so I patched him up, but the bandage went red, and so I spent the flight holding his arm in the air in the hope that it would stop bleeding, which it did until he put it down on arrival.

Several of us had just repeated a refresher first aid course the previous week. I was also on the crew as a French speaker. We found out later that he had gone to hospital in Calais where they had patched him up – giving him 22 stitches.

It wasn't all blood, vomit and panic; there were also the good days and plenty of lighter and amusing moments. In this next tale Di Stewart was involved in an international misunderstanding:

I was once approached by a very good-looking German gentleman who asked in a very strong accent, "Ver can we make a baby?"

I was about to explain that it was only a 30-minute crossing and felt we should get to know each other better first when his wife appeared with a rather smelly infant!

Not my lucky day after all!

And there was, as Katrina Patchette says, always the downright embarrassing:

Di Stewart.

Five minutes after setting off on a flight, one of the girls came to me in the front cabin and said, "I have to stay here for the rest of the trip – I can't go into the main cabin again."

Apparently, she had asked a passenger with a huge nose if he, "...would like any duty-free nose"! She was so embarrassed, she had turned and run.

I let her serve my passengers in the front and took care of hers. Poor guy did have a big nose, though.

By now, if the reader has any existing doubt as to calibre of our cabin staff, this final story from Madeleine Marsden puts any argument firmly to rest:

At the end of one of our three return crossings we were a few minutes away from Ramsgate Hoverport, had finished our usual duties and were about to count the bars for customs on our arrival. I sat in the stewardess' 'jump seat' preparing to take the forms for customs from my folder.

The craft was full and we had been very busy, but throughout the crossing I had noticed a chap sitting next to the window in the forward section of the middle cabin. I just thought he had been a nervous traveller, or had been a bit sea sick.

I had been sitting for a minute or two when he leaned forward and said something very quietly to me – I didn't catch it and replied, "I'm sorry I didn't hear what you said."

I then noticed he had a gun or pistol protruding from his right-hand side underneath his jacket. He then spoke again, still quietly, "I want you to take me to your captain on the flight deck now."

I immediately replied without any hesitation or qualms, "Oh no, no, no – it's far too late now – we are just coming into Ramsgate and I have to count the bars for customs. You should have asked at the beginning of the flight. Please remain seated quietly while I announce our arrival."

He looked completely flabbergasted, slipped the gun back under his jacket and sat there in disbelief. I got up, made the announcement and that was that. It was nipped in the bud so quickly he couldn't believe it and it just seemed instinctively the right thing to do. He didn't believe it and quite frankly neither did I.

I was certainly not going to have any nonsense like that at the end of six busy crossings! I mentioned it to the operations room when I went to sign off.

They informed the local police station and some days later I learnt that they caught up with a man who was using the hovercraft to make a getaway from France into the UK. He was later detained somewhere in northern England.

All in a day's work!

The above is so typically illustrative of our amazing budgies: efficient, dedicated and imperturbable, approaching every situation with *savoir faire* and aplomb.

Madeleine's remark faced by a man with a gun, "I certainly wasn't having any nonsense like that...", and the casualness of, "I mentioned it to the operations room..." are real classics.

She deserves a medal – in fact they all do!

Chapter 18

Keeping Things Moving

The Engineering Department – An Accolade

An engineering team posing for the camera during a winter overhaul.

Emrys Jones – Technical Director

Emrys was born in Llanystumdwy, North Wales in 1920, one of nine children. He was educated at the local primary school and the Portmadog Grammar School. He had hoped to go on to university, but the war intervened. In 1938 he became an engineering apprentice at the Bristol Aircraft Factory. When the war broke out in 1939 the Air Ministry sought aircraft engineers to join the Ministry of Defence. He had hoped to join the RAF, but as he was in a 'Reserved Occupation' he was unable to do so. Emrys was based at Brize Norton (17 miles west of Oxford) and was one of the youngest aircraft engineers to advise and train RAF engineers on the *Spitfire* and later on the Supermarine *Swift*.

He left the Air Ministry in 1951 to join Vickers-Armstrongs. He again trained RAF engineers on the *Swift* and also went with the RAF to Mönchengladbach in Germany and other destinations, as an engineering instructor to the RAF. In 1958 the Air Ministry began to cut back on the manufacture and development of aircraft, as a result of which many employees at Vickers-Armstrongs seemed likely to lose their jobs. Emrys had been working with others on the VA-3 hovercraft and, in 1961, he was the engineer assigned to the hovercraft when it was taken to Rhyl for trials across the Dee Estuary to The Wirral. He stayed at Vickers until 1965 when he was invited by Les Colquhoun to look after the engineering requirements for a new hovercraft venture based at Ramsgate, eventually becoming Technical Director of Hoverlloyd.

His dedication to Hoverlloyd as a viable operation was total. He would arrive at the hoverport very early in the morning. He always went home for lunch and high tea with the children and would invariably return to Pegwell Bay, often not returning home until the early hours of the morning.

He retired in 1981 when Hoverlloyd merged with Hoverspeed, and moved to Criccieth in North Wales in 1985. He died in 1986, unable to enjoy a full retirement in his beloved Wales.

If Clyde Hover Ferries could be considered a founding father of Hoverlloyd then Vickers-Armstrongs could also claim equal right to parenthood. Les Colquhoun had cut his hovering teeth in that company's experimental craft and, when he moved to Hoverlloyd, he took two very valuable assets with him. Most importantly Emrys Jones, who eventually became technical director and, of no less value, Maurice 'Monty' Banks, who joined him as his second-in-command in the post of chief engineer.

Together, these two stalwart gentlemen worked through the initial N6 years and then on to the N4 days at Pegwell to forge a first-class engineering department, one dedicated to ensure that each day and night these hard-to-maintain vehicles departed as near to on time as possible and fully operational. On the many instances where more serious problems meant greater delays, one could guarantee our highly-motivated technical men would be hard at it, not wasting a second. We were lucky to have them.

Recruitment

The lessons learnt from Clyde Hover Ferries, that such a complex machine would want more than 'a man and a boy' to keep it running, was not lost on Emrys. What was also needed was a diversity of skills, and in 1966 a team of four was immediately recruited for the SR.N6 start in Ramsgate. Four engineers representing the key skills joined the company: John Bartlett (Structures), Ted Unsted (Electrical), Jim Weston (Electrical and Radio/Radar) and Dave Parr (Mechanical).

In April 1967 three more engineers joined – Mike Fuller (Structures), Ron Wheeler (Mechanical) and Walter 'Wally' Truslove (Mechanical).

Right from the start, according to Mike Fuller, Hoverlloyd's working ethos was immediately apparent:

> *Although the engineers had designated trades, there was no demarcation and everyone waded in and helped each other out. This attitude ran through engineering during the whole of Hoverlloyd's existence.*

Engineer Dave Parr (left) and Assistant Chief Engineer Monty Banks (centre), in the maintenance area at Ramsgate Harbour working on an SR.N6. Note the frame and chain lift at the stern for lifting the craft. 1966.

Regardless of their background and skills, all found themselves learning the hover business from scratch. Mike Fuller:

> *A lot was learned about operating aircraft-engineered parts in a salt-water environment as well as erosion caused by sand being blasted against light alloy structures.*
>
> *Repairing the skirt was something completely new, which entailed a lot of trial and error. The craft was lifted frequently with two large frames and chain lifts for inspection underneath. This took about an hour or so to set up during the day but temporary repairs would be made with the craft sitting on the pad.*

In 1969, with the inauguration of the N4 service at Pegwell, the engineering department went into overdrive. From maintaining two very small craft on a minimal schedule with very little winter activity to an intensive all-year schedule with the two larger craft meant a huge increase in technical staff to cope.

Most of the new employees came from a service background, mainly the RAF and the Fleet Air Arm, but some came from the commercial airlines, including some based at the nearby Manston airfield.

During this expansion process, three new senior engineer positions were created: Wally Truslove was appointed to supervise the engine, propeller and gearbox maintenance (propulsion). Jim Newbury joined the company from BHC as the structures and hydraulic systems engineer and Bob Taylor, another Vickers man, came on board to cover the electrical systems.

As this was by necessity a 24-hour undertaking, the workforce was organised into three shifts, each made up of engineers from all three trades. Three Shift Crew Chiefs were appointed: Micky Bacon, Mike Fuller and Roy Fuller (not related). Ted Unsted and Ron Wheeler went to the operations department to head up the new car deck crew department.

A stores department was formed, headed by Roger Key from BHC with Les Knight to handle the daily stores requests. So with a couple of ancillary staff the engineering section quickly expanded from a mere half a dozen in Ramsgate Harbour to upwards of 50 at Pegwell Bay.

The Dark Art

Not unlike the flight crew experience, the engineers, although thoroughly familiar with aircraft structures, control systems and engines, were now faced with a very unfamiliar environment and forced to rethink a lot of their assumptions.

One thing that hadn't been thought of was that the skirt system was a technology all of its own and required its own set of experts to manage it. In the hovercraft world there was one more trade to consider and a new word entered the vocabulary – the 'skirtie'.

This new profession, working in appalling conditions, wasn't the most enviable of occupations. Mostly at night at the end of the day's schedule, often crouched inside a soaking wet cavern of rubber, heaving around very heavy inflexible material was not the ideal work environment. Bad enough in summer but in winter, with no means of heating the area, it would have been miserable.

If that wasn't enough, very often at the same time, somewhere above, engines and superstructure were being washed down to get rid of the accumulated salt and sand, all of which had the potential to dribble down to where the 'skirties' were working. It seemed almost akin to the arduous back-breaking job of a stoker on a coal-burning ship in the early part of the twentieth century, except that the 'skirties' suffered the cold and damp rather than the searing heat of the furnaces the stokers were firing.

A 'skirtie' making a temporary skirt repair on a beach. Captain Ted Ruckert is looking on behind the 'skirtie'.

As would be expected, the turnover of staff in this area was quite brisk initially, but gradually a hard core, experienced in the 'dark art', accumulated to become as valuable an asset as any other calling. Those who were originally hired as non-skilled labourers were in due time recognised as skilled technicians in their own right and quite rightly paid accordingly.

The most outstanding of this new breed was Don Nicholls who, through his avid interest and dedication to the new technology, rose to become the skirt maintenance supervisor.

He is remembered as a man who kept himself to himself, very helpful and co-operative on the job but not inclined to continue the conversation. Mike Fuller says:

> *All I knew about Don was that he was ex BHC from Cowes, but he always seemed to keep his distance from people. The skirt*

shop was his life and his main topic of conversation. All his work practices and mods (modifications) came straight from his heart, no drawings, just a lot of aluminium templates he used to make up. He certainly sorted out the rear bags and segment ten, the last bit under the intake caboose[1].

Don was in fact responsible for many other ground-breaking modifications, many of which the manufacturer BHC picked up and used for future projects like the skirt systems for the US air cushion landing craft (LCACs).

The rear corner skirt segment that Don Nicholls and his team did so much to improve.

It is not possible to describe the N4's unique skirt system without mentioning the ubiquitous 'Pegwell Penny'; another new phrase to add to the hovercraft vocabulary. Pegwell pennies were light alloy washers which, in conjunction with 'huckbolts', made up the fasteners joining each skirt segment and fingers. Whether Mike's estimation of 'thousands' is correct or not, it certainly seemed as if there were:

Bloody thousands – each lap joint (where a skirt section joined another one) was huckbolted with a double row all round from the outer hinge line to the inner hinge line. Each huckbolt had a 'penny' on either side of the skirt material. Each finger was held on by probably 30 huckbolts so you have 60 'pennies' there to start with.

[1] Segment ten, the rear corner skirt section along the side and situated under the 'caboose', the name given to the box-like structure housing the engine intake filters.

Discarded 'pennies' scattered all over the maintenance area were a common sight, and it is highly likely that archaeologists in a hundred years time will be digging them up by the dozen.

The Jacking System

Emrys Jones, Technical Director, standing beside one of the jacks, which had just been installed. 1968.

For any maintenance work to be done under the craft there needed to be a system of jacks with sufficient lift to at least raise the skirt configuration clear of the ground. For the SR.N4 this needed to be up to a maximum of 7 feet[2].

The system installed comprised five individual jacks which matched up with the stubby landing pads under the buoyancy raft, known as 'elephant's feet'.

The routine was to put each craft up in turn on the jacks every night to check for skirt and/or structure damage.

The lifting operation was controlled from a panel in the Technical Records Office, which faced the jacking area. Each day, after the craft

[2] Initially a maximum of 7 feet but later skirt modifications lifted the maximum to 9 feet.

had finished operating, the crew would hover the machine around from the arrival pad and position it in the jacking area, so the five elephant's feet under the craft landed as close as possible to the jack heads.

Emrys had devised a set of cables running around the craft through large pulleys, connected to an electric winch in the engineering workshop, which positioned the elephant's feet over the individual jacks.

A hovercraft partially parked near the jacking area. Buildings left to right: Engineering Department; arrival vehicle Customs Hall, in front of which is the duty free bonded store (on one occasion this was partially demolished by an out-of control hovercraft trying to get onto the jacks); main terminal building and offices; vehicle check-in kiosk. 27th June 1969.

At the trials of the first system, where the craft was held in position by a set of bridles front and back, it was discovered that in any sort of cross-wind as the craft was lowered, it would drift off the jack heads.

To solve the problem, Emrys created the modification as illustrated on the following page.

Once the craft was driven to a 'near enough' position, the engineers shackled the front and rear fixed bridles, together with the side-locating cables, which, with the main winch, helped move the machine to the centreline. The locating cable was floating, and fine adjustment, measured against the position marks on the ground, could be made with the Tirfor winches, shown in the diagram.

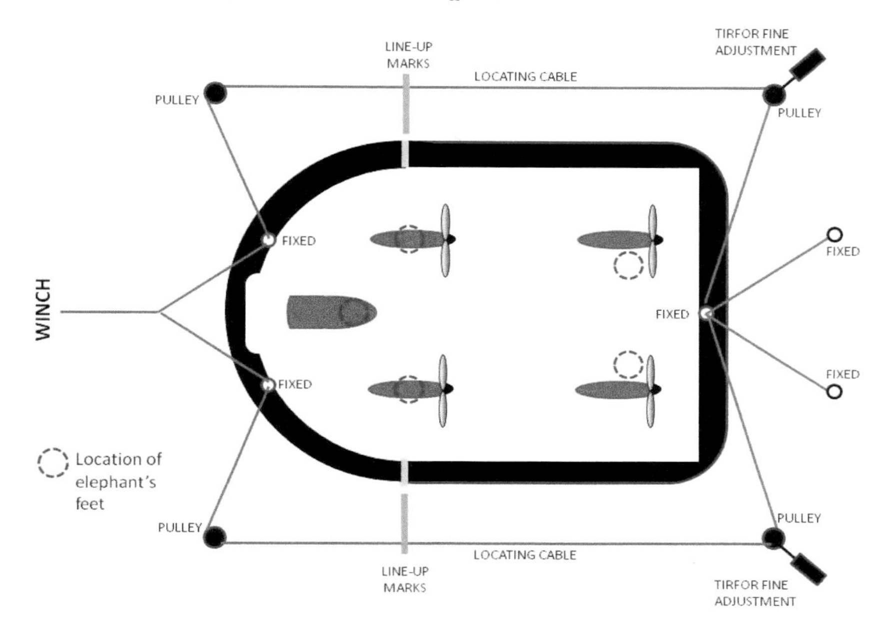

Even with this excellent set-up the captains nevertheless needed to be fairly accurate in their placing of the craft. The system could only handle a metre or so out of initial alignment and, as Mike Fuller says, some captains were better than others:

> *It was the crew chief's duty to advise the flight crew, via a hand-held radio, when to stop, go left or right, and sometimes "back a bit!" It soon became apparent that some skippers mastered the technique of slow-speed manoeuvring in a cross wind better than others, and many times in inclement weather the craft would end up sideways.*

Routine

The working regime differed widely between the summer and winter periods. During the short but intensive summer season, with the primary objective being to ensure that the schedule ran on time with as little interruption as possible, the engineers worked a shift system. Although things quietened down during the winter, this was paradoxically the time of the heaviest workload for the technical staff.

The craft were taken off the route one by one, parked to one side and subjected to thorough overhauls. During the extended overhaul period the majority of the staff engineers reverted to the Monday to Friday pattern and the overhaul season began.

Sir Christopher on jacks during the winter overhaul with the skirt completely removed.

Modelled on the maintenance system he had been used to in BOAC, crew chief Mike Fuller explains the rotation system covering 24 hours a day:

> *During the summer season the three shifts rotated. After seven nights (Friday to Thursday – 11.00pm to 7.00am) you had four days off then started early shift (Tuesday through to Monday – 7.00am to 3.00pm) then three days off and back to nights again. After that night shift the next shift was seven 'lates' (3.00pm to 11.00pm).*

Crew chiefs always came in early to accept the hand-over briefing from the previous team.

Every day, after operations finished, the flight crew handed the list of accrued daily 'snags' to the duty crew chief so that they could be fixed overnight. A post-flight inspection was then carried out to check for damage and all the oils were checked and topped up. In addition, the craft were subjected to a rolling maintenance check schedule, which meant that all critical components and systems were regularly checked or changed according to hours flown.

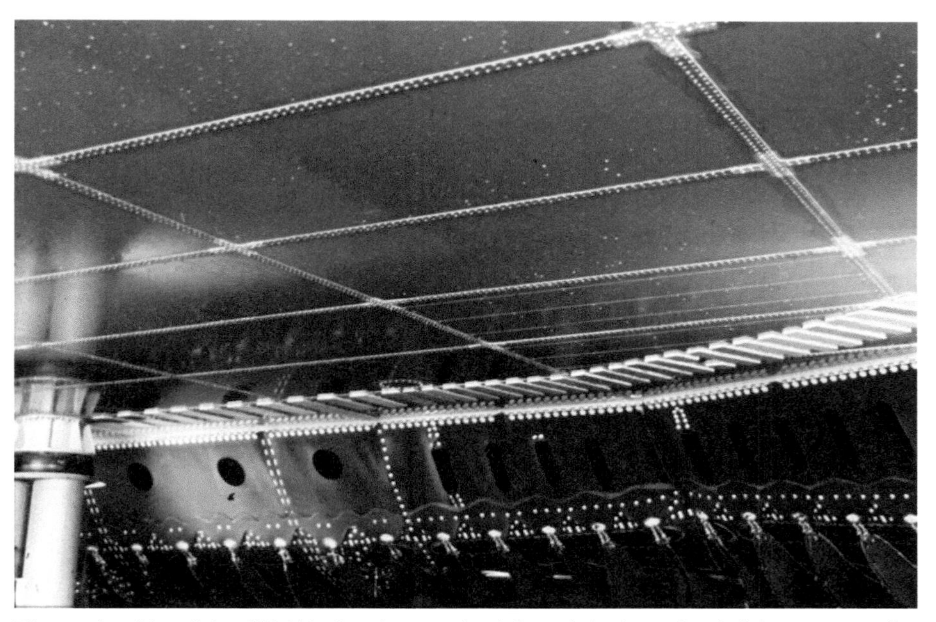

The underside of the SR.N4 showing on the left an 'elephant foot' sitting on top of a jack. Below the air holes of the inner skirt, one can see the top finger attachments to the skirt.

As soon as the summer season ended each craft underwent a thorough overhaul lasting some six weeks.

With a very minimum timetable running in the non-peak months, the daily routine necessitated only two 'duty crews' to keep the operational craft going, working four days on and four off. This was two shifts of about six engineers plus about eight 'skirties'.

Although their rostered time was nominally 11.00am to 11.00pm, if things were going well, they could usually expect to finish by early evening.

In the meantime the overhaul crew worked a standard five-day week, 8.00am to 5.00pm, plus weekend overtime and handled the departure of the first service of the day.

Unlike airline maintenance, the craft was overhauled outside in the open, all through the winter. Large components like engines, propellers, pylons, fins and APUs were removed and taken into the workshops for modification and overhaul. Some components, such as life rafts and gearboxes, were sent to the manufacturer for overhaul.

Mike Fuller, Crew Chief. 1976.

An important part of the craft maintenance was the keeping of meticulous records of every technical event that took place. This closely followed airline practice and was handled by two technical records clerks, John Dickenson and Don Baigent working normal office hours. As it was well before the computer era everything had to be recorded by hand and it was a big task to make sure that all the information was precise down to the last detail.

In addition to recording the craft and engine hours based on the daily flight crew log, the serial numbers of every component put on and taken off the craft were noted. They issued check sheets and worked out schedules so that items such as engines and gearboxes requiring overhaul, due to hours flown, were staggered to avoid the impossible situation of everything having to be dealt with at the same time. They also ensured that the aircraft check system 1 to 4 (Check 1 being a basic check leading to Check 4, the winter overhaul) was adhered to, and recorded the one item exclusive to hovercraft, skirt and finger changes.

The daily snag sheet from each craft ended up in Technical Records and it was to this office that the captains of the first flights out went each day to check the serviceability and sign for their respective machines.

Mike Fuller's last day at Hoverlloyd (picture taken by Mike) with his crew enjoying a celebration banquet. Left to right: Pete Allen, unknown, unknown, Mick Prior, David Ward (another of the same name as the captain), Peter Mathews, Bob Bannon, Keith Mason. 1981.

Moving On

By the time the third SR.N4, *Sir Christopher*, was delivered in June 1972, the engineering section had risen to nearly 100, including the plant maintenance section that looked after the day-to-day running of the hoverport infrastructure.

In addition, the new-found confidence in the company's future saw some management changes and an upgrade of much of the engineering facilities.

In this picture the engines, pylons and fins have been removed for overhaul. The base on which the starboard fin rests is to the right of the open starboard engine hatch. Just in front is the fan air intake and the starboard aft pylon support.

Wally Truslove was appointed to Assistant Chief Engineer and his position of Senior Engineer (Propulsion) was taken by Rupert (Nobby) Clark. New workshops and offices were built to facilitate in-house maintenance of components, including engines, propellers and hydraulic packs and jacks. A large skirt workshop was built where skirt sections could be repaired and manufactured under the supervision of Don Nicholls.

A second set of jacks was installed alongside the first set, as by now a fourth craft was to be added to the fleet.

At this juncture, as Mike Fuller says, the system had settled down to a smooth and well-run machine:

> *This schedule continued to operate successfully; the intricacies of the SR.N4 had been pretty well mastered and generally things ran smoothly. Events like propellers coming off, the 'Lloyds Bank' incident at Calais and the loss of the skirt at Sangatte would come along and stretch the resources of the department, but in true Hoverlloyd spirit everyone rose to the occasion.*

The incidents Mike recalls all occurred during the first difficult two years and are included in Chapter 21: *Hoverlloyd's SR.N4 – The First Years*, with the exception of the 'Lloyd's Bank Incident', which, as it had the distinction of being the first major occurrence that required an engineering team to travel to Calais to effect temporary repairs, is worth recounting here. One assumes that Mike remembers this, not only because it was the first incident of note, but that during the process he escaped serious injury:

> *In March 1970 Captain John Lloyd (the tall one!) was approaching the Calais Hoverport in* Swift *with a strong south-westerly blowing. The craft suddenly drifted to port and hit, side-on, the large bank of sand that was between the two ramps leading onto the port.*
>
> *The impact had crushed the side of the craft but John managed to land on the pad successfully and no one was hurt. The next day I went over on* Sure *with a group of engineers, together with all our tools and boxes of spares to affect a temporary repair.* Swift *certainly looked a bit worse for wear – an 18-foot section of the plenum chamber was crushed inwards. We ripped off all the damaged sheet metal to find that the damage was really quite superficial; the main skirt support structure was un-damaged, as was the internal buoyancy chamber and the main fuel lines that ran alongside the chamber. It became clear that all we had to do was to fill the gap with something, after which* Sure *could return to Pegwell for a permanent repair.*
>
> *Fortunately, the Air Registration Board was very lenient with us regarding matters like this so we decided to patch the damaged area with several large sheets of plywood. A local Calais hardware merchant supplied us with the necessary sheets and we set about cutting them to shape and bolting them into place.*
>
> *I always will remember this operation as I escaped serious injury by a couple of inches. I was inside the plenum chamber holding up one of the plywood sheets waiting for another engineer to drill a quarter of an inch hole through the sheet in a predetermined area. I had my head against the sheet when the drill came crashing through in the wrong place; it grazed past my temple and hit my ear! To this day I shudder at the thought of the drill being two inches over to one side and drilling a neat hole straight into the top of my head.*
>
> *Once all the plywood sheets were bolted into position, we loaded everything onto the car deck and successfully returned to Pegwell with John Lloyd at the controls.*

The large bank of sand was forever known as 'Lloyd's Bank' after the incident, but in 1973 it was removed completely and the area concreted over.

It has to be said that the name Lloyd's Bank was very much an engineer's private joke; it is not certain that even John was aware of it, certainly very few of us in the flight crew were.

Ad Hoc Maintenance

There were one or two perpetual problems, which were of too minor a nature to warrant the time and expense of bringing teams across to Calais to fix.

One continuous 'snag' that remained with us for all of the N.4's life was the 'quill drive'. This was a deliberate weak point in the main engine start system that was meant to fail should there be a problem that might have resulted in burning out the starter motor. Although, because of the trouble-free nature of the engine itself, the times the quill drive did fail were most probably due to old age rather than any major problem.

This was such a fairly common occurrence, but a relatively simple fault to rectify, that quite early on we started to carry spares. After a bit of instruction from the engineers the flight crew could replace the quills themselves rather than return from Calais on three engines, particularly if the weather was marginal. Even so, it took about three-quarters of an hour to do the job, which was usually undertaken by the flight engineer.

Again it is worth noting that we were lucky that, in contrast to the general industrial relations climate in the UK at that time, we had no such thing as demarcation and in no circumstance did anyone object to their particular trade being encroached upon. It is likely, in this case, the engineers, probably having enough on their plate as it was, were only too pleased to let a member of the flight crew get their hands dirty for a change and do the job themselves.

For another *ad hoc* repair, the crew were provided with a suture needle. This was no surgeon's needle for delicate surgery; it was an eight-inch long veterinary needle with a strong wooden handle, ostensibly for sewing up superficial cuts on horses. Threaded with some strong nylon string, this fearsome looking instrument was just the thing for quick repairs on skirt rips, certainly good enough to make the return to Pegwell, where the experts waited to affect a proper repair.

The Peak

The arrival of *The Prince of Wales* in 1976 did not necessitate taking on more engineers, only 'skirties', and the summer season was covered by a three-shift system with about 110 engineers and 'skirties' employed in the direct hands-on operations. This seemingly large number was necessary because of the requirement to jack two craft each night, as well as attending to the daytime operational engineering requirements. In addition there were seven 'shinies'[3], the management men in suits, with overall responsibility for their particular areas of expertise.

The expanded Engineering Department with its extra workshops and second jacking system built on reclaimed land. The proximity of the two jacking systems could be challenging for captains trying to manoeuvre craft on or off in strong winds. 1976. *(Courtesy of Fotoflite.)*

On any one day, two shifts were rostered on and one shift was rostered off. Each of the shifts consisted of eight structures engineers (riggers), eight propulsion engineers (fitters) and four electricians, plus two crew chiefs. One complete team covered the night shift, and one half of the other team covered the early day shift while the other half covered the late shift. The 'skirties' worked in two shifts of four nights on and four off, with about 12 on each shift. The 'skirties' in the skirt shop covered

[3] According to Mike Fuller, 'shinies' is a generic term, which was used both in the RAF and at Hoverlloyd by the blue collars, to describe the white collars in the office. It stems from the seat of office workers' trousers becoming 'shiny' after years of sitting on wooden office chairs.

any daytime rips and, if a daytime jacking job became necessary, 'skirties' on their day off would be called in on overtime.

There were some 12 engineers, including the 'skirties', in the workshops, plus seven 'shinies'.

In the final analysis it was the ARB and, from 1972, the Civil Aviation Authority who had to be satisfied that the craft were being maintained properly, which they did by carrying out regular inspections at the facilities. Despite the unusual conditions, generally cold and wet, under which the staff worked, no time was lost due to failing to satisfy the authorities as to the standard of maintenance required, or for that matter to industrial action; a tribute to the management and personnel alike.

Throughout its brief history, Hoverlloyd's technical department probably had the most difficult job of all without any of the glamour, and were possibly rather taken for granted. But without doubt they played the leading role in making sure that Hoverlloyd's SR.N4 cross-Channel services, when the operation settled down, consistently achieved a reliability record of 98 per cent.

Chapter 19

Making a Profit

Economics

Pegwell in its heyday. Post-1977 when four SR.N4s were in service. Note Manston Airport (now Kent International) in the background.

Without doubt, Christopher Cockerell's invention was an amazing breakthrough in the field of transport. It certainly worked and did what its inventor said it would do. However, the only real measure of success for any mode of transport is simply, can it make money for its operators? Further than that, in a highly competitive market such as the cross-Channel ferry business, can it compete successfully with its rivals?

Speed at sea is costly and it has already been noted that any fast sea-going vessel needs a high volume market willing to pay the higher price it entails. The four Proteus 9 marine engines in the N4, burning some 1,000 gallons of ATK[1] per hour, were expensive to run compared with the slow speed diesel of the average ferry. Another consumable cost unique to the hovercraft, was the continual expense of skirt maintenance, very much akin to changing the tyres of a car on a daily basis.

---

[1] Aviation Turbine Kerosene, also known as Avtur. (Consumption refers to Mk.1 and Mk.2 hovercraft.)

To justify the high operational cost, speed was used to make as many crossings per day as possible, together with as high a load factor as possible. This the N4 achieved in fine style. While a ferry would make four return crossings in 24 hours, the hovercraft, at three times the speed, could achieve eight return crossings in a 16-hour day from the first flight out at 6.00am to the last flight at 8.00pm. Although the N4 could operate in the dark, subject to reduced operating limits, it would not have been practical to operate round the clock during the peak period, not only because of noise disturbance to the local community, but also because of crewing and maintenance considerations too.

Sales and Traffic

These vehicles have checked in and loading is underway. Passengers can access the building by a side entrance to visit the cafeteria and duty free shop. Pegwell Bay Hoverport 1969.

Attending to the need for high volume and high loads was the task of our Marketing and Sales Departments, based in the Broström office in Charles Street, London, but the real job of making sure it was all fine-tuned for maximum profit was firmly in the hands of the traffic personnel in the Pegwell and Calais hoverports.

Hoverlloyd was fortunate in having recruited first-class sales and traffic staff, the majority of whom came from the airline industry. The airline connection was purely fortuitous in that Hoverlloyd was formed shortly after Eagle Airways'[2] demise in 1968, leaving a number of very experienced traffic personnel looking for work. This well-trained and highly professional staff had a significant and lasting influence on the company, making Hoverlloyd not so much another ferry company, but clearly an airline operation at sea. Mike Castleton, one of the original traffic officers, describes the Eagle Airways influence:

> *Simply because Howard, Alan[3] and I knew no other way, we treated the SR.N4 as a large aeroplane and moulded the entire Traffic Department on the lines of an airline operation. Therefore, during that first season the influence of British Eagle could be seen in many areas.*
>
> *The maroon uniform worn by both receptionists and cabin staff was ex-Eagle, purchased as a job-lot from the receivers of the airline by Howard Archdeacon. Similarly, the* Traffic Manual, *which laid down operating procedures for the Traffic Department, was compiled by Alan and me, using the Eagle manual as a basis. Even boarding passes and tickets were direct copies of the equivalent airline documents.*
>
> *Incidentally, the influence of Eagle also extended to the original Sales Department of Hoverlloyd in London. The Sales Manager was Bob Harvey, formally Route Inspector for Eagle, ably assisted by Alan Ashwin, a former Eagle duty officer.*

The conscious decision of Traffic and Sales to treat Hoverlloyd as just another airline made the company so very different to its other high-speed rivals, Seaspeed, and later the Belgian RTM jet-foil service. Both these companies operated their vessels as ferries and therefore with a pronounced maritime flavour. RTM in fact recruited its operational personnel direct from their existing conventional ferry services. With the pervasive influence of the Eagle staff, Hoverlloyd operated as an airline and it wasn't just the methodology; it extended to the terminology used and even to the smart airline uniforms of the flight crew, which were sourced from the British Airways depot at Gatwick.

---

[2] Eagle Airways was formed in 1948 by Harold Bamberg and made its initial profits from the Berlin Airlift. Over the years it tried operating several scheduled passenger services, but, like the other second tier airlines, it was always up against the monopoly of the state-owned airlines, BOAC and BEA, who objected to licences being handed out on the profitable routes. The final nail in the coffin was when Cunard, who had a 60% stake in Eagle Airways to form Cunard Eagle, transferred its allegiance to BOAC.

[3] Howard Archdeacon – Associate Director Traffic; Alan Dewey – Superintendent, Pegwell Bay.

Seaspeed uniforms remained marine in nature, Merchant Navy-style uniforms and braid, together with the distinctive white-topped marine officer's caps. Individually these may seem somewhat trivial differences but the overall effect was a noticeable sense of difference in Hoverlloyd, which, together with a high degree of professionalism, engendered from the outset a pride in the company and its uniform.

The Schedule

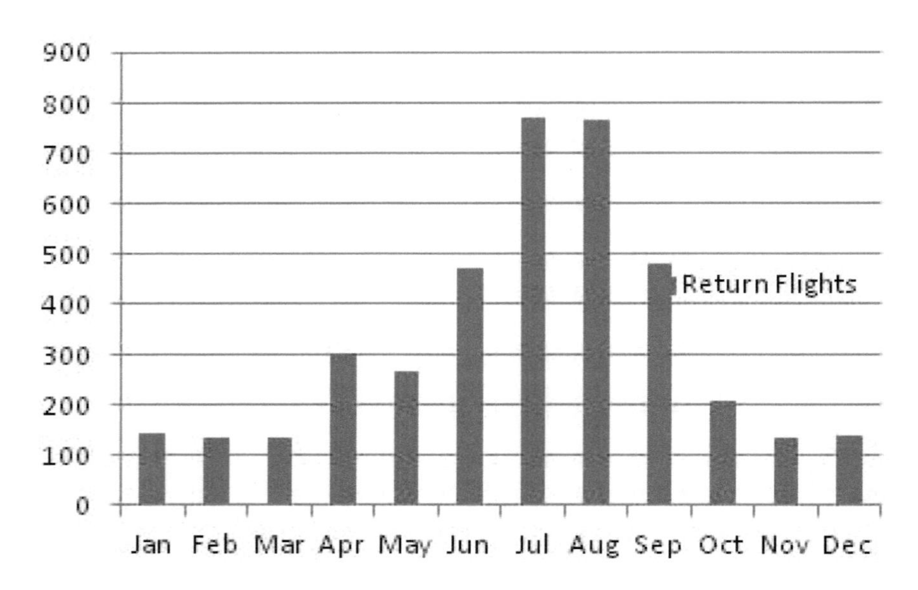

Flights scheduled per month for the first year of four-craft operation.

One of the major problems in the economics of operating hovercraft was the seasonal nature of the cross-Channel market. There was an enormous summer peak coinciding with the summer school holidays from July to early September. The graph above shows the schedule for the first year of the four-craft operation in 1977; the summer high point, including the smaller increase over Easter, accounted for roughly 80 per cent of the annual traffic.

This put Hoverlloyd and the other high-speed interests at a distinct disadvantage compared with the conventional ship ferries. As well as benefiting from their share of the summer holiday frenzy, the ships had a consistent year-round business carrying roll-on roll-off freight, a business that actually expanded enormously over the decade

Hoverlloyd was in operation, eventually reaching its peak with the advent of the 31.4 mile Channel Tunnel in 1994.

In contrast, the high-speed operations, not being of a size to cope with heavy goods vehicles, were wedded exclusively to the holiday market and its seasonal nature.

A car marshaller directing the cars forward for embarkation. 1972.

It must be conceded that the holiday market also experienced a large expansion during this period, hence Hoverlloyd's growth from a two-craft operation in 1969 to four craft in 1977. However, regardless of this welcome increase in business, the market maintained its annual high summer demand profile, an immutable fact of life that the high-speed operators were powerless to change. To the ferry operators the expansion in the holiday market was simply icing on the freight-moving cake, while for the fast craft, it was all they had.

For Hoverlloyd, the effect of this competitive disadvantage put an enormous strain on the need to ensure that every flight went as scheduled and as full as possible. One flight lost because of poor weather or maintenance problems during the summer period was a serious blow against the bottom line, as was any flight leaving under capacity.

The whole company knew that efficiency was everything. We had to leave on time, every time, with as near 100% load factors as possible whenever traffic was waiting to be carried.

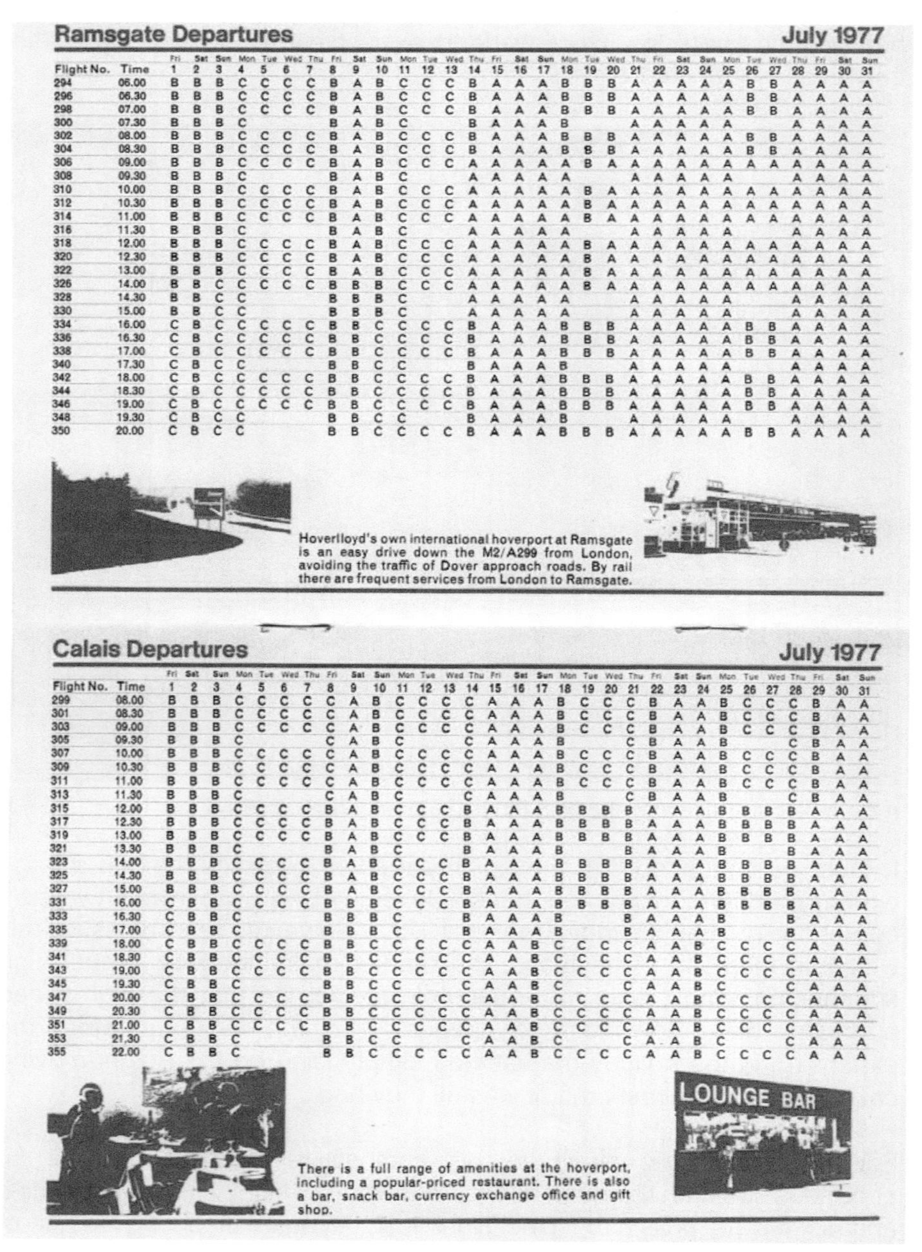

Ramsgate Departures — July 1977

Flight No.	Time	Fri 1	Sat 2	Sun 3	Mon 4	Tue 5	Wed 6	Thu 7	Fri 8	Sat 9	Sun 10	Mon 11	Tue 12	Wed 13	Thu 14	Fri 15	Sat 16	Sun 17	Mon 18	Tue 19	Wed 20	Thu 21	Fri 22	Sat 23	Sun 24	Mon 25	Tue 26	Wed 27	Thu 28	Fri 29	Sat 30	Sun 31
294	06.00	B	B	B	C	C	C	C	B	A	B	C	C	C	B	A	A	A	B	B	B	A	A	A	A	A	B	B	A	A	A	A
296	06.30	B	B	B	C	C	C	C	B	A	B	C	C	C	B	A	A	A	B	B	B	A	A	A	A	A	B	B	A	A	A	A
298	07.00	B	B	B	C	C	C	C	B	A	B	C	C	C	B	A	A	A	B	B	B	A	A	A	A	A	B	B	A	A	A	A
300	07.30	B	B	B	C				B	A	B	C			B	A	A	A	B			A	A	A	A	A			A	A	A	A
302	08.00	B	B	B	C	C	C	C	B	A	B	C	C	C	B	A	A	A	B	B	B	A	A	A	A	A	B	B	A	A	A	A
304	08.30	B	B	B	C	C	C	C	B	A	B	C	C	C	B	A	A	A	B	B	B	A	A	A	A	A	B	B	A	A	A	A
306	09.00	B	B	B	C	C	C	C	B	A	B	C	C	C	A	A	A	A	A	B	A	A	A	A	A	A	A	A	A	A	A	A
308	09.30	B	B	B	C				B	A	B	C			A	A	A	A	A			A	A	A	A	A			A	A	A	A
310	10.00	B	B	B	C	C	C	C	B	A	B	C	C	C	A	A	A	A	A	B	A	A	A	A	A	A	A	A	A	A	A	A
312	10.30	B	B	B	C	C	C	C	B	A	B	C	C	C	A	A	A	A	A	B	A	A	A	A	A	A	A	A	A	A	A	A
314	11.00	B	B	B	C	C	C	C	B	A	B	C	C	C	A	A	A	A	A	B	A	A	A	A	A	A	A	A	A	A	A	A
316	11.30	B	B	B	C				B	A	B	C			A	A	A	A	A			A	A	A	A	A			A	A	A	A
318	12.00	B	B	B	C	C	C	C	B	A	B	C	C	C	A	A	A	A	A	B	A	A	A	A	A	A	A	A	A	A	A	A
320	12.30	B	B	B	C	C	C	C	B	A	B	C	C	C	A	A	A	A	A	B	A	A	A	A	A	A	A	A	A	A	A	A
322	13.00	B	B	B	C	C	C	C	B	A	B	C	C	C	A	A	A	A	A	B	A	A	A	A	A	A	A	A	A	A	A	A
326	14.00	B	B	C	C	C	C	C	B	B	B	C	C	C	A	A	A	A	A	B	A	A	A	A	A	A	A	A	A	A	A	A
328	14.30	B	B	C	C				B	B	B	C			A	A	A	A	A			A	A	A	A	A			A	A	A	A
330	15.00	B	B	C	C				B	B	B	C			A	A	A	A	A			A	A	A	A	A			A	A	A	A
334	16.00	C	B	C	C	C	C	C	B	B	C	C	C	C	B	A	A	A	B	B	B	A	A	A	A	A	B	B	A	A	A	A
336	16.30	C	B	C	C	C	C	C	B	B	C	C	C	C	B	A	A	A	B	B	B	A	A	A	A	A	B	B	A	A	A	A
338	17.00	C	B	C	C	C	C	C	B	B	C	C	C	C	B	A	A	A	B	B	B	A	A	A	A	A	B	B	A	A	A	A
340	17.30	C	B	C	C				B	B	C	C			B	A	A	A	B			A	A	A	A	A			A	A	A	A
342	18.00	C	B	C	C	C	C	C	B	B	C	C	C	C	B	A	A	A	B	B	B	A	A	A	A	A	B	B	A	A	A	A
344	18.30	C	B	C	C	C	C	C	B	B	C	C	C	C	B	A	A	A	B	B	B	A	A	A	A	A	B	B	A	A	A	A
346	19.00	C	B	C	C	C	C	C	B	B	C	C	C	C	B	A	A	A	B	B	B	A	A	A	A	A	B	B	A	A	A	A
348	19.30	C	B	C	C				B	B	C	C			B	A	A	A	B			A	A	A	A	A			A	A	A	A
350	20.00	C	B	C	C				B	B	C	C	C	C	B	A	A	A	B	B	B	A	A	A	A	A	B	B	A	A	A	A

Hoverlloyd's own international hoverport at Ramsgate is an easy drive down the M2/A299 from London, avoiding the traffic of Dover approach roads. By rail there are frequent services from London to Ramsgate.

Calais Departures — July 1977

Flight No.	Time	Fri 1	Sat 2	Sun 3	Mon 4	Tue 5	Wed 6	Thu 7	Fri 8	Sat 9	Sun 10	Mon 11	Tue 12	Wed 13	Thu 14	Fri 15	Sat 16	Sun 17	Mon 18	Tue 19	Wed 20	Thu 21	Fri 22	Sat 23	Sun 24	Mon 25	Tue 26	Wed 27	Thu 28	Fri 29	Sat 30	Sun 31
299	08.00	B	B	B	C	C	C	C	C	A	B	C	C	C	C	A	A	A	B	C	C	C	B	A	A	B	C	C	C	B	A	A
301	08.30	B	B	B	C	C	C	C	C	A	B	C	C	C	C	A	A	A	B	C	C	C	B	A	A	B	C	C	C	B	A	A
303	09.00	B	B	B	C	C	C	C	C	A	B	C	C	C	C	A	A	A	B	C	C	C	B	A	A	B	C	C	C	B	A	A
305	09.30	B	B	B	C				C	A	B	C			C	A	A	A	B			C	B	A	A	B			C	B	A	A
307	10.00	B	B	B	C	C	C	C	C	A	B	C	C	C	C	A	A	A	B	C	C	C	B	A	A	B	C	C	C	B	A	A
309	10.30	B	B	B	C	C	C	C	C	A	B	C	C	C	C	A	A	A	B	C	C	C	B	A	A	B	C	C	C	B	A	A
311	11.00	B	B	B	C	C	C	C	C	A	B	C	C	C	C	A	A	A	B	C	C	C	B	A	A	B	C	C	C	B	A	A
313	11.30	B	B	B	C				C	A	B	C			C	A	A	A	B			C	B	A	A	B			C	B	A	A
315	12.00	B	B	B	C	C	C	C	B	A	B	C	C	C	B	A	A	A	B	B	B	B	A	A	A	B	B	B	B	A	A	A
317	12.30	B	B	B	C	C	C	C	B	A	B	C	C	C	B	A	A	A	B	B	B	B	A	A	A	B	B	B	B	A	A	A
319	13.00	B	B	B	C	C	C	C	B	A	B	C	C	C	B	A	A	A	B	B	B	B	A	A	A	B	B	B	B	A	A	A
321	13.30	B	B	B	C				B	A	B	C			B	A	A	A	B			B	A	A	A	B			B	A	A	A
323	14.00	B	B	B	C	C	C	C	B	A	B	C	C	C	B	A	A	A	B	B	B	B	A	A	A	B	B	B	B	A	A	A
325	14.30	B	B	B	C	C	C	C	B	A	B	C	C	C	B	A	A	A	B	B	B	B	A	A	A	B	B	B	B	A	A	A
327	15.00	B	B	B	C	C	C	C	B	A	B	C	C	C	B	A	A	A	B	B	B	B	A	A	A	B	B	B	B	A	A	A
331	16.00	C	B	B	C	C	C	C	B	B	B	C	C	C	B	A	A	A	B	B	B	B	A	A	A	B	B	B	B	A	A	A
333	16.30	C	B	B	C				B	B	B	C			B	A	A	A	B			B	A	A	A	B			B	A	A	A
335	17.00	C	B	B	C				B	B	B	C			B	A	A	A	B			B	A	A	A	B			B	A	A	A
339	18.00	C	B	B	C	C	C	C	B	B	C	C	C	C	C	A	A	B	C	C	C	C	A	A	B	C	C	C	C	A	A	A
341	18.30	C	B	B	C	C	C	C	B	B	C	C	C	C	C	A	A	B	C	C	C	C	A	A	B	C	C	C	C	A	A	A
343	19.00	C	B	B	C	C	C	C	B	B	C	C	C	C	C	A	A	B	C	C	C	C	A	A	B	C	C	C	C	A	A	A
345	19.30	C	B	B	C				B	B	C	C			C	A	A	B	C			C	A	A	B	C			C	A	A	A
347	20.00	C	B	B	C	C	C	C	B	B	C	C	C	C	C	A	A	B	C	C	C	C	A	A	B	C	C	C	C	A	A	A
349	20.30	C	B	B	C	C	C	C	B	B	C	C	C	C	C	A	A	B	C	C	C	C	A	A	B	C	C	C	C	A	A	A
351	21.00	C	B	B	C	C	C	C	B	B	C	C	C	C	C	A	A	B	C	C	C	C	A	A	B	C	C	C	C	A	A	A
353	21.30	C	B	B	C				B	B	C	C			C	A	A	B	C			C	A	A	B	C			C	A	A	A
355	22.00	C	B	B	C				B	B	C	C	C	C	C	A	A	B	C	C	C	C	A	A	B	C	C	C	C	A	A	A

There is a full range of amenities at the hoverport, including a popular-priced restaurant. There is also a bar, snack bar, currency exchange office and gift shop.

The July 1977 timetable. The letters refer to the vehicle tariff for a particular departure time. The reduced schedule for the off-peak days of Tuesday and Wednesdays allowed one craft to be taken out of service for maintenance.

One of the quirks that affected the average load factor was that peak times for traffic going to Calais often meant that it was off peak for the return and vice versa. Even so, the car deck and cabin crews worked like demons to move one load off and on, within the very meagre 20 minute turn-around time allotted. Engineers were always on hand to clear any operational snags as quickly as possible, to ensure the next departure left on schedule. However, the real masters of the situation were the traffic people. They were the ones who controlled the figures, making sure that loads were maximised where possible and time was not lost.

The hovercraft, despite its limitations, had one thing in its favour. Leaving hourly on the hour – in the latter days with four machines, every half hour – meant that if passengers were delayed in any way, they were usually on their way within a reasonable period of time. Something similar to the old city bus saying, "There'll be another one along in a minute".

This our Traffic Department used to maximum advantage, either by combining some scheduled departures into one, putting some passengers on an earlier flight while delaying others no more than an hour to take the later one. This was facilitated by the knowledge that human nature of the average passenger resulted in people turning up early for their reserved departure, 'to be on the safe side'. It was often possible to put these early birds on the preceding flight, thus upping the load factor[4]. If this was done on a consistent basis over the morning, sufficient passengers had been 'pulled forward' to end up by early afternoon with an empty departure, which was promptly cancelled. The result; expensive fuel saved on one return crossing and maximum earnings on the others.

This ability to manipulate loads in such a manner was a particular bonus on days of bad weather when our usual 40-minute scheduled crossing could fall apart and expand in some circumstances to an hour and a half. The ability for Traffic to 'lose flights' in the afternoon allowed a certain amount of catch-up time, ensuring that our last flights were not returning in the small hours of the next day, although it couldn't always be avoided.

Dealing with Humanity

Freight never complains. It is impervious to the weather, is not concerned with delays, is quite happy to sit anywhere, is never seasick and never

---

[4] There were many occasions when a car would turn up at check-in and be rushed through, minutes before a craft was departing. On a calm day it was possible for that vehicle to be on the road in France, having cleared Customs and Immigration, within 35 minutes of check-in at Pegwell Bay.

writes disparaging letters afterwards. People are different. They indulge in all of the above and more besides. In addition to our long-suffering stewardesses it was the stoic Traffic Department who bore most of the brunt.

Foot passenger check-in at Pegwell Bay. The uniforms indicate that this picture was taken post-1974.

Trying to juggle flight departures, while keeping passengers informed and happy in worsening weather conditions, must have been a nightmare, particularly in the early days when we were still learning to cope with the vagaries of the Channel weather. Add to the weather delays, the unpredictability of accidental damage and the teething trouble breakdowns that occurred almost daily, the traffic staff were continually trying to cope with a volatile situation that could change by the hour.

Juggling numbers and trying to combine flights included another problem that our experienced ex-airline staff hadn't had to cope with before. Passenger numbers booked for flights were not homogeneous. There were car passengers and foot passengers and it was fatal to mix up the two. Put too many foot passengers onboard and there may not be enough seats for those pouring off the car deck. What follows is Mike Castleton's account of one such muddle:

During the early period, while I was Duty Officer at Pegwell, an incident occurred which was less dramatic than some but equally stressful as far as I was concerned. It happened on an early shift during the summer of 1969 and involved one of the senior receptionists.

Upon starting the day we looked at the loads reserved on each flight and I was depressed to see a major over-booking on the departure scheduled for 10.00am. We had managed to book a total of 313 passengers onto a craft with 254 seats. The flights either side were fully booked, offering no possibility of a lifeline.

My brief to the check-in desk was specific: "We can't bring passengers forward or roll them over because everything is full so we must be prepared to reroute via Dover any surplus above 254. Start with 254 boarding cards and safeguard the vehicle allocation. We'll offload day-trip passengers as required. Let me know when you start to approach the critical figure and I will come to reception and between us we'll control it from there."

Weather was good, both craft serviceable and everything going smoothly. The 09.00hrs departure left on time and was full. The departure lounge began to fill with the 10.00hrs passengers as the closure time for the overbooked flight, 09.30hrs, approached. I enquired from check-in how things were looking and was told by the girl, "No problem, we're going to be full but no more than 254." Sighs of relief all round.

Vehicle and foot passenger loading started at about 0940. I remember thinking after about ten minutes that foot passenger loading appeared to be slow. At that moment the telephone rang. In tremulous tones the reception supervisor said, "We have a problem – I forgot to include the vehicle passengers. We have accepted 303 total." In other words 49 more passengers than seats. By this time chaos reigned on board as people were left standing all over the place with Tom Wilson, the captain, champing at the bit to leave. I can't precisely remember how, but with the assistance of the cabin staff, I managed to remove the surplus load but can recall that it was both time-consuming and difficult to put it mildly.

Eventually the craft was at sea, surplus passengers put on a coach to Dover and peace reigned once more. At this stage I invited a trembling receptionist for a convivial chat, which ended with words to the effect that I would wait until the shift ended before killing her.

Howard Archdeacon, my boss, had the last word on the incident – having witnessed my efforts, but mercifully not intervening, he said to me, " Well done Mike, that was a tricky situation... Dewey would have handled it better."

To prise passengers out of the hovercraft once boarded and seated would have taken tact and diplomacy of the highest order – another situation defused by our excellent staff, but, as Mike hints, there were other incidents far worse, such as this alarming story from Manu Heatley, one of the original traffic receptionists in Calais:

I had only been in the job a few weeks. I was on duty at the ticket desk, before the last departure of the day. Everything was pretty quiet by then and so I was by myself at the desk.

A young Frenchman and his girlfriend arrived and had a heated argument in front of the desk. She checked in and took herself off through the Immigration and Custom controls. After a few minutes he came to the desk to buy a ticket. When it came to paying for it he wanted to pay with some beads. I realised then that he was high on drugs and should have summoned help from the back, but young and inexperienced as I was in dealing with this kind of behaviour, I carried on explaining that he had to pay with money. I was holding his ticket on top of the desk when he produced a penknife, grabbed my wrist and pressed the blade of the knife on the inside of my wrist! I was too petrified to scream and stood there for what seemed to be an eternity.

There was a walkway leading to the upstairs offices, which overlooked the ticket and check-in hall and I suddenly heard a door opening. I looked up and saw Terry Halfacre, the Traffic Superintendent, coming out of one of the offices with someone else. He looked down, saw what was happening and signalled me to keep quiet.

Within seconds he and the other person had rushed downstairs, came quietly behind the young man and grabbed his arms. He dropped the knife, I ran to the back Duty Office and someone called the police. He was taken away and I was given a large brandy by Terry to steady my nerves.

A pretty frightening experience, which taught me that not all passengers were nice.

Luckily for us all, experiences like Manu's were rare if not unique. By and large, passengers behaved well and at times showed a remarkable degree of tolerance and goodwill. Roger relates a particularly heart-warming occasion:

> *It always amazed me how tolerant the British travelling public were, particularly in the first year or so when we used to bugger them about something rotten with breakdowns and delays.*
>
> *I remember one particular day we stuffed around one passenger load, which should have left Ramsgate around mid-day. With the usual APU and hydraulic trouble we had them on and off the craft virtually all afternoon until we finally departed around five pm.*
>
> *That turned out to be not the happy ending it should have been. In the meantime the weather had really deteriorated. We got to the edge of the Goodwins, took one look at the raging ocean off the edge and decided, quite rightly, to return.*
>
> *We got back into Pegwell, shunted most of the passengers off to Dover on the ferry, then retired to the bar to lick our wounds. In those days we were a couple of million pounds over budget, so every bad day like that we had the feeling the receivers would be in the following day. While we were sitting there, pretty po-faced, a couple of the passengers strolled up and commiserated with us, thanked us for trying and bought us a beer. Bless 'em.*

Les Colquhoun and 'Big' Jim Hodgson

As the anecdote suggests, Hoverlloyd's financial position in the first two years was precarious. As previously explained the rumoured budget blow-out was in the region of £2 million, principally due to the construction cost of the Pegwell Hoverport. Although the service was immediately popular with the travelling public and load factors were approaching maximum right from the start, the revenue from the two Mk.1 hovercraft was insufficient to cover the debt.

Things came to a head in 1971, resulting in a management shake-up, which saw the original Managing Director, Les Colquhoun, who was more of an operational expert, replaced by James Hodgson, a consummate businessman, who was also Executive Vice Chairman of Hoverlloyd. He already had a long-standing association with Hoverlloyd, as his London-based company, Bonner Hodgson in Charles Street, carried out the early recruiting of the SR.N6 personnel in 1966 and, right from the beginning, was responsible for Hoverlloyd's PR and marketing.

As personalities go, Les and 'Big Jim', as he came to be known, were chalk and cheese. Les Colquhoun, who had set up Hoverlloyd, was wholly responsible for engendering the spirit of infectious enthusiasm amongst the staff, together with the 'can do' culture. He was warm-hearted and approachable; in modern parlance 'a people person'.

Jim Hodgson and Les Colquhoun view the construction of the Pegwell Hoverport. 1969.

By contrast James Hodgson was an arch-typical boardroom sophisticate, cool and aloof in manner, someone it would be difficult to imagine with his tie off and his sleeves rolled up.

Nevertheless, Big Jim's business acumen was clearly what was needed at this time. He very quickly proved his worth by persuading the Swedes, who by this stage must have been very near to pulling out of the whole venture, to take the opposite tack and invest a lot more money. They accepted his proposal to buy a third craft – *Sir Christopher*[5] was put into service for the 1972 season – and then to embark on a programme of 'jumboizing' all three machines by removing the inner cabins and widening the craft to incorporate extra seating for passengers.

The increase in car and passenger volume stemming from the Mk.2 expansion and the addition of a third hovercraft[6] marked a significant

---

[5] The *Sir Christopher* was the only hovercraft to be named after the inventor, Sir Christopher Cockerell.

[6] Described in more detail in Chapter 22: *Developments 1972 – 1977.*

increase in profitability and stood the company in good stead right through to the end of the decade.

Humour Always Helps

The first years of Hoverlloyd may have been tough and uncertain, but the high spirits of the largely youthful staff kept any sense of doom and gloom well in abeyance. Captain Ted Ruckert's recollection is typical of the times:

> *On one occasion, in the early days of the SR.N4 operation, we were watching the hovercraft arrive from Calais with some VIPs on board. A brass band had been laid on and started playing as the engines shut down. At that moment Les Colquhoun, the Managing Director, came into the Radio Room with the Swedish Lloyd directors, who had come to the hoverport to find out why everything was way over budget. One of them asked, "Why is the band playing?" Before Les could open his mouth, Dave Luff, our cheeky little 2nd officer said, "We always have a band playing when the hovercraft arrives on time!"*

Fortunately Les and the directors thought that was extremely funny.

In retrospect it might be said that Hoverlloyd was fortunate to have both Les Colquhoun and James Hodgson respectively at the helm. Les' easy-going manner, enthusiasm and determination to see the company develop as a properly-supported professional operation, undoubtedly created from the very start the professionalism and company image that everyone involved took such a great pride in.

The hoverport at Pegwell may have been almost catastrophically over budget, but the result was a modern operational masterpiece with a magnificent building to boot. In marketing terms and in terms of staff morale, it resulted in significant benefits.

Certainly the travelling public were of that opinion and flocked to Ramsgate. Even today Hoverlloyd as a brand image is still fresh in many minds. This interesting comment came from the BBC's South East local interest programme *Inside Out* in 2006:

> *The cross-Channel service from Dover was run by Seaspeed, a subsidiary of British Rail. But hot on their heels was Hoverlloyd. It built a hoverport in Pegwell Bay near Ramsgate in 1969. Hoverlloyd was a brand new Swedish company.*

In terms of enterprise, staff motivation and sheer sexiness, Hoverlloyd won hands down.

To that already well-polished image, wholly attributable to Les Colquhoun, Big Jim applied his business acumen and expertise in financial management to turn Hoverlloyd into a profitable company. Sadly both men are no longer with us, but both deserve our thanks.

Calais

So far only one part of the traffic equation has been mentioned. The other essential element for ensuring Hoverlloyd's profitability was the part played by those in Calais. While Pegwell was wholly owned by Hoverlloyd, the Calais hoverport was independently owned and run by the Calais Chamber of Industry and Commerce (CCIC).

The complex set-up is best explained by 'our man on the ground', Mike Castleton, who, having started as a traffic officer at Pegwell Bay, was appointed Hoverlloyd's Traffic Superintendent in Calais:

The position of Hoverlloyd in Calais was totally different from that in Ramsgate. The hoverport in Calais was built in the winter of 1968-69 by the Calais Chamber of Industry and Commerce who were the owners of the building. Hoverlloyd was there as a tenant. Jean Louf of the CCIC, with whom I was to work closely in the '70s and '80s, was appointed manager of the hoverport.

For a ferry company to operate into a French port it is necessary to appoint a broker to represent the operator. Such brokers are known as Courtiers Maritime. Our broker was Claude Foissey, located next to the CCIC in Boulevard des Allies. The primary function of the broker is to calculate port taxes payable to the French government, pay these every ten days to French Customs and then invoice the operator. A separate and lower rate tax is payable directly to the CCIC by the operator.

The function of the broker in the case of Hoverlloyd in 1969 was more far-reaching and complex. The manager of Claude Foissey's agency was Yves Lefebvre and it was Yves who was the true architect of Hoverlloyd in Calais. Under his supervision and guidance a traffic, reservations and accounts structure was put together to be in place for the commencement of SR.N4 operations in April 1969. Yves drew upon local sources to appoint staff. A gentleman named Jean de Mouliniere was employed to manage and develop the Traffic Department. Three duty officers formed the next layer of the structure. They remained with Hoverlloyd for many years: Francis Balloy, Marcel Hamy and Jacques Bodhuin. Marcel later joined the Sales Department and was replaced by Jean-Claude Romana in 1972.

In those formative years Yves Levebvre, who liaised closely with Les Colquhoun, was very much Mr. Hoverlloyd in Calais and as such wielded great influence.

The first Englishman in Calais, Terry Halfacre, took over from Jean de Mouliniere as Station Superintendent in 1971.

Calais Hoverport in the late '70s. A unique photograph showing, from left to right: A Hoverlloyd SR.N4 Mk.2, the newly jumboized Seaspeed SR.N4 Mk.3 and the ill-fated Sedam N-500.

After a thousand years of French/British antagonism, in stark contrast to the norm, Mike's description of Franco-British cross-Channel relationships suggest they were convivial in the extreme. Whether Les Colquhoun's legendary people skills were again in evidence is a matter of conjecture, but certainly they wouldn't have been a hindrance.

Perhaps even more telling is the manner in which Mike Castleton took on the position after Terry's departure:

Yves was hastily summoned to Pegwell Bay by Les Colquhoun and Howard Archdeacon to advise upon a replacement for

Terry. His advice was typically shrewd and concise: "Les, if you appoint a Frenchman to replace Terry you will never be quite sure what is happening in Calais. His replacement must be English."

For a Frenchman to suggest an English person as a replacement for reasons of good management indicates a mature, non-partisan relationship of the highest order, and a tribute to all those involved. A mutual respect and understanding at the top that ensured the same relationship between all levels of the company throughout its lifetime.

Turning to Michael himself, who claims at the time his spoken French to be 'non-existent', the initial stages of settling into a foreign country must have been very stressful. However, yet again, French goodwill came to the fore:

Being no linguist, my French was never fluent unlike that of Pat, my wife. I did however struggle and managed to get by. Reasonable fluency in English was a prerequisite for all traffic and reservations personnel, but not a requirement for the accounts department.

My first weeks in Calais were to open my eyes to the enormity of the step I had taken. Although so few miles separate our two countries the respective ways of life are poles apart. I began to wonder if I had made the right move. Here I must mention the total support of Francis Balloy, Jacques Bodhuin and Jean-Claude Romana. Their loyalty to me never wavered over the years and without that support I doubt that I would have survived those first weeks.

Outside the operational sphere I had to develop a general understanding of local practice and customs in Calais. Two people were particularly helpful; Yves Lefebvre and my bi-lingual secretary, Marthe Evrard. Yves made a point of introducing me to all the people who I needed to know and was endlessly patient. Madame Evrard was invaluable in helping me to understand the day-to-day running of the hoverport as a whole. It is noteworthy and so typical of French protocol that never once in 16 years did she address me by my Christian name.

Both these friends are unable to accept my thanks today. I only hope that I expressed my gratitude adequately at the time.

Even when relationships were a little strained at the start, it is a credit to all parties that things were resolved in an amicable and beneficial manner:

Francis Balloy and Rod Lake, as 1st officer, walking on the apron as *Sir Christopher* makes an approach up Calais' eastern ramp. This was before the eastern and western ramps were joined to enable craft to drive up directly from the beach onto the apron. Circa 1973.

The caption probably has to be: 'To ensure a happy holiday, it pays to wait until the rear unloading ramp arrives before driving off.' The most likely explanation, however, is that for some reason the truck behind unfortunately shunted into the transit van, pushing it over the edge.

Francis Balloy and Jacques Bodhuin assisting Stirling Moss (later Sir Stirling), legendary Formula 1 racing driver at Calais Hoverport. 1970.

Cars waiting to embark in Calais on an SR.N4 Mk 2 Hoverlloyd craft. 1973.

Working within the constraints imposed by the CCIC was often difficult. Jean Louf, the hoverport manager, and a man of many talents, regarded the hoverport as his personal fiefdom. Jean had to be approached with tact and diplomacy but over the years we had a good working relationship. He was a qualified pilot and I spent many hours in a variety of light aircraft flying with him.

Calais Duty Free

Prior to the removal of customs controls with the further integration of the European Union, the sale of duty free goods was a lucrative market for all of the cross-Channel operators. For the high-speed hovercraft with their smaller market and higher costs, it was essential.

For example, Hoverlloyd's operating profit for 1979 was £2,137,520, including profit on duty free goods, estimated at 30 per cent of the total, which alone generated a turnover of around £2,250,000 for that year. Without income from duty free a reasonable profit would have been turned into a small loss.

Seaspeed started an experimental service from Dover to Calais in 1970, which became a permanent service in 1971. The Seaspeed craft is on the left. Cars are embarking at the stern of the Hoverlloyd craft. 1970.

It was fortunate for the company having a first foot on the ground in Calais as it meant it had a hand in setting up the duty free sales in the hoverport. Again, because of the independent management of the facilities, Hoverlloyd was not the sole beneficiary. This was shared with the French interests in the port, CCIC and the franchisee of the sales outlet. What was a bonus, and what must have been difficult for Seaspeed to swallow, was that when they eventually started services form Dover to Calais, their passengers would also be patronising the shop and therefore passing the revenue to their rivals.

Although Hoverlloyd was just one shareholder of the revenue, it still contributed excellent profits, as Mike Castleton explains:

> *There was a formal agreement between the Chamber of Commerce and Hoverlloyd for the distribution of duty free revenue, which was divided three ways between the franchisee, Leon Ghesquiers, the Chamber of Commerce and Hoverlloyd. Most of the stock was provided by Hoverlloyd Ramsgate, although Leon was free to purchase stock locally when this was not available from Pegwell Bay. I don't know the precise structure of the financial agreement but I do recall a comment by the Finance Director in about 1976, to the effect that, "the money is coming in so fast that I don't know what to do with it." The profit margins on duty free goods were very high and enabled the three-way agreement to be beneficial to all parties.*

Coach Services

The previous story concerning the overflow of foot passengers causing a severe headache highlights how popular daytrips from both sides were right from the start. The British flocked to Calais to pick up cheap beer, French cheese and other delicacies from the hypermarkets, with a relatively strong movement from Calais, mostly composed of French housewives, to buy clothes in Britain. Marks and Spencer's ladies underwear did particularly well.

Added to this brisk business over the succeeding years was the development of through coach services. Routes from London to the major European centres became an increasingly significant market in which the Calais staff played a major role in the expansion.

The first route in 1969 was a daily service through to Paris, licensed on the French side to two French companies, STRV of Calais and the Toulouse-based Voyages Fram – the Hoverlloyd terminal in the capital was conveniently situated close by the Gare du Nord. The London to Ramsgate sector, starting from Victoria Coach Station, was operated by the East Kent Bus & Coach Company.

The UK currency at the time was pounds, shillings and pence – decimalisation was not introduced until 1971 – and the cost of a single fare from London to Paris was just £4, or if you joined at Ramsgate, £3.50. There was also a coach service to Ostend for the grand sum of £3.38, or, if joining at Ramsgate, £2.78[7]. By 1977 the London to Paris fare had risen to £10.50, but was still considered good value.

An East Kent Bus Company vehicle, in Hoverlloyd livery, which operated the London to Pegwell Bay part of the coach/hovercraft through Continental services, alongside *Swift*. This was a publicity picture as the passengers arrived at the main entrance and proceeded through check-in as foot passengers.

The service appealed to the cheap end of the market and, without continuous motorways, or dual carriageways – yet to be built – from London to Ramsgate, or from Calais to Paris, the journey was somewhat tedious, taking a total time of around eight hours to complete. Nevertheless it was extremely popular and load factors were high.

This first successful enterprise was followed in 1973 by the opening of the London to Brussels route, run by Royal Tours of Ostend, which also proved to be a lucrative market. Royal Tours went on to initiate a route to Amsterdam, thus tying up for Hoverlloyd the three major government and business centres in North Western Europe.

---

[7] This, originally written as £2 17s 6d, and the preceding sums are the modern decimal equivalents of the old 'pounds, shillings and pence' currency, which changed to decimal in 1971.

An attempt by the same company to move the network further afield to Cologne in Germany was unfortunately not so successful after objections raised by German Railways were upheld.

Emboldened by the runaway success of these various enterprises, steps were taken later to expand the coach system even further. However, despite initial good fortune, the tide turned:

In the late seventies a number of new coach destinations were introduced. These consisted of a service to the Côte d'Azur terminating in Nice with multiple traffic stops en-route. Another service was to Lourdes and Tarbes via Bayonne and Pau. The final long-haul service to be introduced was to the principality of Andorra. Planning these services took up a large part of my time and that of Gordon Gillon, an amiable Scotsman in charge of the accounts department, for several months.

The operation of long-haul routes moved us into a completely different sphere. Unlike the existing services, which simply went from A to B with one stop, the new ventures had multiple commercial stops all of which had to be strictly controlled. Each service required a team of two drivers in both directions and, in order to maintain the optimum utilisation of seats and an adequate care of passengers, it was necessary on all the new routes to carry a bi-lingual courier. These were mainly sourced from the Traffic Department in Calais with some input from Pegwell Bay. Laurette Wacogne and Catherine Dannel were particularly effective in this role. In keeping with parliamentary tradition I also utilised my wife Patricia in this capacity.

The commercial success of the new routes was immediate and load factors were high.

As the summer passed however, it became increasingly clear that the operational problems were intensifying and complaint levels rising. Problem areas included coach reliability, the indifferent quality of a minority of drivers and, above all, the impossibility of observing schedules in the summer traffic density of southern France.

Before I could begin to work out remedial measures the problem was solved for me and a different one created. The Sales Director, Bob Harvey, bore overall responsibility for the commercial and legal aspects of the coach network. I was consulting with Bob on the operating problems when it was brought to our notice that the entire operation was threatened by serious licensing irregularities, which had been concealed from Hoverlloyd by the coach operators for their own ends. This could have resulted in action by the French Ministry of Transport, thereby placing the entire Hoverlloyd operation in jeopardy. Bob and I conferred with Howard Archdeacon and it was decided to pull the plug on the long-haul services with immediate effect. This I did, resulting in passengers stranded all over southern France. A rescue package was devised involving the transfer of booked passengers to the SNCF rail network at Hoverlloyd's expense. This was a logistical nightmare whereby traffic staff had to be despatched from Calais with bags of cash, but it proved successful. Not a happy chapter in the history of Hoverlloyd.

As Michael says, it may not have been one of the high points of the Hoverlloyd story, but nevertheless, it still remains a clear demonstration of the company's major asset, its personnel; a management capable of making quick well-considered responses to problems as they arose, plus a capable, highly dedicated and motivated staff ready to put decisions into action with a high level of efficiency.

Industrial Relations

Although in strictly legal terms Hoverlloyd in England and Hoverlloyd in France were two separate entities, the staff considered themselves one company and behaved that way. The loyalty and love for the company as a 'brand' was indivisible and, if the turn out at the various reunions from both sides of the Channel is any indication, it remains as strong as ever to this day.

One major difference between the two was in the matter of industrial relations.

France has always had a strong socialist approach to workforce organisation, and union participation in companies like Hoverlloyd had strong government and legislative support.

In contrast, despite the 1970s in the UK being the pinnacle of labour unrest, Hoverlloyd, because of an enlightened management approach, was an island of tranquillity amongst the surrounding chaos.

The decade was notorious for industrial disputes on a massive and continuous scale. A succession of poor governments, weak industrial management and a union movement out of control, all combined to lead to an industrial relations nightmare and a catastrophic decline in Britain's economy.

In one of the many attempts to placate union power, the Labour government of the time, following very much the well-established French pattern, passed an act making union membership compulsory if a sufficient vote of the workforce required it.

Up until that point Hoverlloyd had very little union involvement, but after a small group demanded a vote in accordance with the new legislation, a ballot was duly taken.

The result was something around 95% against union representation.

With this background in mind, it must have come as something of a shock for Mike Castleton to find a totally different working environment on his arrival in Calais:

> *It is a legal requirement in France for an employer the size of Hoverlloyd to allow the formation of a works council to represent staff. This is known as the Comite d'Entreprise (Works Committee) and members are elected by the staff themselves. It was incumbent on me to arrange such an election and to meet with this committee monthly and to minute the agenda and its conclusions.*
>
> *Having observed this process, I was taught another lesson in French industrial law. A letter addressed to me was received from the most militant of France's trades union, the CGT[8], informing me that a member of my staff had joined the union and had been nominated by the CGT as its representative in Hoverlloyd.*
>
> *By law I had no choice but to accept this development and to allow this representative to attend works committee meetings. If this was not time-consuming enough I was also legally obliged to allow the election of four more staff representatives known as 'Delegues du Personnel'.*
>
> *I never really saw the logic of such dual staff representation, but I do recall that the entire procedure was irksome in the extreme, consuming disproportionate amounts of management time. It became a constant thorn in my side.*

Considering the CGT was an extreme radical organisation which, at the time, enjoyed close links with the French Communist Party, the potential for trouble was very real. It says a lot for the Calais staff and, to his

[8] Confédération Générale du Travail (General Confederation of Labour).

credit, most probably Mike Castleton's quiet management style, that industrial relations in Calais, like its counterpart in Pegwell, remained an oasis of calm throughout the company's short life.

It is easy to say this was pure luck, but luck, like most things beneficial, has to be worked for. Looking back on it all, it can be acknowledged now that Hoverlloyd – Swedish, English and French – worked very hard at it.

Entente not so Cordiale

Utopia it may have been, but where humanity is involved nothing is all sweetness and light. One constant cross-Channel niggle was the noticeable difference in passenger complaints on the French side. Of the total complaints received, Calais accounted for 70% of them.

Unfortunately, the British invariably jumped to the easy conclusion; it was typical of the French to be rude, particularly when it came to the old enemy across the water.

Hoverlloyd Traffic staff at Calais Hoverport. Left to right: Jacques Bodhuin (duty officer); Jean Philippe Pollet (car marshaller); Alain Tassart (seasonal staff); unknown. Behind is the hoverport control tower. 1970.

It has been alluded to before that written criticism is a very English trait; so it is not surprising to note that 90 per cent of written complaints came from *Les Anglais* in Calais.

As Mike says:

> *Continentals are not given to writing, they just shout at the time.*

He believes there is a more plausible explanation:

> *The statistics show that the great majority of motorists from Ramsgate arrived for their booked flights and were accommodated on that flight or soon after. In other words their expectations were met. Also they were travelling with a holiday to look forward to.*
>
> *Taking the reverse scenario the passenger mindset is quite different. Their holiday is over, the holiday budget has been exceeded, the kids in the back have squawked a hundred times, "are we nearly there?" They have travelled hundreds of kilometres with only work to look forward to in the weeks ahead and are hopelessly late for their booked flight. Then, someone with a French accent tells them something they don't want to hear.*
>
> *Allegations of rudeness were always investigated and very rarely found to be true. I thought it was summed up very well during a meeting by one of my supervisors, Jean-Phillipe Jaze, when he said, "Sir, as soon as you tell an Englishman something he doesn't want to hear you are being rude."*
>
> *There is a lot of truth in that observation.*

Departures

Another anomaly between each side of *La Manche* must be acknowledged, namely the odd fact that the departing English were always more feted in France than they were in their own country.

There were many other occasions, but two of Mike's illustrations will have to suffice:

> *Upon the occasion of Tom Wilson's very last flight we decided to give him a little send-off.*
>
> *Francis Balloy and I climbed the ladder and I presented Tom with a couple of bottles of champagne. I said to him, "Enjoy*

these in Ramsgate." He replied, "Why wait until then?" and corks began popping. Francis then presented him with a huge pair of white woollen socks[9], which Tom tied to the vertical mast in front of the flight-deck windscreens. The craft left and, minutes later, came a message from the radio room, "We've reached 75 knots and the socks are holding fast!"

Luckily for us all, no one has ever been booked for 'driving a hovercraft under the influence'. Even if there had been such a thing as the 'Channel Police' they wouldn't have stood a chance of catching the old so-and-so at that speed:

The other little tale relates to none other than Robin 'Bonjour les oiseaux'[10] Paine. I think it was Robin's last flight as captain. We hatched a plan in Calais to replace him with another captain travelling incognito to take the flight back to Ramsgate. We could and did, then hijack Robin and took him to my flat in Calais where Pat and I, with many others, gave him a liquid send-off. Details thereafter are sketchy but I believe Robin disembarked in Ramsgate on a baggage trolley followed by a lap of honour of the pad. And, to this day, I still have his epaulettes as a souvenir!

We could never have done that in the stagnant environment of Hoverspeed.

If Mike's last remark sounds overly critical of the later company, a touch bitter even, it has to be acknowledged that any set-up after Hoverlloyd was bound to be a disappointment. Companies that engender that level of love and loyalty are few and far between; those rarities that do exist, people write books about them.

This volume is a case in point.

Without exception, the atmosphere prevalent throughout Hoverlloyd's short life was repaid a thousand fold on both sides of the Channel by a level of dedication hard to imagine now, even by those of us who experienced it.

All of those who worked in the traffic department, ensuring a high degree of customer service, therefore ensured in return a high degree of customer loyalty, all deserve our thanks.

---

[9] Tom was renowned for his white socks.

[10] Robin would always arrive in the Calais Traffic office with this greeting to the girls.

Mike Castleton and Laurette Wacogne at the 40-year reunion, May 2009. After all that time the uniform still fits!

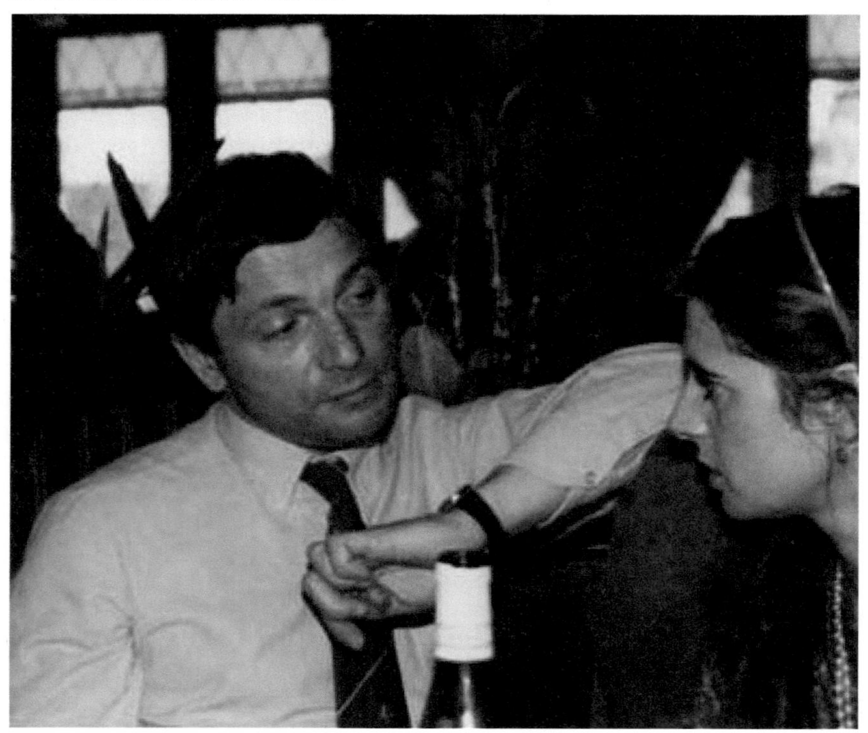

Calais 1973 – Francis Balloy with Laurette.

Laurette Wacogne in 1974.

It's hard to leave out the equally deserving, but Mike singles out two in particular in Calais who were indeed exceptional. None of us who knew them would argue:

> *Such was the spirit of Hoverlloyd that, for many of us, it was not just a job but a way of life. Two people who symbolised that were Francis Balloy and Laurette Wacogne.*
>
> *Francis lived for Hoverlloyd and would have worked seven days a week if I had asked.*
>
> *Of all the devotees of Hoverlloyd there was never one more fervent than Laurette. Volatile and incredibly stubborn, she and I had many disagreements but they never lasted and her loyalty to both Hoverlloyd and to me never wavered.*

The X Factor

So much for the personnel who worked their utmost to make Hoverlloyd a success, but there was a hidden factor, an inanimate 'X Factor', which also contributed to put Hoverlloyd ahead of its rivals.

Much was made earlier of the company's emphasis on procuring the most advantageous route, and commensurate with that the right bases from which to operate.

Once operations commenced in earnest that dedication to getting it right was proven to be amply justified. The chosen Ramsgate to Calais route proved to be the perfect scenario for a successful hovercraft service. A route providing not only shelter from the prevailing weather for much of its length over the Goodwin Sands, but also a means of choice in adjusting the day-to-day passage to take the best advantage of wind and tide at any time of the day.

It could be said that once the Mk.2 modifications were delivered in 1973, Hoverlloyd held all four aces – good management and staff, excellent well-sheltered bases of operation, a craft that could produce a return on investment and a hovercraft-friendly route.

After an uncertain start the feeling in the company was that we were set for a bright and prosperous future.

PART FOUR

The Hoverlloyd Years

Chapter 20

Ramsgate to Calais

Basic Economics

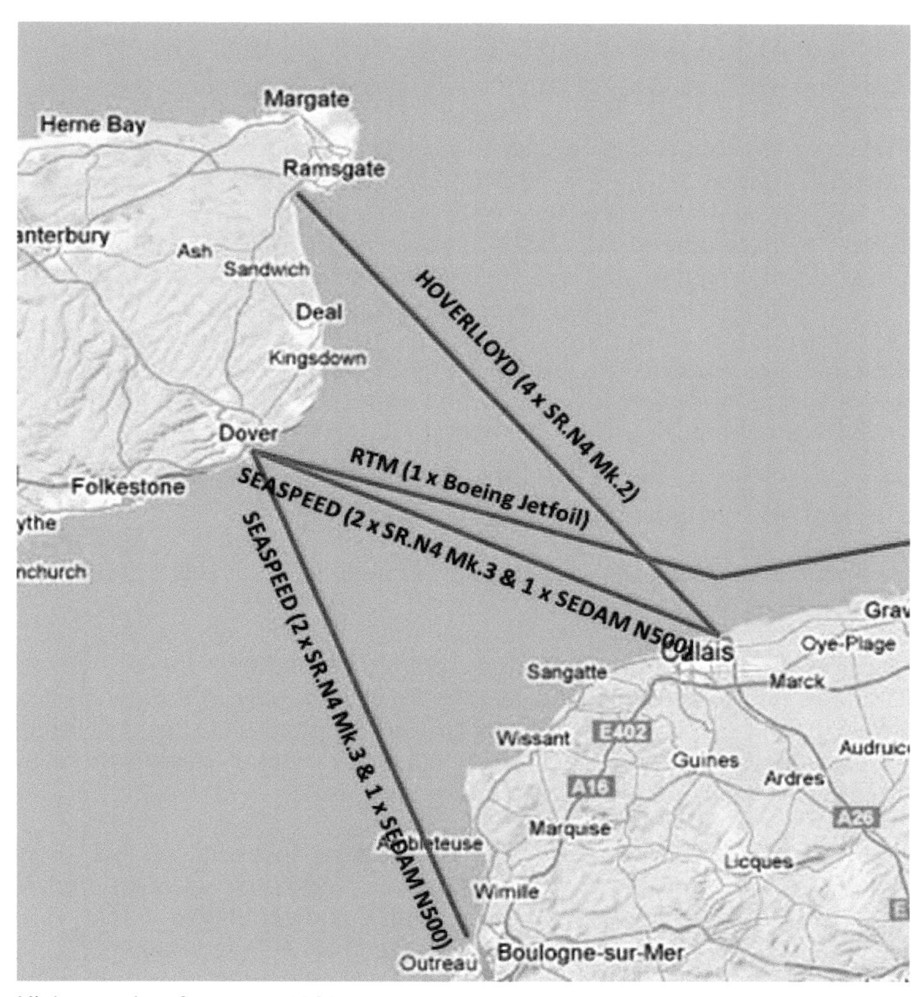

High speed craft routes. 1980.

One of the pitfalls associated with any new invention is the tendency for some entrepreneurs to desperately seek a reason to use it.

Someone will come up with an idea for a hovercraft service and it very quickly becomes apparent that the whole concept is based on pure infatuation, rather than any hard-headed evaluation of the actual need. Time after time, people are seduced by the sheer excitement and thrill of something that glides over any sort of surface with breathtaking speed and ease. After all, above all else, the hovercraft *looks* a lot of fun.

By the mid '60s enthusiasm for this amazing new concept was at fever pitch and ideas as to how this incredible machine could be employed abounded, many of them wild in the extreme.

What you were going to do with a hover-caravan is anyone's guess!

Putting sheer fantasy aside, any project evaluation boils down to three simple questions:

- Is there a need for a fast transit?
- Is there a need for an amphibious machine?
- Is the public able and prepared to pay for speed?

If there is not a pressing requirement for the first two of these attributes, and a willing market for the third, then there can be no valid reason to invest; the economics will simply not add up.

The Hoverlloyd Situation

Using the criteria above, was there a genuine need for a cross-Channel hovercraft service in 1969?

Considering the vast number of passengers carried over the company's lifetime it has to be conceded as affirmative, but only in the sense that Swedish Lloyd and Swedish America had a commercial need to enter an expanding and lucrative market, a market hitherto denied them by the existing ferry operators in Dover and Calais.

Also, it has to be said that if any service was to be considered from Ramsgate, then only an amphibious vehicle could provide a straight line route over the extensive shoal areas of the Goodwin Sands.

Although not necessarily a need, it has to be accepted also that a fast connection to the Continent was attractive to the passengers. An average ferry crossing was around one and a half hours but the comparative slowness of loading and discharging, meant that the road to road time for a car could be two and a half hours or more. By contrast with Hoverlloyd's scheduled turn-round time of a mere 20 minutes, cars arriving at Pegwell, say, at 06:30 could find themselves on the road in Calais by 07:40.

Another crucial aspect of any high-speed service, regardless of whether it is a hovercraft, a hydrofoil or, the latest craze, the wing-in-ground (WIG) craft, is the cost of operation. Moving vessels at sea above conventional ship speeds is always expensive on fuel, so there has to be a reasonable return, engendered by both the passenger volume and the price of a ticket. It follows that such potential markets can only be those having large populations with a high standard of living; the UK cross-Channel market is one of the few that complies.

Year in year out, where else in the world is there a cold wet island with a 60 million population of comparatively high income, all desperate to go south each summer to guaranteed sunshine on the other side of the water? From today's perspective, accepting the fact that no other SR.N4 routes ever came to fruition anywhere in the world, the answer to that question is sadly in the negative.

It has to be concluded that the SR.N4 was a unique vehicle operating in unique circumstances. It may also have something to do with the fact that we ended up with some unique people.

The Route Structure

From the navigational viewpoint the supreme advantage of the hovercraft over conventional vessels is its amphibious nature. No need for a planned voyage around dangerous shoals ending in a circuitous route from port to port; the hovercraft needs only a straight line from A to B. Obviously there are occasions when there may be hard structures such as navigation beacons and exposed wrecks to avoid, but in the Hoverlloyd case, from our fairway buoy, placed due south of Ramsgate Harbour, marking the approach to Pegwell, a straight line of clear passage could be drawn directly to the hoverport at Calais. Any ordinary vessel would have had to alter either north or south around the Goodwin Sands, adding extra miles to the port to port distance.

The distance from the Pegwell approach buoy ('First Buoy') to the CA2 Buoy off the Calais Hoverport, some 28 nautical miles, was used to measure our cruise speed for the trip. At our average cruise speed of 60 knots, which meant 28 minutes buoy to buoy, allowing for a gradual slowdown on the approach, the 40-minute schedule advertised to passengers was well within the capabilities.

But of course that was on days with reasonable weather. The English Channel is notorious for bad weather, which is by no means confined to the winter months. It was amazing how 'Murphy's Law' invariably dictated Force 8 and 9 gales on Easter and other Bank Holiday weekends.

Comparative Crossing Times		
Cross-Channel Car Ferries	**Company**	**Time**
Felixstowe - Antwerp	Transport Ferry Service	14 hrs
Harwich – Ostend	Belgian State Marine	5 hrs
Tilbury – Antwerp	Transport Ferry Service	16 hrs
Southend – Ostend	British Air Ferries	45 mins
Southend – Calais	British Air Ferries	30 mins
Ramsgate – Calais	***Hoverlloyd***	***40 mins***
Dover – Zeebrugge	Townsend	3½ hrs
Dover – Ostend	Belgian State Maine	4 hrs
Lydd – Ostend	British Air Ferries	45 mins
Dover – Dunkirk	British Rail	3½ hrs
Dover – Calais	French Rail/Townsend	1½ hrs
Dover – Boulogne	British Rail	1½ hrs
Dover – Boulogne	***British Rail Seaspeed***	***35 mins***
Lydd – Le Touquet	British Air Ferries	20 mins
Lydd – Deauville	British Air Ferries	45 mins
Newhaven – Dieppe	British Rail	3½ hrs

Comparative Crossing Times (*Air-Cushion Vehicle, April 1969*).

The direction of the wind and its strength was important in determining whether or not the hovercraft could operate. For example, winds of a certain strength in the quadrant between north-west and east would curtail operations sooner than if a comparable wind was coming from the south-west. The strong tides also played an important part, as when the wind was against the tide the sea state could be quite ferocious.

It is interesting to note now how much more advantageous Hoverlloyd's route was with regards to weather than the Dover/Boulogne route initially chosen by Seaspeed. Although much the same length, the Hoverlloyd route provided flexibility to choose the most advantageous course, whatever the direction or strength of the wind. By contrast the course from Dover to Boulogne offered no choice at all as the whole route was exposed to the prevailing south-westerly gales. The hoverport at Boulogne was particularly vulnerable, facing as it did, due west into the teeth of it.

Captain John Syring, of Seaspeed, recalls:

> *Getting out of Dover through the Eastern Entrance was not a problem in a south-westerly. The real problem in a south-westerly was Boulogne. You've got the seas breaking on the beach and hitting Boulogne Harbour wall. The waves then*

come straight back out again so that up to three miles out you would get random pyramidal big seas.

We used to battle to Boulogne for an hour or more. We'd come out of the Eastern Entrance at Dover and go inshore two thirds of the way to Dungeness, so we could turn to port and put the sea on our quarter and run down to Boulogne. We were battering the craft, battering the skirt and frightening the passengers. I believe our cancellations were due more to the weather at Boulogne than at Dover.

The Hoverlloyd route gave a comparatively sheltered ten miles out to the edge of the Goodwin Sands, time in which to gauge the weather conditions and decide in perfect safety whether or not it was wise to continue on. On several occasions one look at the heavy surf over the edge of the shallows was enough for discretion rather than valour to advise a return to base.

On their return to Dover, Seaspeed pilots also had to deal with negotiating the hard, stone-walled harbour entrances, something bordering on the 'risky' for a free-moving vehicle like the N4. Although the Seaspeed captains never regarded the entrances as a hazard, the sheer fact that the craft had to be negotiated through two solid walls still presented an obstacle.

A potentially dangerous situation, which ultimately proved to be a reality, was when in difficult conditions *The Princess Margaret*, by then part of the merged company Hoverspeed, hit the downwind wall of the Western Entrance in March 1985, killing four passengers and injuring many more. The only tragedy to mar an otherwise unblemished safety record of over 30 years of cross-Channel hovercraft operations.

It is little wonder that Seaspeed could never achieve Hoverlloyd's record for reliability, which, after our first couple of years of initiation, was consistently in excess of 98 per cent of scheduled crossings. This is by no means a reflection on the skill and dedication of our fellow hovernauts at Dover. They coped with these difficulties as manfully as they could, difficulties that we were just lucky to be without.

The Development of Weather Routing

The Ramsgate/Calais track, with the benefit of such a wealth of options open to us to cope with the vagaries of the weather, allowed us to pass quickly through the bottom of the learning curve and eventually become experts in judging the conditions to a high degree of accuracy when choosing a route to the best advantage.

Swift on her way to Calais in mid-Channel in probably a south-westerly Force 7.

Dealing with the persistent south-westerlies was the major issue. Although north-easterly weather was less common, they could in many ways be more severe – in Robin's opinion:

> *The north-east and east winds were utterly ghastly as they seemed to throw up 'holes' all over the Channel and the ride was extremely uncomfortable. The south-west winds were generally a much better bet and, even with wind against tide, it was more of a roller-coaster ride rather than the dreaded easterly punch-bag effect.*

The pilot operation from 1966 to 1968, with the much smaller SR.N6, gave some background knowledge of what to expect once the N4 service commenced, but because of the vast difference in size and manoeuvrability, it was inevitably nothing more than a best guess. The route map opposite, published in *Hovercraft World* in 1969, is a case in point. Based on the N6 experience the initial thoughts suggested bad weather routes north and south of the Goodwins. The actual N4 experience developed a different approach, particularly on the return. It was common practice in really bad south-westerly gales to pull to windward as far down the coast, toward Cap Gris Nez as was necessary to put the wind abaft the beam and then shape up for the Goodwins, joining our normal track line on the south-eastern edge. If this sounds somewhat reminiscent of old sailing parlance, the fact is, because the machine was so much subject to the wind, we were in fact 'sailing' the craft to best advantage in any situation. Fascinating to think we, in our very modern machine, were relearning skills that our sailor ancestors took for granted some hundred years before.

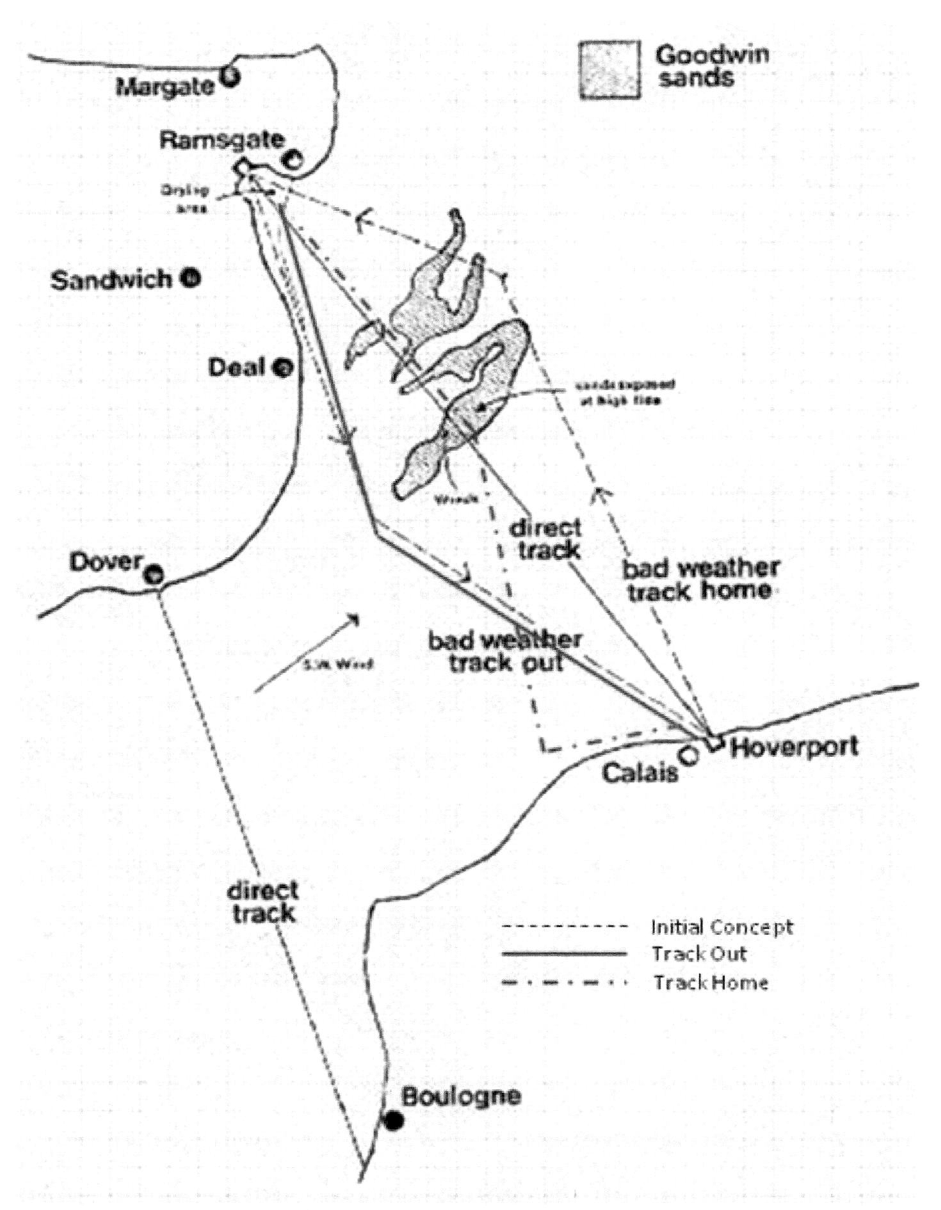

Bad Weather routes conceived prior to N4 commencement in 1969. Coloured lines show routes developed with the benefit of experience. (*Original diagram – HovercraftWorld March/April 1969)*

Also, part of our education was coming to understand how much the Dover Strait tidal streams, at times as much as 4 knots, affected the sea-state in Force 8 winds and above.

We were limited, by our safety certificate, to a significant wave height of 2.4 metres (7.8 feet), but in reality the potential danger lay not so much in the height of the waves but their wavelength – that is, the period between the crests.

With wind and tide in the same direction, the N4 could handle the comparatively gentle slopes of the waves with ease, but as soon as the tide turned against the wind, the length of the waves shortened up dramatically and with it the gradient of the slope. Sometimes so short and steep, it was very much like negotiating a series of vertical walls. These were the sort of conditions that often resulted in smashed forward structure and broken windows and contributed greatly to the seasickness among passengers (and sometimes crew).

It is very difficult to gauge the exact height of a wave to know whether or not one was within the operating limit. There were definitely occasions when pilots were probably outside the limit, but there were other factors to take into account, which only experience could dictate. In very severe weather the usual process was to cancel the operation while the tide was against the wind. As soon as the tide turned the schedule would be resumed in weather that may well have been producing seas in excess of three metres, but the period between the crests and the troughs was sufficient for it not to be an issue.

To sum up, we at Hoverlloyd, after a few short years of experience, had become masters of our environment in all its manifestations. Another important factor was that we had learned to appreciate the 70 square miles of the Goodwin Sands as being 'home'. In any poor weather we knew once we were over the edge of the sands we were in an area of calm with no other vessel traffic to bother us. This was our territory and, after surviving the ordeal of the Dover Strait sea and swell, we were now safe and sound with nothing more to do but to avoid the wrecks and point towards Pegwell.

The Great Ship Swallower

Without doubt Hoverlloyd's situation was unique. After more than two thousand years of seafarers looking upon the Goodwins as a much feared source of death and destruction, we were the first and only mariners to look upon the dreaded sands as a home and a haven.

During that long period from pre-Roman times to the present day, conservative estimates suggest the sands are a graveyard to some 1,500 ships and who knows how many lost lives[1]. In times past the area was known to seafarers as 'The Great Ship Swallower', a name it richly deserves.

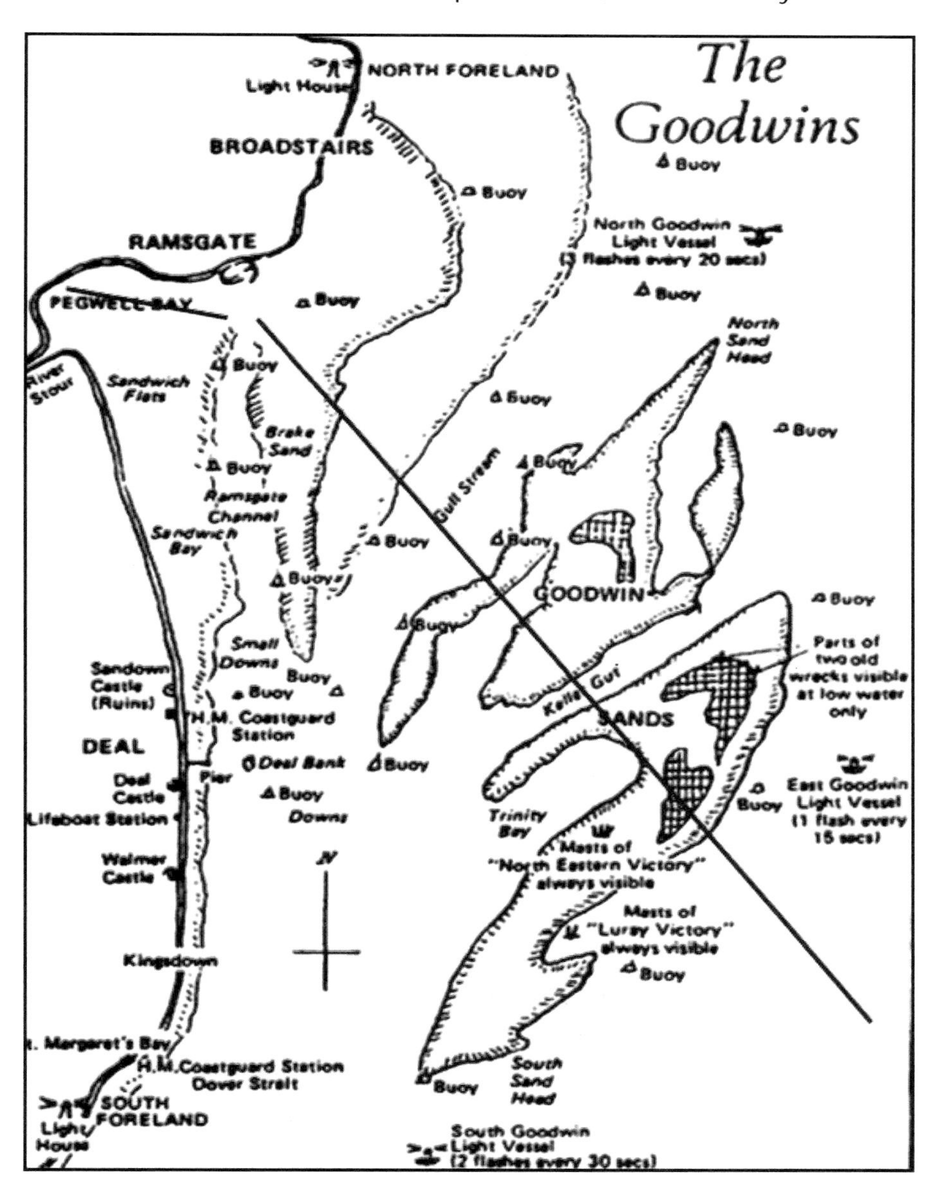

[1] R. Larn (1977), *Goodwin Sands Ship Wrecks*, David & Charles.

Although several scientific studies have been conducted over the years, primarily to try and understand the dynamic nature of the shoals that have constantly changed their shape and location, none of these investigations would have matched the day-to-day scrutiny of over a decade by Hoverlloyd's flight crews. At the same time it couldn't be said that we came to know the area like the back of our hands because the hand kept changing. What we did learn was how amazingly fluid the sands were, constantly ebbing and flowing mostly in a north-east, south-west direction in what was apparently an annual migration. The sand tended to pile up to the south in the summer months only to be pushed back up north, presumably by the autumn and winter gales.

Traversing the Goodwin Sands at low tide at 60 knots, (69 mph). The square black and yellow plate, seen under the strut, was attached by a stiffened flexible wire to the top of the bow door. This was the 'hi-tech' solution for letting the captain know when the bow door was closed.

Although the direction and timing of this movement appeared consistent, the extent was certainly not. Some years moved more sand than others, changing the topography considerably, sometimes in unexpected manners. The first extreme low tides of each summer would show the degree to which the sand had piled up each year.

As can be seen from the map on the previous page, the sands are in fact not one entity but two diamond-shaped sections pointing north east, separated by a narrow channel known as the Kellet Gut.

Ted Ruckert had an interesting experience:

> *Towards the end of that first summer of 1969 we were crossing the Goodwin Sands towards Ramsgate after dark at low tide on a clear evening, when the navigator said he had a target on the radar in the middle of the sands, which shouldn't be there. At the same time we saw a flashing light ahead of us from that position. Robin Paine was the 1st officer, but I can't remember who the navigator was. I slowed down and, as we got closer, we could make out the outline of a yacht aground on a falling tide in the middle of a big pool in the sands. In the distance were the bright lights of Ramsgate. I thought we had better stop and see if the occupants required any help. We pulled up alongside and I sent the navigator down to see what the situation was.*
>
> *Eventually he called us from the telephone by the main passenger door chuckling away. Apparently it was a French yacht en route from the Medway towns to Calais. When the navigator asked to see their chart they produced an Esso Road Map with a line drawn from North Foreland to Calais showing the whole area across the Goodwin Sands and the English Channel as all blue! With the weather worsening I suggested they dropped their anchor and came back to Pegwell with us, but they declined. We informed Deal coastguard and I believe they were later towed off.*

Where most sand tends to pile up is on the eastern side of the outer section, to seaward of the Kellet. On these particularly low tide days the hovercraft would be skimming over sand interspersed with shallow pools to the north of the Kellet, in other words sand that was only just barely covered by water. Once across the narrow Kellet channel the sand to the south and east was always well dried out and firm.

What was not noticeable however was that, as the craft accelerated, which it always did on a hard surface free from the drag of the water, the gradient of the sand bank was rising slowly but surely and our height above sea level was increasing. The big shock came when the outer edge was reached, an edge that fell away dramatically to reveal a drop of some two or three metres at low water springs. By this time the 200-ton machine might have increased to maybe 70 knots and with that sort of momentum must have been airborne for a second or so before hitting the Channel surf on the other side.

All very exciting but alarming nonetheless. However, we needn't have worried. Again the incredibly robust nature of the N4 structure took it in its stride. The skirt – this time really acting as a cushion – absorbed the fall with hardly a bump or tremor and carried on serenely. It's to be wondered if the passengers actually noticed.

On the outward journey, leaping into the void on the edge was one experience, meeting the wall of sand on the way back was quite another, as Robin remembers:

> *When a new trainee captain had amassed a certain number of hours, and achieved a basic level of competency under a training captain, the trainee was allowed to drive the machine from the left-hand (captain's) seat in certain circumstances under the supervision of the craft captain sitting in the right-hand seat. This had to be because it was not practical to manoeuvre the craft from the right-hand seat as a result of having to stretch across the throttle levers to reach the all-important raised propeller pitch levers, which were on the captain's right.*
>
> *I recall an incident when I allowed one such trainee on a calm day to drive from the left-hand seat. It was flat calm and we sped across the Channel from Calais to Pegwell at a very generous 60 knots. It was low water spring tides and, from some way off I could see what I thought was an enormous bank of sand on the edge of the Goodwins – let us say at least two metres high. Hitting that at 60 knots was not going to be a good idea, so I suggested to the trainee that he started to take some pitch off the props to reduce speed and at the same time give us maximum hover height.*
>
> *Nothing happened, so I repeated the request.*
>
> *Still nothing happened and as I looked at the trainee I saw he had frozen with fright in his seat. I was now quite cross and said, "For goodness sake, get that bloody pitch off", at the same time as my hands went across to the pitch levers to do the job myself.*
>
> *Unfortunately, instead of hitting the pitch levers, as intended, I closed the two forward engine throttle levers, which meant that the bow started to drop, and even more quickly as a result of the pitch levers still being forward. The bow drop started just before we hit the bank, but the incredible thing was that with the drag of the skirt, by then on the sand, we simply came to an undignified stop in the middle of nowhere with no damage done.*

The Wrecks

There were, and still are, only two wrecks on the Goodwins, which are always visible regardless of the height of the tide. Both vessels, the *North Eastern Victory* and *Luray Victory,* which went aground in 1946 within months of each other, lie close south of the Hoverlloyd track. The

North Eastern Victory is the most visible showing three large masts with, back then, cargo gear still attached. The *Luray Victory* showed nothing more than a thin spike of a topmast, a very poor target on the radar in bad weather. Records show there to be three other large vessels, all from the same period, within close range of these prominent two, and three or four more ranging northward along the edge of the sand towards the East Goodwin Light Vessel.

The *North Eastern Victory*, as she was in 1969 (*Provenance unknown*)

Today the topmasts and cargo gear have been swept away, leaving only the stumps of the main masts. (*Dover Lifeboat J. Miell 2008*)

The brief reappearance in 1973 of *U-48* sunk by British patrol vessels in 1917. (*Peter Powell, Broadstairs.*)

Where are they now? For that matter, where is the remaining evidence of the 1,500 vessels that litter the sands as a whole?

The fact is the two Victory ships, which appear to be sitting on an underlying outcrop of chalk, are the exceptions that prove the rule. The 'Great Ship Swallower' does, in effect, do just that. Most wrecks sink into oblivion within a matter of months of going aground, not necessarily however, never to be seen again.

Every now and then a small chunk of metal will appear, either because of a larger than usual shift of sand to the south-west, or, another possibility, the moving quicksand rolls the long dead vessels around in their graves, so that a part of the structure may pierce the surface occasionally.

Whatever the reason for their existence, any sightings of new pieces of metal were identified as to their size and position, and the rest of the hover fleet alerted. Normally this was not of particular concern as these new appearances were never of a size that could not be cleared easily by the skirt height. Nonetheless, sharp bits of corroded metal could still do damage to the skirt fingers adding unnecessary expense to our maintenance costs. It was prudent to keep clear.

Roger claims bragging rights for being the first to report the slow reappearance of the German submarine *U-48* in 1972:

In the summer of 1972, I reported sighting a piece of cylindrical shaped metal on the northern edge of the sands, about a mile north of our normal inbound track. That's all there was at the time, just a small kitchen bin sized object sticking out of the sand. A year later it revealed itself to be the remains of the periscope of the U-Boat, U-48, *sunk by British patrol vessels in 1917. This time the sand had scoured away around the vessel to reveal not just the periscope but the complete hull and conning tower. A ghostly return from the grave, which lasted most of that year, only to vanish once more into the depths, never to be seen again. It's more than likely it will reappear at some future date, but, without the hovercraft, who will be there to notice?*

Wrecks East of Calais

From a point half a mile east of the eastern edge of Calais Hoverport, as shown in the sketch map on the following page, the beach towards Dunkirk was strewn with World War Two debris, not just shipwrecks but all the remains of the battlefield: tank traps, other built-in obstructions and even more dangerous, areas of dumped ammunition. Although not directly involved in the Dunkirk evacuation in 1940, the location would still have seen enough action in the German capture of Calais at the time, and later, with its subsequent recapture after the D-Day landings in 1944.

This sketch map was compiled by Geoff Riches and Ken Mair in 1980. According to Ken, just walking around the place was obviously hazardous:

I seem to remember that we and others wanted to see if we could run the craft towards the east on the sand for a couple of miles when we had an easterly blow. I went to Calais and I remember driving with one of the Calais guys for about two or three miles towards Dunkerque. As I recall, my guide picked up several small bombs and rubbish, and I was quite relieved when we finished. There was a lot of stuff just lying on the sand and I imagine there would be much more underneath.

Why the map was sketched so late in Hoverlloyd's history is not clear. Prior to that, we all were vaguely aware that it was not a good place for a hovercraft to be, particularly at low tide, and therefore kept well clear.

Sometimes out of desperation, to claw up wind in a bad north-easterly, some braver than others – or more foolish – had a go and learned a thing or two. Roger remembers his own salutary lesson:

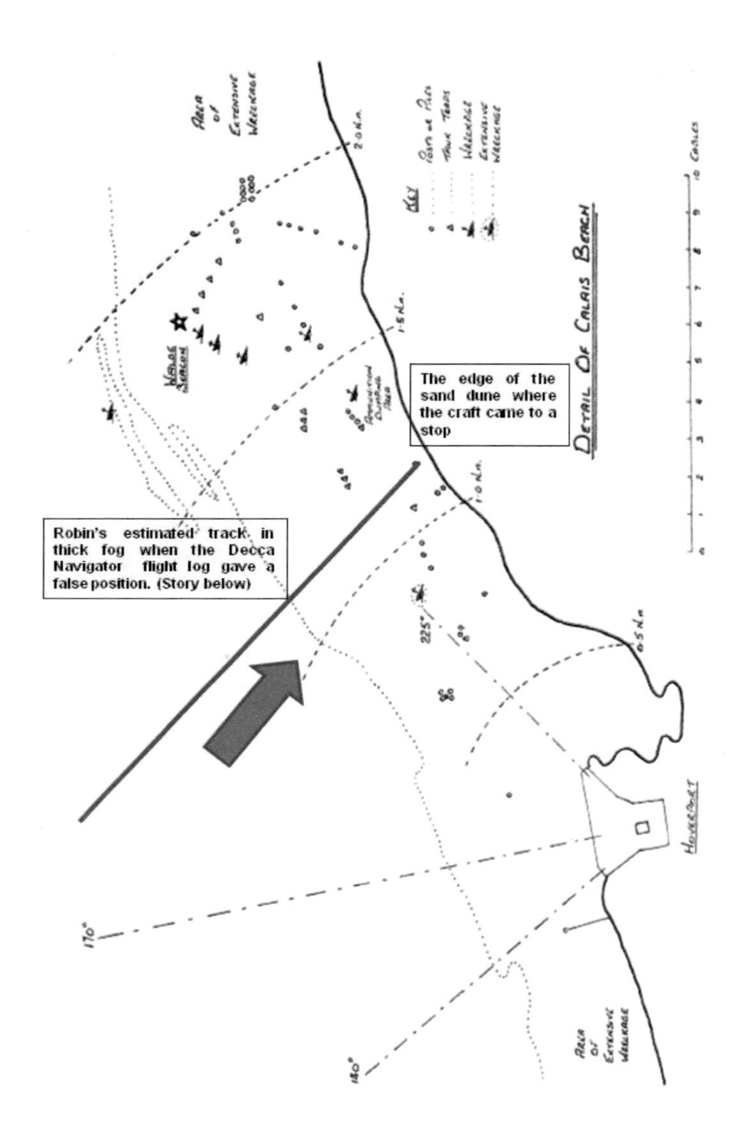

My own frightener was leaving at night trying to go up the beach. My intention was to head up to the Walde Beacon then turn off for home with the strong north-east wind behind me a little.

It was overcast, no moon and very dark, so the only visibility was courtesy of our bow searchlights, which only showed a limited arc ahead. I had no idea what I was going to encounter.

I think I had some mistaken idea that most of the problems were on the far side of the beacon. That is until the first large piece of metal flashed past on my seaward side and I was starting to get an inkling that I definitely shouldn't be there. Problem was, when I tried to turn out to sea, I was faced with other bits of debris.

I eventually found a hole that I got through without mishap and headed out to sea with one very big sigh of relief.

Looking at Geoff and Ken's chart now I am fairly sure I passed south of the three poles and the wreck shown at the end of the 225° bearing line and probably continued south of the three other poles and through the gap between them and the tank trap.

It was a lesson learnt and I never ventured into that area again.

Probably without exception most captains of that period can all relate similar 'hairy' experiences. One redeeming feature of the N4 was the solid 13,500 shaft horse power of the Proteus engines, which could get you out of trouble almost as quickly as you got into it. Robin has cause to be particularly thankful:

In common with all accidents, or near accidents, as mercifully this one was, its cause was a combination of factors rather than just one.

The major culprit was a malfunction in our primary navigation instrument, the Decca Navigator, but although a serious failure in itself, it would not have necessarily culminated in near disaster if it hadn't been supported by three other elements. A crew that was working together for the first time, and therefore not as yet fully functional as a team; a language problem between myself as captain and a newly trained navigator, a young Hong Kong Chinese, whose English was perfectly fluent but strongly accented, all of which could have been dealt with quite happily if it hadn't also occurred in zero visibility at 6 o'clock in the morning.

We have explained before that the relationship between captain and navigator on the SR.N4 was essentially one of complete trust, a trust that could only be built up over a period of time. In Y.K. Lung's case I hadn't had a chance to work with him prior to this day and in fact had not even met him except on the Safety Training Course, of which I was in charge. The idea of venturing off in thick fog with such an unknown quantity was not something to relish. So, in order to make life easier for him and make me feel more comfortable, I told YK that I would follow our track line on the Decca Navigator, which would leave him to concentrate wholly on anti-collision.

The Decca Navigator decometers with the flight log below them can be seen at the top of the captain's windscreen. In the diagram below the crossing numbered lines can just be made out. The pointer on the arm moves across the chart while the chart rolls up or down in response to numbers on the decometers, which in turn should show the craft's accurate position.

We eventually launched forth into the gloom at high speed with my eyes constantly moving between the Decca Navigator track and the gyro compass in front of me, with the occasional cursory glance at the other instrumentation. My ears were listening to YK's patter, which I initially found rather difficult to understand.

Luckily there turned out be a minimal amount of ship traffic through the Strait that day and there was little for us to worry about, that is until we neared the French coast, which according to the Decca Navigator indicated we were passing the RCN buoy a few miles off the beach.

I noticed that YK's voice was becoming rather agitated, but I hadn't got a clue what he was talking about. All of a sudden I saw my 1st officer, Martin Childs, trying to climb out of his seat backwards with a look of horror on his face. I was about to ask him what the hell he thought he was doing, when I looked down over the bow of the hovercraft through the gloom to see that we had made the transition from land to water at 60 knots at low water spring tide, when according to the Decca Navigator we had just over three miles to go.

I slammed all four pitch levers into full reverse and ordered full power. One of the early lessons in driving the hovercraft was that the props would give a lot less thrust in reverse. I had no clue where we were except that I suspected we were well to the east of where we should have been. The visibility was no more than 50 yards and I was terrified that we might encounter the World War Two wreckage that littered that area.

The speed began to drop and we eventually came to a halt in the hover just before the large sand dunes, too steep and high for the craft to clear. From memory we were about a mile to the east. We gingerly made our way along the sand to the hoverport and once on the pad and the engines had been shut down we had a 'discussion' as to what had gone wrong.

Obviously the Decca Navigator, still showing our position as at least a mile off track, had malfunctioned. It had been perfectly accurate when leaving Pegwell but had somehow 'slipped a lane'[2] on our way across.

---

[2] On a ship there is time to regularly check that the 'slave decometers', and on an aircraft the Decca flight log too, are showing the correct numbers against a 'master' dial. This is not the case on the hovercraft where the captain's workload is such that it is not possible to do so other than before leaving the hoverport.

When I asked YK why he hadn't said anything, he said he had. He said, "I tell you Cap'n, Calais to the light, but a rong rong way to the light!" My Chinese obviously wasn't up to scratch.

As the season progressed he turned out to be an extremely competent navigator and we had a very happy time[3] together. It was just bad luck that we should have encountered these difficult conditions on our first trip.

The Bait Diggers and 'The Battle of the Bait'

There was no wreckage to bother us on the home end of the route. Instead we were presented with a very different problem, a set of obstacles, which varied in number and distribution on a day-to-day basis. On any given day, at low tide, there could be anything from half a dozen to fifty or more people, hard at work digging into the flat mud of Pegwell Bay, all in pursuit of the lugworm.

At Low Water Springs, the tide would go out from the bottom of the ramp for as much as 1¼ miles towards Ramsgate Harbour. Pegwell Bay was a popular spot for bait diggers to dig for the lugworm, as witnessed at the two public inquiries, which they would either use for bait themselves or sell to the local fishing tackle shops for a tidy little sum.

From the beginning it was clear that there was going to be a conflict between the hovercraft and the bait diggers. It was also clear that some form of boundary extending from the north to the south of the hoverport and out towards the sea was necessary to define a 'flight path' in which the hovercraft was expected to operate. Accordingly, Hoverlloyd positioned its own full-scale buoy 1½ miles to seaward from the hoverport, which marked the 'entrance' to the flight path. On either side to the north and south, at a generous distance apart, smaller inflatable marker buoys marked the boundaries of hovercraft operations all the way to the Ramsgate and Sandwich ramps.

Unlike an airport, where it is a criminal offence to be on a runway unless authorised, Hoverlloyd was confronted with the issue of foreshore rights. The claim that the Crown is the owner of the foreshore was confirmed by the House of Lords in 1891. The ruling said:

---

[3] A nice little footnote to this story is that on the last day of the season YK came into the Radio Room smoking a large cigar, almost as big as him, and offered Robin one too. Robin enquired as to the occasion for this lavish gesture, to which YK replied, "My wife have baby". "Congratulations," says Robin. "Ah, yes," said YK, "and I call him Lobin!" Sadly, he has never seen YK since, nor met up with his sort of namesake.

It is beyond dispute that the Crown is prima facie entitled to every part of the foreshore between high and low-water mark, and that a subject can only establish a title to any part of that foreshore, either by proving an express grant thereof from the Crown, or by giving evidence from which such a grant, though not capable of being produced, will be presumed.

The approach to Pegwell Bay Hoverport at low tide, as viewed from the flight deck, on this occasion without bait diggers.

There was added confusion when the Ramsgate Town Clerk said that, contrary to what the bait diggers thought, Pegwell Bay was not common land. It belonged to the Corporation and any interference with the navigation of the hovercraft could result in a High Court injunction. He said:

We have allowed the bait diggers to use the land for many years but we will not tolerate any unreasonable behaviour.

The flight path did take up a large area of what was previously the sole preserve of the bait diggers and there was resentment that they were now going to have to share this area with the hovercraft. On occasions, particularly at weekends, there were numerous diggers scattered all over the area, which from the flight deck looked like being confronted by a rather tricky obstacle course. Apparently one of the attractions of digging in the flight path was that the vibration of the machine brought the worms nearer the surface.

Two of the red buoys marking the flight path can be seen behind *Swift* as she leaves Pegwell Bay Hoverport at high water.

Les Colquhoun decided that there was a better way of dealing with the situation other than confrontation, possibly leading to expensive legal battles, over an issue involving foreshore rights. Using his considerable people skills, he had a meeting with representatives of the bait diggers to try to find a solution, which he duly did shortly before the wife of the then Prime Minister, Mary Wilson, named *Swift* at the ceremony on 23rd January 1969. It was agreed that the bait diggers could dig in the flight path, but would not intentionally obstruct the hovercraft. Furthermore, Hoverlloyd would investigate ways of warning bait diggers of the impending arrival of a hovercraft and issue them with luminous jackets. Nothing in fact happened with investigating a warning system or issuing luminous jackets, but an uneasy truce ensued, which didn't resolve anything except to temporarily take the steam out of the situation and allow the bait diggers to continue to dig in the flight path. This less than satisfactory arrangement led to the inevitable incidents later on.

At the time, however, the local press hailed the result of the negotiations as 'It's Peace on Pegwell Mud Flats', and published a wonderful picture of a Mrs. Beer, who was sent from the launching ceremony of *Swift* by Les Colquhoun with champagne and glasses for the bait diggers, to celebrate the new-found peace.

When there was a head wind the hovercraft was much easier to control than with a tail wind making it less difficult to manoeuvre around them.

Making peace with the bait diggers. Waitress Mrs. Nell Beer serving champagne to three bait diggers on Pegwell Bay mud flats during the inauguration ceremony of *Swift* on 23rd January 1969. *(East Kent Times.)*

The idea was to line up on the ramp some way out, or in the case of the Sandwich ramp, a point just to the south of it. Depending on the wind direction, the drift angle could be 30° or more, which in simple terms means the hovercraft is heading in one direction, but in order to do so must point 30° in another direction. With bait diggers in the line of the approach this was not always possible. Some bait diggers were considerate and moved out of the way to let the hovercraft pass while others stood their ground. David Ward recalls two who came off second best:

> *I seem to remember arriving for work as duty captain early one morning and being confronted by two very bedraggled looking characters sitting in the entrance to the hoverport. I thought they were a couple of vagrants and asked them what they were doing. They answered that they had just had a close shave with one of the hovercraft and it was pretty obvious that it had been a very close shave.*
>
> *In the course of our conversation they told me that they were sure one of our captains had it in for them and tried to go straight for them. I assured them that none of our chaps would do that.*

At that point I knew exactly who the culprit was, but I omitted to enlighten them about the points system he had in force.

I do believe that on another occasion one of the bait diggers threw a bucket of water at someone in the Radio Room. Luckily it was not me this time and I managed to convince them to go home after I bought them a cup of coffee.

We cannot verify David's mention of the bucket of water in the radio room but there were eyewitnesses to Monty Banks, the senior engineer, getting a bucketful over him one day.

According to engineer Bernie Chilvers, who was with Peter James:

At about 8.30 on a cold autumn morning with dense fog covering the bay, the craft, on its return run from Calais, could be heard approaching but not seen. We were standing on the pad awaiting its arrival. Suddenly, out of the thick murk a vague apparition morphed into the familiar shape of the hovercraft, which came up the pad and settled in position. We were about to move the craft steps into position to allow the passengers to alight when we observed a male figure also appearing out of the fog walking up the pad, carrying a bucket. As he came closer we realised that he was soaking wet and covered in sand and mud.

He reached our position and demanded to know who was in charge of the operation and, as Monty Banks happened to be nearby, he was pointed out as the man. The bait digger then lifted the bucket and threw the contents over Monty, soaking him in water and lugworms.

He had been out bait digging with a friend when the craft had arrived, but, due to the density of the fog, the captain had not seen him and had driven right over him and his friend.

Naturally being upset he had proceeded to tell Monty what he thought of it all in a forceful way. The situation could have got really nasty but for Monty bursting out laughing at his predicament and was more concerned that the man and his friend were alright.

No doubt the bait digger was quite naturally upset!

This definitely happened and many others have verified it, although it is highly unlikely the two bait-diggers were actually run over. If either had been hit they would have sustained injuries, or at least a severe bruising.

Although the skirt was rubber, it was a very hard material and at full hover the air pressure would have given it the rigidity of a car tyre. Hitting someone at the approach speed of 25mph would have had dire results.

An example of the damaging force the N4 skirt could exert, even at slow speed, was amply illustrated when a small fibreglass boat was hit just at the foot of the Ramsgate ramp. At that point the hovercraft would have been doing a little more than 15 knots (17mph). The state of the boat after it had been hit was amazing. Much of the GRP had delaminated and it looked as though it had been picked up and screwed around like a wet piece of washing.

This is the only occasion where a small boat was struck in this manner and the circumstances are worth relating. Bob Adams was captain at the time and Dennis Ford was the navigator:

> *My memory may be playing tricks on me, but as far as I recall we were happily plodding towards the Ramsgate ramp, with this boat on the port bow, when one target suddenly (and very close) became two. I don't recall whether Bob could see by that stage, or he rightly surmised that a second target, which we were on course to hit, could be the occupants taking fright, but either way I think we actually drifted left to avoid the people and flattened their boat instead.*
>
> *Good call on Bob's part, but they might have preferred it if we had got them instead of their boat. Later, as captain, I ran over a dog that emerged looking somewhat bemused, but otherwise none the worse for the experience.*

Sight-seers and Sights to See

Watching the hovercraft arrive and depart from Pegwell became one of the major tourist attractions in East Kent. In the early days people would watch from the car park by the replica Viking Ship above the hoverport, or try to glimpse round the fence by the departure car park. Eventually a viewing platform was erected on the roof of the hoverport and people were charged a 50p entrance fee. It proved a huge success and subsequently a bridge was built from the Viking Green over the hoverport car park area to enable visitors to access the viewing platform more easily.

According to David Ward there were other sights to see too:

> *One very fine and sunny day I was making an approach to the Ramsgate ramp and noticed a couple quite close to the approach and close to the ramp. I was wondering who on*

earth had given them permission to stand there when just as we got near them the girl suddenly whipped off her coat and the chap proceeded to take lots of photos of this gorgeous girl in her 'altogether' – gave me quite a turn and it resulted in my approach not being the best I ever made!

Calais Hoverport had a similar public viewing platform, which proved popular too.

Surprise Visitors and Visitors Surprised

Bait diggers and holiday makers were not the only visitors to Pegwell Bay. One of the objections at the public inquiry to the hoverport being built in Pegwell Bay was that it would be detrimental to the wildlife in the sanctuary to the south of the hoverport. The exact opposite was the case and more birds than ever descended on the area. One summer a pair of flamingos arrived and stayed for a couple of weeks, much to the delight of everyone[4].

---

[4] Sad to say although the pair, thought to have escaped from a sanctuary in Holland, made a temporary nest very close to the Sandwich ramp no photographs have been traced. In this day and age with cameras on mobile phones there would have been hundreds to choose from.

On odd occasions seagulls, flying in close proximity to the hovercraft, got caught in the air stream of the propellers, which resulted in a messy end to their flying careers. It is likely that Darwin would have been happy with our conclusion – excuse? – that only the slow and less agile were the principal casualties, thus improving the species.

As with anything new, the challenge of conquering the obstacles with which one is confronted is not only exciting, but, when solutions are found, provides a great sense of achievement. The SR.N4 provided just that, but it took some six years to reach a point where the level of experience and expertise among the flight crews was such that they could say with some confidence that most flights were largely routine. The one exception, however, was fog, and in particular the approach to Pegwell Bay at low water.

Chapter 21

SR.N4 Hoverlloyd – The First Years

An Aircraft at Sea

Swift and *Sure* in the Mk.1 configuration at sea between Ramsgate and Calais. 1969.

It is indicative of what a new departure for the makers it was, that the build notification SR.N stands for Saunders-Roe 'Nautical'. The Isle of Wight-based company, now reborn as the British Hovercraft Corporation, having spent the previous 50 years building aircraft and boats, obviously felt the need to emphasise that this particular vehicle was destined for the sea.

Although the company had made its name as a builder of flying boats as early as 1913, the accent and intention lay in the first word, 'flying', not the second. They landed and took off on water but spent the rest of their journey a few thousand feet in the air, safely out of harm's way. Their take-off run had never before stretched as far as the 31 nautical miles from Ramsgate to Calais.

Despite BHC's long experience with aircraft on the water, this was an entirely new departure and the reality of an aircraft structure in the severe weather conditions of the Dover Strait were yet to be recorded.

We were about to find out.

It has to be confessed that we viewed the whole idea with a certain degree of scepticism. To us seafaring types, used to vessels built of mild steel, whose top strakes of plating were measured in inches of thickness, the prospect of taking a painfully thin structure, not a whole lot different to a collection of aluminium beer cans, out into a Force 8 gale in the English Channel seemed a somewhat risky business, to say the least.

Damage like this could be repaired in a few hours.

What added to the uncertainty was the very brief period of prototype development prior to the introduction of the passenger service proper[1]. Nonetheless, as already mentioned, lack of adequate finance on the part of the manufacturer and Hoverlloyd's blow-out in infrastructure costs, demanded an operational start in the summer of 1969 to commence a revenue flow, regardless of the development status of the machine.

[1] Although the very first passenger service was operated by Seaspeed from Dover to Boulogne in 1968, the brief period of operation before system failures forced 001 back to Cowes for modification, can only, in retrospect, be considered as an integral part of prototype development. When both cross-Channel companies operated a full summer season in 1969, could this be construed as the start of passenger services.

Hoverlloyd's first year was very much a question of learning to trust such innovative and untried technology. However, as the experience of one Channel gale after another increased our appreciation of what this amazingly new and very different form of transport was capable of, so grew in turn our expertise and confidence.

By the end of the second year of operation we were no longer sceptics but enthusiastic converts. What we had previously considered absurdly delicate proved in operation to be amazingly robust. And it has to be emphasised this was not a result of timidity on our part, or of any canny expertise in dealing with the vagaries of the weather – that came later. Sheer ignorance and a strong desire to prove ourselves and our company impelled us to grit our teeth and drive the machines to their permitted operational limits, and many times, unwittingly beyond. It has to be admitted that this often resulted in severe damage to the structure and the skirt, but finding out what we could and couldn't do was all part of the initial learning curve.

What impressed us and inspired the most confidence was how well the N4 could survive extensive damage, to skirt and structure, plus loss of engines and control systems and still continue to operate. Time and again craft limped into Pegwell, maybe on three engines, skirt segments in tatters and large holes in the structure. If a comparable loss of structure and systems had occurred to a conventional ferry the result would, at the very least, have been a couple of weeks in a dry dock.

Not so the intrepid SR.N4. Our engineers were immediately on the job as soon as we landed. Pop riveters began replacing sections of plating, the skirt gang replacing and repairing the rubber, engine and control people sorting out the other problems. Depending on the severity, the very worst that could have happened might have seen the stricken craft being put on the jacks for a couple of days, but more often than not the machine was ready for service again in a matter of hours.

Cutting edge of technology?

In 1969 we may have thought ourselves as pilots of the future, but to our youthful perceptions, already fired up by the moon landings, the reality of the SR.N4 hovercraft as a piece of modern technology fell a long way short of futuristic. It was probably very unfair, but from our perspective as operators, the concept may have been brand new and innovative, but our impression was that it had been put together with the technological equivalent of, 'What have we got at the back of the garage?'

Navigational instrumentation, such as the Sperry G4B compass and Decca Navigator emanated from the immediate post-war period. The

Sperry compass was originally developed for the 'V Bomber' programme and saw service in the Avro Vulcan in the mid-50s.

The instrumentation, for example, was by and large derived from the drawing boards of previous failed projects, such as the *Princess* flying boat on which the original design work started in 1945.

The engine chosen to provide the N4's power had a similar history; design work on the Proteus started in 1944 and was first tested in 1946 by attaching it to the bomb bay of an *Avro Lincoln*. There were many problems to be rectified, so it underwent a redesign and was again tested in 1947. The Mk. 600 Proteus could deliver 3,780bhp and, in addition to the *Princess* flying boats, it was designated to be used on the *Bristol Brabazon*, which was another failed project.

Later versions of the Proteus were used on the *Bristol Britannia* and marine versions were subsequently developed for use by the Royal Navy *Brave* and *Bold* class fast patrol boats and the SR.N4, BH7 and VT-2 hovercraft.

Such were the historic connections of the SR.N4, that when old World War Two veterans visited our flight deck, their remarks on the lines of, "It looks just like my old Lancaster I flew in the war!" weren't far off the mark.

It tended to deflate the image somewhat.

1969 – Mk.1, Initial problems

In 1969 *Swift* and *Sure*, numbers 002 and 003[2], as delivered to Hoverlloyd for the first summer season were, like Seaspeed's 001, Mk.1 craft. Undoubtedly, Hoverlloyd benefitted from the first operational experience with Seaspeed in 1968 and the modifications that ensued, but the two craft, delivered to us a year later, were still far from operationally reliable, as we were soon to find out.

The first and foremost concern was the skirt system which, unlike the structure, control systems and instrumentation, had no history of development and therefore no solid base of expertise to deal with the problems of maintenance. The skirt segments frequently tore, sometimes ripping into long tails of rubber, which in turn lashed at the body of the craft, tearing large holes in the flimsy structure. At times, the long tentacles of rubber reached up far enough to slam at the cabin windows, like a monster of the deep in some low-budget movie. It must have been a frightening experience for the passengers.

---

[2] These yard numbers should not be confused with the registration numbers, which for *Swift* and *Sure* respectively, were GH-2004 and GH-2005. When used as call signs, the crews used to abbreviate these to 'zero four' and 'zero five'.

Initially, faced with these frequent episodes of skirt failure, there was little our engineers could do but repair the damage and sit back and wait for the next occurrence.

Paradoxically, the most serious skirt damage occurred some two years into Hoverlloyd's operation when it could reasonably be expected that the maintenance teams would have accumulated some expertise, and the routine maintenance measures should have been more effective.

The Sangatte Incident

On 10th November 1971 *Swift*, with David Wise as captain, Ian Philip as 1st officer and Roger Syms as 2nd officer, suffered severe skirt damage at the start of its return trip from Calais, so severe that it was impossible to hold a course into a strong north-easterly wind and make a return to Calais. The decision had to be made to beach the craft some miles to the west of the hoverport at Sangatte[3].

Later inspection after beaching revealed a total loss of the skirt system from the bow and all the way down the starboard side and partially from the bow, back along the port side.

The immediate effects of the damage were a pronounced starboard list and a decrease in speed initially from 30 knots down to 10 knots. Any adjustment of course further to the east to head for the hoverport resulted in a complete drop off in speed, enough to bring the craft to a standstill.

A secondary problem was the amount of spray thrown over the craft from the escaping air no longer contained by the skirt system. It reduced visibility in the control cabin, but the most serious effect was the seepage from such a large volume of water, which found its way into the intercom, radio and public address systems. Very shortly after the initial incident the control cabin intercom failed, followed by both VHF radios, leaving the crew with no means of communication, both internal and external.

Luckily the Seaspeed N4, *The Princess Anne,* was in the immediate vicinity and had been asked, while VHF was still usable, to stand by and escort the damaged craft to safety. Once VHF contact was lost, communication was 'back to the '50s', and we were reduced to signalling by Morse code on the Aldis Lamp between the two machines. So those signal exams we took for our Certificates of Competency were of use after all.

The following description of the approach, landing and eventual evacuation of the 145 passengers is taken verbatim from Captain David Wise's report:

---

[3] Now the location of the French terminal for the Channel Tunnel.

Swift beached at Sangatte – most probably at the point of trying to move further towards the beach line. November 1971.

A couple of life rafts on the beach used to evacuate some passengers from the starboard side of *Swift*. The craft is still trying to move further up the beach with the incoming tide. November 1971.

At 1345, it was judged all lift had gone from the starboard side and round the bow section with little lift remaining at the rear. Speed was estimated at 5 knots, longitudinal trim was approx. level, with a substantial starboard list.

Slight slamming was occurring as presumably the underside of the forward buoyancy tanks were impacting on the wave crests.

The port rear end of the craft (weather side) was also being pounded by occasional waves, which tended to lift the port rear end and swing the craft nose round to port more into wind.

Considering the craft was now virtually a displacement vessel, steerage remained good, using 30° of positive pitch on all propellers to achieve the best possible speed. The craft rode the sea state well.

During this time Car Deck Crewman R. Wheeler, who was acting as messenger, made several visits below to the cabin staff to ascertain the passengers were in good shape, and that no fresh problems had developed in the passenger cabins or on the car deck.

The craft was off the beach at Sangatte (approx. 3.5 miles SW of Calais) at 1410. As the water shallowed to approximately 10 feet, forward speed virtually stopped. The conditions off the beach were estimated as:

Wind NNE x 30 knots with 4-foot short surf conditions. Slow progress was made towards the beach each time incoming waves lifted the craft and carried it forward.

A position between two rows of wooden stakes leading down to the waters edge about 100 yards apart was chosen as a suitable landing area.

A further attempt was then made to beach the craft, full power was used on all engines and the craft moved forward approximately 15-20 feet until the bow section went aground. The craft was then turned approximately 30° to the left to put the wind on the port beam; this gave a good lee on the starboard side.

The main engines were left running at idle with maximum forward propeller pitch to keep the craft in position.

The craft was approx 150 feet from the water's edge at the bow end and 200 feet at the stern end. The depth of water at the starboard main passenger door was estimated at 6 feet.

The Princess Anne, *who was now close on our port side, was called on the Aldis lamp and told that* Swift *was OK.* The Princess Anne *then departed for Calais Hoverport.*

Evacuation:

At approx 1415 the captain went below and talked to car deck and cabin staff. Although no immediate danger existed, the captain decided to evacuate the passengers from the craft. The reasons behind this decision were:

1. *The port rear side of the craft was taking quite a hammering from the incoming surf so structural damage was feared, especially if weather conditions deteriorated.*

2. *The tide was rising, high water was not expected until approx 1730 (H.W. Calais 1718) so it would probably be 1830 before the tide had receded sufficiently for passengers to walk off without getting wet – a wait of a further four hours for the passengers after a rather harrowing experience.*

3. *145 passengers off loaded would lighten the craft by approx 10 tons. This would assist in moving the craft further up the beach.*

4. *Added advantage of daylight, only 2½ hours of daylight remaining.*

5. *With the wind on the port side, a good lee was provided on the starboard side with reduced surfing effect. Passengers could be evacuated smoothly and unhurriedly by life raft with little risk of panic, injury or getting wet.*

The captain informed cabin and car deck personnel that an orderly evacuation would take place, using life rafts on the starboard side. The passengers on the port side were to remain seated until the starboard side passengers had been evacuated. As the public address system was inoperative, passengers were informed of the evacuation procedure by the stewardesses.

The evacuation went smoothly, although not as fast as the theory suggested. There always tends to be a large discrepancy between how emergency procedures are supposed to work, compared with how they work in the real situation, with real people. Our evacuation drills were predicated on a 15-minute evacuation for a full passenger load. In this case the whole operation to land, what was little more than a half load of passengers, took exactly one hour from start to finish.

The combination of a north-easterly wind and an easterly running flood tide was the principal difficulty. Passengers could only be taken off easily from the starboard lee-side, and the problem for the captain was keeping the craft in that position, against the tide. Engines had to be kept running to maintain the craft in position to create the lee, but it was soon discovered that in so doing the airflow pushed the life rafts away from the side and made them impossible to load.

The only solution was to run the engines for 5 or 10 minutes to turn the craft to port to make the lee, shut the engines down and load a life raft as quickly as possible before the flood tide swung the craft back through the wind and the lee was lost. This procedure was repeated three times and three life rafts with 79 passengers got away without incident.

At this point it was noticed that each time the engines were restarted the craft repositioned itself, as the rising tide had the effect of allowing the craft to ground nearer and nearer to the beach line. It was therefore decided to curtail the evacuation for the moment and use full power on all engines to force the machine as far up the beach as possible. This manoeuvre resulted in *Swift* grounding firmly within 20 feet of water's edge in approximately 1 foot depth at the bow.

The bow ramp was lowered and all the remaining passengers were able to paddle ashore, some of our chivalrous car deck men giving the ladies a piggy-back.

So ended an incident that could have been a lot worse. Not the best of publicity for the still comparatively new hovercraft, but not a complete public relations disaster for Hoverlloyd, if the subsequent reactions of the ship-wrecked passengers are anything to go by. When some hours later, after securing the craft on the beach, the tired and bedraggled crew eventually turned up at George V Hotel, the passengers staying there gave them a standing ovation.

Sangatte – the Recovery Operation

If the actual beaching of *Swift* at Sangatte was a drama, then the ten days taken to recover *Swift* from the beach and return it safely 'home' to Pegwell was equal to it. A saga all of its own, as Mike Fuller, the crew chief in charge, relates:

Swift at Sangatte – recovery operation well under way. November 1971.

The skirt on an SR.N4 is a complicated structure that hangs underneath in a half loop and runs under the whole craft perimeter, and weighs over 12 tons in total. The initial report back from Swift *was bad; they had lost nearly all of the skirt, apart from one rear section. So we had a real problem on our hands, and Les (Colquhoun) called in a favour by getting the RAF crash rescue team involved.*

The main problem was that the craft would have to be jacked up so that we could hang the new skirt onto the structure. I was to be sent over the next day on a Sealink ferry from Dover with every skirt engineer we employed, plus half a dozen engineers and a complete skirt. The RAF crew would be on the same ferry and I was told to liaise with the officer in charge. I met up with him on the ferry and he immediately tried to put me at ease saying that he knew all about the SR.N4 and produced the instructions from an 'Airfix' model kit!

We arrived at Sangatte in a convoy of vans carrying the skirt and tools to find Swift *was beached bow first on the high tide mark up against the sea wall. The RAF plan was to lift the craft using inflatable airbags pushed underneath the flat bottom of the craft. Immediately we saw this would not be possible, as the tides had settled* Swift *flat on the beach. Fortunately, a very capable Chief Tech. was in the RAF party and he made all of his men available to do whatever I requested.*

The local Hoverlloyd rep was a good English guy by the name of Terry Halfacre and he was a godsend. He organised a large amount of long-handled shovels and a JCB digger, as by now we had realised that we had a lot of digging to do. So over the next six days, working 12 hours a day we dug out sand, dragged in sections of skirt and hung it onto Swift. *The French JCB driver was magnificent; he could push his large bucket right under the outer structure and pull back the sand much quicker than we could dig.*

Every day Terry organised large quantities of baguettes filled with sliced meat and lots of hot soup, as the weather was very cold and windy – it was November after all. After a few days Terry asked me if the lads were enjoying the rolls, which everyone thought was the best beef they had eaten. He explained that beef was very expensive and that we had all been eating 'viande de cheval' (horse-meat) rolls. We decided to keep quiet and let everyone enjoy the daily feast. We stayed in a local hotel every night and after this conversation I became suspicious of the steak we had every night with a huge pile of French fries.

Finally we had got all the sections of skirt hung on but could not join the sections together as this did require the craft to be jacked up. So we put all our equipment and tools on the car deck. Dave Wise (returned from Ramsgate) and crew gently and slowly hovered Swift *back along the beach at low tide with the skirt blowing through all the open joints. We managed to get back to the Calais Hoverport to find a team from British Hovercraft Corporation, the craft manufacturers, waiting with a portable jacking system.*

That night, a sou'westerly gale roared up the Channel and had Swift *still been against the sea wall it would have been badly damaged. After four days the skirt was all buttoned up and on the 20th November we loaded everything back onto the car deck and returned safely to Pegwell Bay.*

Such events as the Sangatte beaching, while unwelcome at the time, nevertheless, as can be seen in Mike's story, served as accumulated experience and added to the learning process. In a comparatively short space of time, through trial and error, the in-house knowledge and expertise grew and in very few years Hoverlloyd could boast a skirt reliability record second to none, although it cannot be emphasised enough what a drain on resources, both in manpower and material, the skirt was to the bottom-line profit.

As a by-product of this growing fund of in-service knowledge and skill by Hoverlloyd's team of skirt technicians, led by the indomitable Don

Nicholls, suggestions for improvement were passed back to the manufacturer.

BHC's own expertise in the design and manufacture of skirt systems had also developed enormously during this period, particularly with new bonding techniques and a better understanding of the dynamics of the air flow. The latter led eventually to significant modifications to the skirt configuration, whose major benefit was a more positive balance of pressure towards the forward end, rendering plough-in virtually a thing of the past.

Swift finally departing the beach at Sangatte after a new skirt had been fitted, but the segments not joined. November 1971.

Mechanical Failures

In the list of early-year problems, although the skirt was the major concern, mechanical failures were not far behind. The two Rover gas turbine auxiliary power units (APUs), which provided AC/DC electrical power to the control systems and other craft services, were a constant source of breakdown, almost disastrously in one case.

Although the incident, which became known eventually as the 'Pegwell Waltz' was initiated by the failure of both APUs, the real cause of what could have been a major catastrophe, with most likely dozens of casualties, was a combination of inexperience, a poorly designed and labelled switch, and lack of departmental liaison.

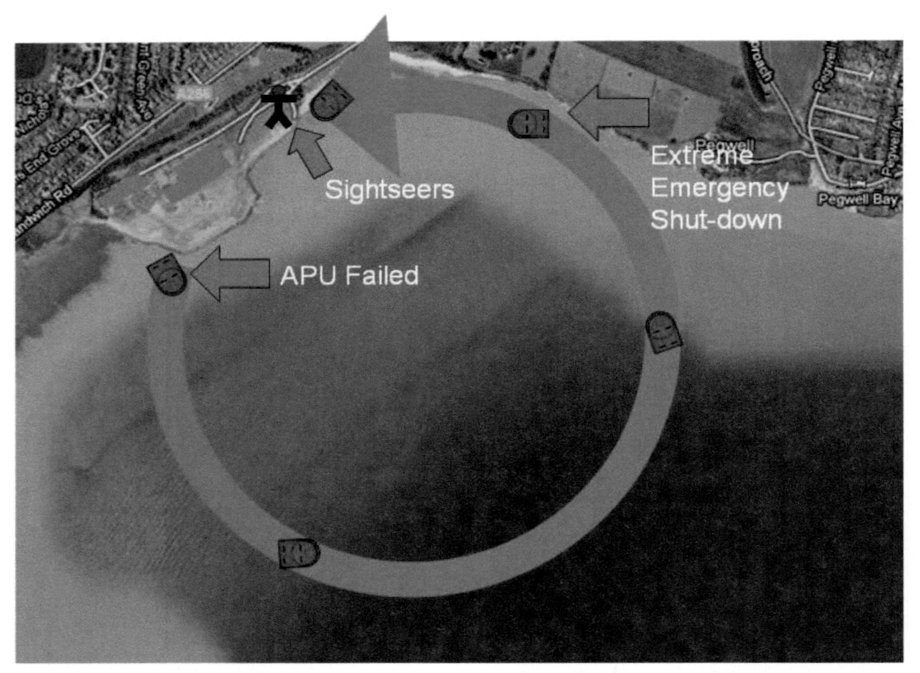

A schematic of the 'Pegwell Waltz' track. From the point of the APU failure all control of engines and craft direction was lost.

The event occurred during the first few months of the N4 operation in 1969 when no one in the flight crew had enough hours in their respective seats to be considered fully competent. It all boiled down to a lack of understanding as to the function of one small, insignificant switch and the lack of clarity in the way it was labelled.

There was also, at this early stage, a lack of formal checklist procedures on the flight deck; we were after all, still learning to be aircraft pilots. As Robin Paine who was 1st officer/flight engineer that day confirms:

> *Unlike an aircraft there was no formal checklist, although the 1st officers did check the basics, such as fuel pumps 'on', APUs running satisfactorily etc. Sometime later checklist cards were placed above each position, but even then there was no formal double check system. Although it was not particularly taxing to master the 'whichery of switchery', it is only with experience that one is able to deal quickly with a problem when it arose, and experience in all three seats on the flight deck was in very short supply in 1969.*

When the second APU failed the craft was left without power, which in turn left the craft completely without any control of the engines and even worse, any directional control of the hovercraft. It was impossible to shut the engines down and having no control of the pylon direction, which remained locked in a port turn, the craft described a perfect circle off the Sandwich ramp, continuing at speed around towards a point just north of the Ramsgate ramp, an area much frequented by crowds of sightseers and beyond that the departure car park. If the crew hadn't found a solution to the problem, in the very short time available, the out-of-control machine would have cut a swathe through the sightseers and continued on, smashing through the waiting cars and probably on into the departure lounge. The resulting carnage is unthinkable.

Sure departing down the Sandwich ramp, requiring a sharp turn to port at the bottom. It was about this point that the second APU failed leaving the captain without any means of control with the engines still sucking fuel.

Ted Ruckert, one the original N6 captains, tells the story:

> *I boarded* Swift *for the 0800 departure with Robin Paine as my 1st officer and Mike Rowland-Hill as 2nd officer sitting behind us manning the radars. We were running a bit late as the second APU wouldn't start. The weather was fine, but there was a hint of mist/fog and that was why I was dubious about leaving with only one APU on line. However, the craft was loaded and, as usual, the pressure was on 'to get the show on the road', so I*

decided to depart on one APU with an engineer on board to work on the other en route.

In the absence of any proper checklist, and all of us being relatively inexperienced at that stage, one only paid attention to the immediate necessities. In the training we were told that the DC switch on the overhead panel was always left on, and perhaps its overall importance was probably lost in the maze of all the other information that had been thrust upon us. It was not marked or highlighted in any special way – it was just another switch.

For departure I asked Robin to set the compressor settings to 9,500 revs on the front and 10,500 revs on the rear engines. It was low tide in Pegwell Bay when we launched ourselves down the Sandwich ramp at the bottom of which I put on hard left rudder. I then ordered 11,300 revs on all four engines. Robin just looked and smiled and tried to say something to me, but there was no sound through the head set, so I shouted, "I said eleven threes all round!" "Sorry!" said Robin, shouting back, and duly tried to increase the power, but nothing happened.

Mike Rowland-Hill then came from behind his radars to announce his radars had failed and he couldn't hear or say anything – not a promising start to the day if you are charged with navigating a hovercraft at 60 knots across the busiest shipping lanes in the world. I then said, "Put her down", but the throttles would not respond. We closed the HP cocks, the fuel pump switches and every other switch we could find, but still nothing happened because, although we had lost all power – AC and DC – the engines had the ability to suck fuel even when the fuel pumps were not working. The pylon controls and pitch levers were locked over in a port turn and would not respond either. We were now continuing to turn to the left and move in an arc out of control, which would land us in the departure car park. It is difficult to estimate the time from when the APU and subsequently all DC power failed, and when we finally came to rest, but it was probably no more than 60 seconds.

Luckily, during that very brief minute, BHC's excellent training came to the fore, and Robin remembered the 'emergency shut-down' switches. These switches worked off the emergency batteries, which were in total isolation to the rest of the control systems, and were there for no other purpose. If there ever was a time to use them, with the craft only a few hundred metres off the unsuspecting watchers on the beach, it was now.

The four 'Extreme Emergency Shut-Down' switches, also known as 'the Jesus Christ' switches, that may have saved the giant hovercraft industry from prematurely being written off.

Mike Rowland-Hill, the 2nd officer/navigator, without radar, radio or intercom was reduced to mere observer:

> *We lifted off, and proceeded down the Sandwich ramp, and executed a slow left turn to line up with the southern flight path.*
>
> *About halfway round this turn, my radars etc. failed, and I reported this to Ted. I immediately realised that he could not hear me, as power to the intercom had gone also, so I opened the curtains, and saw the back of the two heads in front of me in animated conversation. I recall Ted shouting to Robin, over the din of the propellers which were in reasonably fine pitch, and enquiring as to what the problem was. (Shades of Dick Dastardly, "Dooooooooo something, Muttley!" come to mind!)*
>
> *All A/C or D/C power (or so we thought) had failed, and so all controls were out of action. Only the main engines, which were 'self sustainable' and able to feed themselves, remained in action.*
>
> *Such was the efficiency of Ted's turn, the craft seemed to me, to be totally in balance, and was moving sweetly sideways,*

while describing a pleasing arc around the bay. Less pleasing was the fact that this arc would have meant we would have ended up in a crowded departure car park, had it continued.

It must have been only seconds, but felt rather longer, before at almost the same time, we remembered our engineering training, and all eyes, and most importantly, Robin's left hand, targeted a small console behind the throttles, proudly announcing 'Extreme Emergency Shut-Down'.

Somewhat shakily, but thankfully, Robin lifted and pushed the innocent looking switches. Slowly, all too slowly, the engines stopped, and air was removed from under the craft.

This decay, however, seemed far too slow, but with luck on our side we finally came to rest in a muddied heap, with the starboard side close to the car park fence.

It was a huge relief to be down, at which point Ted asked me to check the starboard side for any damage, as we had landed on the rock rubble on the edge of the car park.

Professional to the last, I obeyed. I descended the flight deck ladder to the car deck, crossed to the starboard main passenger cabin, opened the door, and stepped down. The side was covered with Pegwell mud, and I slipped. Fearing a swallow dive, face down in the mud, I chose a slightly less embarrassing entry, and jumped, ending ankle deep in the slime.

Of course, even from this most advantageous of positions, I could not see much apart from a muddy skirt, so I collected my thoughts, and, with two shoes full of mud, I returned to the flight deck to report.

I had to run the gauntlet of several passengers 'enquiring' what was going on, but I recall the cabin crew protecting me from most of it.

'It's an ill wind...' of course, because Bill Williamson, the operations manager, told me to go home and get changed, so I missed the initial inquiry as to what happened.

When everyone had composed themselves after the shock from the incident and an engineer came on board, it was soon clear what had happened. There had certainly been an AC power failure, but this had not resulted in a DC power failure. It was simply that the DC switch on the overhead panel was in the 'off' position. Neither the main batteries, nor the reserve batteries, which served to handle all the DC services, principally the control systems and communications, were connected.

The three position DC switch (circled) with the 'Off' position not marked, located far right above the flight engineer's seat.

So why was it left in the off position for the first flight of the day when it should have been permanently on?

What wasn't understood by the flight crew was that the engineers' usual procedure at night was to connect hovercraft to an external Ground Power Unit (GPU). However, because this turned out to be another unreliable piece of machinery, the engineers had decided to turn the DC switch off on the craft. If the batteries remained connected, when a GPU failure occurred, the craft switched to the battery power and drained them.

Part of the chain of errors in this incident was that the engineers had failed to inform the Operations Department.

With the limited experience of the crews, inadequate procedures and the breakdown in communication on this occasion between the Engineering and Operations Departments, it was only a question of time before this was going to happen. There is always a feeling of guilt that something so obvious wasn't spotted, but like all incidents in the world of aviation, shipping and hovercraft, it is usually a combination of more than one factor that is the cause.

The propensity for error was further aggravated by the fact that the DC switch had three positions: 'Main Battery', 'Off' and 'Reserve Battery', but only the battery positions were labelled. In fact the circuit in BHC's Operating Manual also failed to identify the centre position as 'Off'. Up until this debacle it is highly likely most of the flight crew assumed the switch to be a two position one, not three.

In this case the solution was to highlight the DC switch with a bright orange 'day glo' ring around it and inform all the flight crews that this was an item that should be double checked, and after this very near disaster no one ever forgot.

Aero Systems in a Maritime Environment

An overall problem, and probably something that could not have been envisaged by the manufacturer, was the filthy environment that Hoverlloyd, particularly, was operating in. A vast expanse of mud flat at low water on the approach to Pegwell, not to mention the miles of often exposed sand over the Goodwins, presented an environment for which aircraft systems were simply not designed.

A good example of the spray being thrown up from under the hovercraft. Not only were there problems with the salt water, mud and sand getting through and clogging up the filters, but the visibility for passengers was not particularly good.

Apart from looking embarrassingly out of date, all this transferred aero technology had more intrinsic problems. It was designed to operate several thousand feet up, where the air is comparatively clear and clean, not at sea level amid the salt, mud, seaweed and dead seagulls. This played particular havoc with the hydraulic systems in the propeller gearboxes. It wasn't so much a matter of initially causing a problem, so much as once a problem occurred, trying to rectify it in such a dirty atmosphere invariably perpetuated it.

If a problem eventuated with the prop pitch control, the mere opening up of a system in that environment, a system that relied on two micron filters for smooth operation, inevitably meant more microscopic particles entering and causing further problems down the line. The usual pattern, once a prop pitch snag had occurred, was almost a guarantee to produce a series of repetitions over several days before the works were sufficiently cleaned up and sealed again.

Despite not having the same levels of mud and sand that Hoverlloyd had to cope with, John Syring says Seaspeed had much the same problem. The somewhat uninformed advice given to the flight crew was typical of the time:

> *That first season we had lots of hydraulic failures. We used to carry a spanner so that when we were on the pad in Boulogne when a pylon was stuck over and the propeller pitch was full astern, there was this silly idea whereby we could whip out a ceramic filter with our spanner, wipe it on our clean hankie to get the sand off, put it back and it was supposed to be alright, which it wasn't for any length of time. We didn't always carry an engineer with us, but it wasn't until we got a hydraulics flushing rig in a Land Rover that the hydraulic problems got better.*

The Newhaven Mine

It was inevitable that Hoverlloyd's engineers, dealing as they were with the everyday problems of the dirty environment, would learn quickly the importance of cleanliness and develop strategies to cope with it. In contrast, whenever craft were returned to Cowes, for whatever reason, the operational inexperience of the BHC staff was very evident.

Bill Williamson, Hoverlloyd's operations manager at the time, confirms the problem in this tale of the Newhaven mine:

> *We had to go to Cowes to bring back the craft after a 'refit'. As on other occasions when a craft left Cowes it was nothing but trouble, usually hydraulic.*
>
> *On this occasion, after leaving Cowes, things started to go wrong and we gradually lost systems and engines. By the time we arrived off Newhaven we were down to two engines and it was obvious we were not going to reach Ramsgate.*

Swift at Newhaven – no sign of the mine!

There was only going to be one approach to this arrival and the craft staggered as far up the beach as we could get it. The back half of the craft was still in the water. The BHC staff commenced repairs and, if I remember correctly, had to get spares from Cowes.

During the day repairs continued and gale warnings were issued for Dover Straits and the Channel. Then a bloody mine appeared in the surf, right astern of our position. It was black, shiny with holes in it, but no prongs.

There was a local policeman on duty at the location and through him we tried to contact Bomb Disposal for advice. There was nothing doing and it was either take off over the mine, or wait for the gale to severely damage the craft[4]*.*

The policeman advised me that the mine had probably been rolling about on the beach since World War Two and the chances of it being active were about nil.

---

4 To add some historical perspective to the incident, the hovercraft, at this location, was probably well out of VHF range of both Cowes and Ramsgate. In those days, long before the advent of mobile phones, all communication that day would have necessitated trudging up the beach to find a local telephone box.

We decided to go when repairs were complete and shortly afterwards departed Newhaven for Ramsgate, fortunately with no further problems.

Not an enviable decision for any captain to be faced with, and the betting is, despite his reassurances, the local bobby stayed well back when the craft departed.

The Proteus Gas Turbine

The only pieces of machinery that worked well in this environment were the four main engines, the Rolls Royce Proteus gas turbines. By the time they were installed in the N4 this particular engine already had a history as a power plant in marine craft. By 1969 it was considered to be reliably 'marinised'.

Nevertheless, not surprisingly, the extreme atmosphere and the volume of salt water that the cushion of air threw up presented a greater challenge to engine maintenance and life than had previously been experienced on other vessels. This required a new development phase of its own to bring it up to scratch.

Even with a very complex filtration system to the intake air, salt accumulation on the turbine blades was a continuing headache. Throughout the N4's life, keeping the engines clean and free of salt was a continuing chore for the engineers, which never let up.

The remedies tended to follow the established pattern of N4 life, primitive but effective.

Mike Fuller, crew chief:

Each night the required scheduled check items would be carried out, along with washing the structures and the engines. The Rolls Royce Proteus turbine engine did not take too kindly to ingesting large quantities of salt water, so a method of washing the salt from the hundreds of turbine blades inside of the engine was devised.

The usual routine at the end of the day (and if the JPTs[5] were all within limits), was to wait for an hour or more and then squirt water directly through the engine intake mesh (not the 'Knitmesh' filters in the caboose) during a dry cycle. All four motors would be done and this would soften the salt on the

---

[5] JPT (Jet Pipe Temperature) was expected to remain at 450°C at normal cruise operation. Prominently placed on the front engine parameter panel, it was the flight engineer's principal indicator of engine health.

compressor blades; technically it was called a 'Compressor Wash'. After another hour the engines would be dry cycled to blow out the residue.

When flight crew reported engine temps creeping up we did some extra washing, which was to wait an hour, dry cycle the motors and pour about a gallon of kerosene through the engine that was running hot. Wait another hour, dry cycle the motor again and spray water through to wash out the kero and salt, and then after another hour dry cycle the motor. If the temps were really high we would carry out this procedure and in the morning we would 'Carbo Blast' the engine with a product called 'Turco' which was ground up walnut shells. The actual airflow through an axial flow compressor is straight through, so the Turco had to be evenly distributed all round the engine intakes. A pound in weight (half a spew bag!) was the maximum we could use and the engineer would stand between the two motors and pour the contents of the bag into the intake screens at the back of the engine (reverse flow compressor don't forget), while it was running at about 10,500 rpm. What Health and Safety would make of this today is anybody's guess.

The problem that caused the most breakdowns and delays at one stage was not the Proteus engine itself but the Rotax start system.

Electrician Mick Prior, points out yet again how superannuated the technology was:

In 1956, my first posting in the RAF was to 601 City of London Auxiliary Squadron, flying Meteor Eights. The aircraft were fitted with twin Rolls Royce Derwent engines, which also used the Rotax start system. In the event of a failure to start, being the youngest one present, I had to open a panel at the rear, underneath the cockpit and strike the Rotax timer hard. The result was that 9 out of 10 times it would start, the pilot would fly away and we would change it when he landed.

When I arrived at Hoverlloyd I was quite surprised to find this old technology still in existence. The timer consisted of a solenoid, which wound up a clockwork mechanism. It was a three stage starter, as the clockwork ran down it closed three contacts.

I well remember an incident with Captain Bob Middleton. As per usual I banged the timer and all engines started. I said, "Hang on I will get a timer and change it going over." However, he didn't wait, and not surprisingly found that he couldn't start the engines for his return trip. In the meantime I had caught the next craft over and was able to change the timer for him. I thought he looked a bit sheepish.

The opened up Rotax Panel.

Robin Paine remembers inventing a new maintenance tool:

> *When we tried to depart Calais one of the engines wouldn't light up. Having recently been told the trick of hitting the Rotax, I went down to the port electrical bay with my headset. I remember my weapon to deliver the thump was my right shoe. The 1st officer commenced the start sequence and when he called out that the orange light was on (or whatever light it was), that was the signal for my shoe to make contact with the Rotax Panel. It worked perfectly. What appealed to my sense of humour was the ability to start a jet engine with one's right shoe!*

Does this suggest for some reason, a left shoe would not have been so effective?

In today's terms, in this digital world of ours, Mick's description of the Rotax system, the solenoid winding up clockwork, sounds positively Neanderthal, but as Mick said, he considered the system well out of date even in 1969.

The Day the Propeller Fell Off

In all this chapter of accidents and incidents the strangest and most unexpected of the lot was the day the propeller fell off. Duty Officer Mike Castleton was in the departure lounge at the time and, with the waiting passengers, had a grandstand view:

> *I was duty officer on a Saturday morning in November 1969. One of the craft,* Swift *I think, was sitting outside the duty office and* Sure *was inbound from Calais. During the night there had been a very heavy snowfall and* Swift *was covered by a white blanket. Captain Tom Wilson boarded* Swift *and the engines were started to move the craft to the jacking area. Because of the snow the visual effects were spectacular as* Swift *rose and completed a 90° turn.*
>
> *The departure lounge was full of passengers waiting to embark and many were at the windows to watch the craft manoeuvring. As* Swift *moved towards the engineering base there was a noise like a giant rubber band snapping and Car Marshaller Peter Price, who was in the duty office shouted, "Christ! The prop's come off!" I jumped up in time to see the craft sinking stern first, minus the number four propeller, of which there was no sign. Smoke was pouring from the starboard engine bay and debris was scattered everywhere.*
>
> *After the craft had settled there was a moment of wry humour when Tom Wilson emerged from the side door of the flight deck, trade-mark white socks clearly on display. He walked across the roof of the craft and stood looking in bewilderment at the propeller-less number four pylon. He took his cap off, carried on looking at the pylon for a few moments while scratching his head. Replacing his cap he returned to the flight deck, a picture of confusion!*
>
> *In fact it was a near miracle that no one was killed or injured. Had the propeller detached a few seconds earlier it would have hurtled into a packed departure lounge with horrendous consequences. It emerged that the runaway prop had wrecked the starboard engine bay before cart-wheeling across the pad and lodging itself in the building. It ended, twisted and deformed, draped like a giant spider on the desktop of the Customs Crew Inspection office. Fortunately, the office was unoccupied.*

The possibility of the huge 19-foot piece of flailing metal smashing into the crowded departure lounge doesn't bear thinking about now, but it could have been a lot worse.

Following an investigation of the four existing N4s by the Air Registration Board, both at Pegwell and Dover, four of the remaining 15 propellers were found to be on the verge of detaching, raising the spectre of the possibility of a forward prop falling and carving its way through a full passenger cabin. Like the Pegwell Waltz incident, this was the sort of disaster that would, in all probability, have finished the hovercraft industry in its infancy.

It didn't take long to discover the reason for the failure, Mike Fuller explains:

> *The shaft on the prop box was a lot longer than on the Britannia aircraft engine and the original torque did not stretch the BHC-designed shaft enough to hold the nut in place.*
>
> *To rectify the matter the torque was almost doubled, and the engine guys had to check the prop very frequently to ensure all was well with the nut and locking mechanism. If my memory serves me right the nut had unscrewed through 23 revs to come off completely.*

It is also likely that the way the pitch was used on the N4 was a contributing factor. Reverse pitch was used constantly back and forth to control the hovercraft arriving and departing, whereas it would have been used less frequently in the aircraft situation, putting less stress on the shaft.

One more lesson learned.

Job Satisfaction

It must be said, in retrospect, that although the preceding incidents present the impression of a succession of disasters, they are best reviewed as a set of milestones in a steadily improving situation.

In between failures the majority of the flights ran comparatively smoothly, lessons were learnt, equipment modified and procedures improved. In fact, for all the worry about the viability of the operation and the security of our jobs, most of us who were privileged to take part in those early formative years, derived enormous satisfaction from being part of a team that achieved so much.

With no previous experience to speak of, and no one else to help us, we took an unknown, untried concept and slowly but surely made it work. There is no satisfaction greater than that.

Sure passing under Tower Bridge on 7th October 1970. The craft took 200 delegates attending an international mechanical handling conference to visit the Port of London's £35 million container port at Tilbury. Hoverlloyd also announced that if the third London Airport was built at Foulness in the Thames Estuary they would investigate a service into London.

Chapter 22

Developments 1972 – 1977

1972 – Mk.2, the Beginnings of Success

Hoverlloyd's third craft, *Sir Christopher*, leaving BHC's Columbine slipway as an SR.N4 Mk.1 on 5th June 1972. After trials at Cowes it arrived at Pegwell Bay on 14th June and entered service on 3rd July.

By as early as 1971 Hoverlloyd could claim to have surmounted the majority of the early teething troubles and could now take some time to review progress and plan for a steadier future.

'Big Jim' Hodgson had taken over as managing director from Les Colquhoun at the end of that year and almost immediately instituted the development phase, which saw the introduction of a third hovercraft to the fleet followed closely by a two-year programme to modify all three N4s to Mk.2 status.

The new addition to the fleet, at a cost of £1.75 to £2 million, arrived for the 1972 season, named *Sir Christopher*[1] and registered as GH-2008.

---

[1] In honour of Sir Christopher Cockerell, the hovercraft's inventor, Lady Cockerell performed the naming ceremony.

The business case for the proposal was clear. An increase in operational costs by the addition of the third machine, nominally 50%, was offset by the increase in carrying capacity of the three new Mk.2 SR.N4s from 508 passengers to 840 at the completion of the programme; a rise in potential revenue in excess of 65%.

Lady Cockerell performing the naming ceremony of *Sir Christopher* on 29th June 1972.

It is likely that the increase in operational expenditure was significantly less than 50%. Obviously fuel and maintenance would have increased commensurately, but administrative costs would have remained the same. One would have also assumed that operational budgets would have risen accordingly, but in fact a comparatively small increase in staff was required and then mostly in short term casuals to meet the summer peak. Overall, it suggests that a three-machine operation was inherently more efficient.

Some personnel were taken on permanently, but they mostly replaced staff members who had left. Other vacancies were covered by temporary employees, which the company had little difficulty in recruiting. Even comparatively expensive flight crew positions were easily filled. Hovercraft continued to fascinate officers at sea and many were happy to join us, even for a few months experience and a 'toe-in-the-door'. Many came back each summer on a regular basis until opportunities for permanency arose.

By the time of the third craft purchase in 1972, after several years of hovercraft activity since the mid '60s, it was to be expected that people with previous experience would come knocking at the door. Peter Gray, whose time with Seaspeed driving the side-wall HM2 has already been mentioned, joined us in 1972 together with Dennis Ford. Dennis' sojourn in the Arabian Gulf was altogether a bit more exotic than dashing around the Solent:

In 1970 BHC had sold the Saudi Arabian Coast Guard & Frontier Force a package consisting of eight SR.N6 hovercraft, plus all the spares, training, and the bases to operate from – to be split between Jeddah on the Red Sea and Dammam on the Arabian Gulf.

I was posted to Dammam, just opposite Bahrain, with an English base commander, an experienced BHC pilot, and two gentlemen from the 9th 12th Lancers – a tank regiment. The idea being that we had enough experience between us to cover all aspects of hovercraft operations, quasi-military aspects, and marine navigation and seamanship.

The craft were to patrol the Gulf, tracking down potential smuggling operations and investigating other vessels passing through Saudi territorial waters.

As potential instructors we were taken through the usual courses on the Gnome engine, craft systems, and then driver training, hopefully to the point where we could teach others. This somewhat ambitious programme started in June and lasted for about five months, before we flew out to Saudi in November.

Dennis Ford as an SR.N6 instructor in Saudi.

It is interesting to note the length of the training period. Obviously lessons had been learnt since the early days when training, even for personnel who were destined to teach others, rarely amounted to more than a few hours.

From this juncture at Hoverlloyd a pattern of employment developed which saw a cadre of highly experienced staff kept as permanent employees, whose principal purpose was to train new seasonal recruits each year to cope with the peak season. In flight crew terms this resulted in a split of approximately two-thirds permanent staff to one-third seasonal.

There were further economies possible within this structure because the permanent staff was also divided into permanent and seasonal ranks. The latter, although retaining their nominal rank of captain or 1st officer throughout the year, were only paid as such once they were required and rostered in that position as the peak season approached.

It is likely that the car deck and cabin staff, without the constraints of a three-rank hierarchy were able to cut back even more over the winter period where the average day could be as little as four scheduled departures.

Sure (left) and *Sir Christopher* on the pad at Pegwell Bay with *Swift* approaching the Sandwich ramp as the last craft in at the end of a day's operations. All three craft were SR.N4 Mk.1s. Richborough power station is in the background. The inbound craft was probably destined for the jacks after disembarkation. 1972. Richborough power station was operational between 1962 and 1996 and the cooling towers were finally demolished on 11th March 2012.

Growing Up

Reviewing this transition in Hoverlloyd's development, it is noticeable that it clearly marked a coming of age. In contrast to the initial pioneer years during which, just to keep things running, it necessitated a certain amount of devil-may-care attitude, now that things were settling down to a reliable and relatively trouble-free existence, the pioneer spirit also settled into a relatively more formalised and responsible attitude to match. In many ways, it wasn't half as much fun, but as compensation it was replaced by an equally rewarding pride in a growing professionalism.

In parallel, changes in personnel also occurred, which in their own way had an effect in altering the company atmosphere. Two of whom (it may be said) typified the pioneer era, moved on – Bill Williamson left in 1973, followed a year later by Tom Wilson. The two ex-Clyde Hover Ferries men, who did so much to help build the nucleus of Hoverlloyd in 1966 and imbue it with its 'can do' approach, decided it was time for a change – Bill to pursue other local business interests and Tom to take up a new career in the offshore oil industry, joining Noble Denton, a company specialising in the supervision of rig movement worldwide, closely followed by John Lloyd.

Bill was replaced by David Wise at the point where 'Big Jim' Hodgson was also reorganising the Hoverlloyd management structure into four departments headed up by associate directors. David was appointed Associate Director of Operations, a position he maintained for the rest of Hoverlloyd's existence and then as operations manager of Hoverspeed in 1982.

For the permanent flight crew members, the prior years had been a set of steep learning curves, which meant that it was no longer a question of the blind leading the blind. From now on, new flight crew were treated to a more formal induction and training regime presented by top class experts in their field. At every level from 2nd officer/navigator to captain, training programmes became more codified and structured. Programmes were run by staff, who had not only 'been there and done that', but who had taken an above average interest in their chosen subject. They not only had practical experience but studied its theory as well. This was professionalism at its best.

Tony Quaife.

Tony Quaife and John 'Biggles' Lloyd were typical examples of the expertise involved, supported by a remarkable degree of dedication. This outcome was not just 'feel good' professionalism, but, in addition, led to significant cost savings. Hoverlloyd was now in a position to run its own training schemes and avoid the expense of sending trainees on courses to BHC and Rolls Royce, the latter facilitated by our engineering department's own developing expertise, to the point where they obtained licences from Rolls Royce to conduct a large part of their own engine overhauls.

Tony Quaife:

In March 1977 I went through captain's check out and was promoted permanent 1st officer. In June I did my first P1[2]*. In October I shifted from safety training to flight crew engineering.*

Bob Middleton had set up a good foundation for engineering training but the requirements were now expanding.

I attended a Peter Habens course for new recruits (at BHC) as a refresher and that year I believe was the last time the guys went on a Rolls Royce course at Ansty, so John and I had to put together suitable training and added RR engines in the final examination papers, which we mostly rewrote.

By arrangement I went around to each of the engineering departments and had in-depth discussions, the idea being to up flight crews' general level of understanding and ability to solve problems, especially on the pad at Calais Hoverport.

As we had no engineering staff at Calais, a broken craft would have to wait for assistance from Pegwell. I remember an incident when I arrived at Calais where another craft couldn't start engines using the normal sequence, and being able to advise a work-around for them. That cannot have been the only such time a lot of passenger inconvenience was avoided in that way.

Another part of the thinking was to encourage good practice on board, for example to avoid premature wear of expensive components. (Practices later found to be generally absent among Seaspeed personnel.)

John[3] *and I went down to BHC at our own expense and were shown around by Peter Habens. One of the Seaspeed craft had been cut apart to convert to a Mk.3 so we learned a lot about construction.*

---

[2] P1 – Flight crew logbook abbreviation for pilot in command.

[3] John 'Biggles' Lloyd eventually commanded Hoverspeed's first SeaCat, *Great Britain*, across the Atlantic, winning the Hales Trophy, the Blue Riband, for the fastest passenger ship crossing of the North Atlantic since 1952. See Chapter 25: *1980 The Beginning of the End*.

We drew up charts of craft idiosyncrasies and build differences together with long-standing problems. (Much of this was then copied by engineering management without anyone ever mentioning it to us.)

As I had recently been up for Master's, a lot of basic technical stuff was fresh in my mind, especially electronics, so that was added to courses – e.g. 3-phase theory, capacitance/inductance etc.

This was of particular use for refresher courses, a new development in the late 70s when I would hold forth to a portion of flight crew for a few days. Crew would also learn from each other's experiences and any unresolved questions would be immediately referred to engineers.

Of course, John's enthusiasm was invaluable in developing all this.

I understand that Hoverlloyd had been given dispensation by the BoT for this training so that people checked-out were in effect licensed by the government. I believe we were ever conscious of that and enthusiastic about creating a high standard.

John 'Biggles' Lloyd with some of his crew. Left to right: Sally Plackett, unknown, unknown, Norman Ferguson, Chris Lovesey, Sue Titcombe.

What Tony doesn't mention was that he and Biggles crawled around on their hands and knees in the plenum area noting the scouring in the paint in order to map the pattern of the air-flow. This invaluable information was passed back to BHC who would no doubt have made good use of it had they ever embarked on a successor to the SR.N4.

Bouncing off the Pillbox

But even the best of experts make errors. In the following anecdote it's difficult to fathom why a really experienced and thoroughly competent navigator like David Luff with, at least at that time, a thousand hours radar experience, should have made such a potentially disastrous mistake. One possible explanation is that the radar had detuned[4], unbeknown to him, and the resultant picture was almost undecipherable.

Nevertheless, in the end, sheer experience saved the day. Roger was there when it happened:

> *In the early days there were two approach ramps to the Calais Hoverport, leaving a substantial sand dune between them. Slightly closer to the western ramp in the dune, leaning over at a bit of an angle, was the concrete remains of a World War Two German gun emplacement.*
>
> *One very thick foggy day we were approaching Calais. I was in the right-hand seat and 'Big' John Lloyd was driving. David Luff was navigator and, despite his experience, he had obviously lost the plot completely. We came onto the beach probably doing at least 20 knots as it was necessary to keep the speed up, otherwise the machine would stall halfway up the comparatively steep ramps. First thing we in front knew of David's total disorientation was the sudden appearance out of the fog of this massive chunk of concrete a few metres dead ahead.*
>
> *There was no way we would avoid hitting the pillbox. John, being the experienced and cool-headed captain he was, reversed pitch all around, but crucially, kept the machine on full hover. With the bow skirt fully inflated, we bounced like an over-sized beach ball into the concrete and watched the pillbox disappear quickly into the fog ahead as we rebounded back down the beach for another go.*
>
> *I remember one word from our intrepid navigator, "Sorry!"*

---

[4] Although amazingly reliable considering their age, the Decca radars did need constant 'tweaking' to maintain a clear picture. In this pre-solid state electronics age this often meant adjustment to components, which meant the navigator was unable to adjust things *in situ.*

The SR.N4 Mk.2

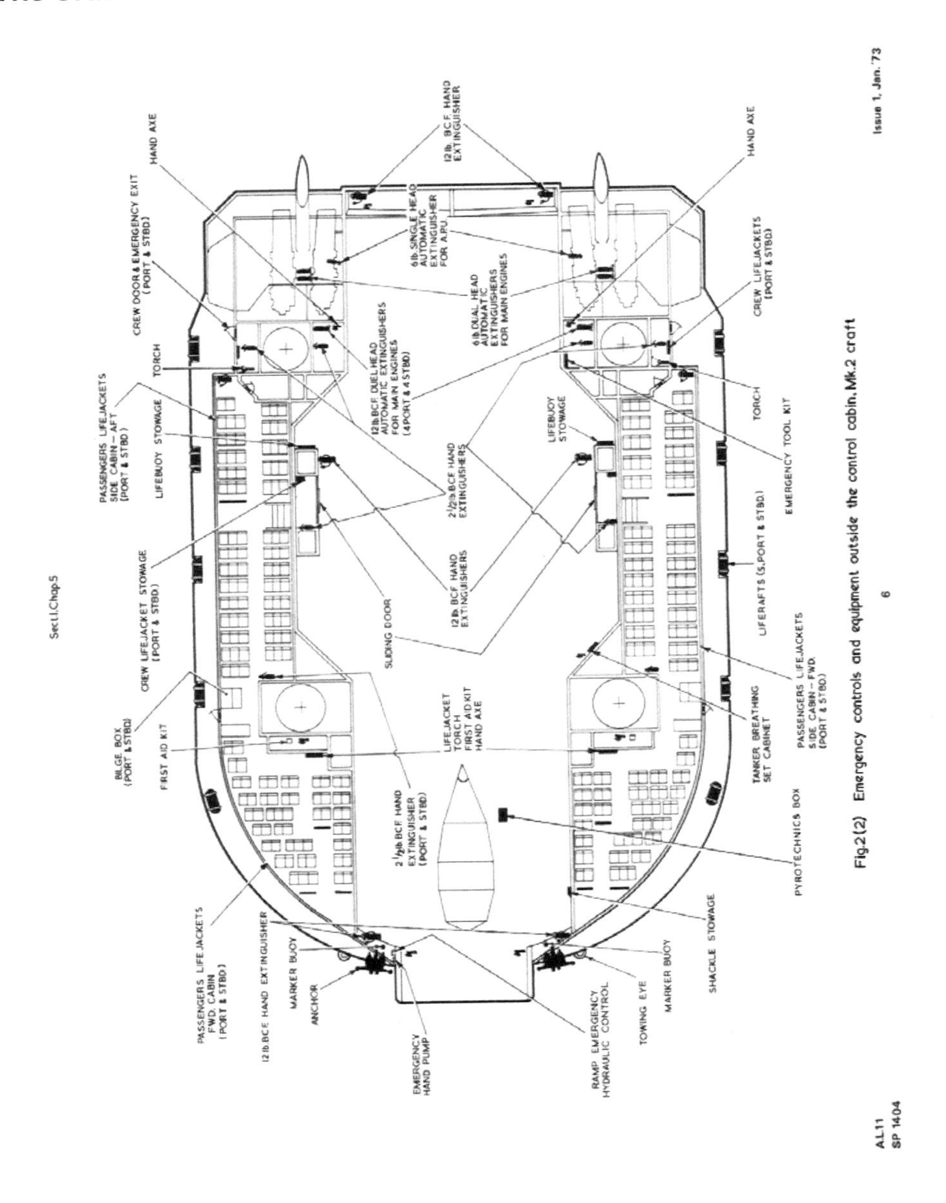

Mk.2 Cabin Configuration.

By the end of 1971, the lessons learned over the first two pioneering years gave Hoverlloyd a clearer insight into what needed to be done to make the company not just operationally reliable, but profitable.

Swift undergoing conversion from a Mk.1 to a Mk.2 during the winter of 1972-73 in the Columbine hangar at Cowes, with a BH7 under construction in the foreground.

A Mk.2 version of the N4 was proposed, which increased the passenger and car deck loads significantly, the priority of which was to dispose of the deeply unpopular Mk.1 inner cabins. The 40-seat enclosed areas, positioned on each side of the car deck, were windowless, claustrophobic and, understandably, universally loathed by the passengers.

No one contemplating an exciting flight on a hovercraft could appreciate being confined to a small windowless space with nothing much else to do but watch fellow passengers being sick. Nevertheless, if they were to be removed, then somehow the loss of 80 seats had to be compensated for.

BHC's solution was to widen the superstructure of the craft, thereby widening the outer passenger cabins sufficiently enough, not only to allow for the inner cabin removal, but to actually increase the overall passenger and car capacity. The resulting SR.N4 Mk.2 increased the Mk.1 carrying capacity from 254 to 280 passengers and from 30 to 36 cars. The extra three cars on each side would fit into the space vacated by the inner cabins.

During the winter off-peak months of 1972 and 1973, the three craft already in service, *Swift*, *Sure* and *Sir Christopher* returned to Cowes in turn to be transformed from Mk.1s to Mk.2 SR.N4s. *Swift* was the first and left for Cowes in September, returning in January.

The new Mk.2 cabin layout transformed the whole feel of the passenger areas; spacious compared with the first model, with every seat facing forward and every passenger with a view out the window; although it has to be admitted the latter was more a question of light rather than visibility – the usual thick cloud of spray and the resulting abrasive effect in conjunction with mud and sand meant that, regardless of the size of the window, visibility was always somewhat limited.

Swift on completion of her conversion to a Mk.2 arriving at Ramsgate.

The new look Mk. 2 forward starboard cabin.

The new Mk.2 interior. The starboard side main cabin. Widening the cabins resulted in a spacious and attractive seating area and, with the inner cabins removed, provided six extra car spaces on the vehicle deck.

Sir Christopher as a Mk.2 craft undergoing an extensive winter overhaul at Pegwell. It shows clearly the structural changes in order to increase the carrying capacity. The original Mk.1 cabin sides have been moved outward to the hinge line of the skirt and a 'V' extension added to provide a new evacuation walkway.

Skirt Improvements

At the same time, advantage was taken during these major upgrades to improve the skirt design. It was obvious from the many failures of the skirt that the whole configuration needed major modifications. From 1969 on, in order to reduce the incidence of plough-in, the established recommended procedure was to operate the machine with a 1° to $1^1/_2$° bow-up attitude. By and large this was effective, but with the skirt hemline also at the same angle, it resulted in the rear end being dragged through the water and an unacceptable increase in skirt damage in that area.

The solution was to taper the skirt height in such a way that the top line of the skirt attached to the hinge line of the craft structure was angled at $1^1/_2$°, while the hemline at the bottom, where all the damage occurred, remained parallel and level with the water[5].

The craft is on jacks. Although this is a tapered skirt configuration, the tattered remnants of the rear fingers show that wear at the back end remained an insurmountable problem. In this instance the rear 'cone' fingers would appear to have lost at least 12 inches off their original shape. Captain Ted Ruckert with his son, Simon.

Although structurally still a Mk.1 design, *Sir Christopher* was delivered with the new tapered skirt. *Swift* and *Sure* were fitted as part of their

---

[5] The upgrade was designed to hopefully reduce finger wear. This particular problem, unforeseen by the N4's designers, which although improved upon by this and future modifications, remained until the end. The large cost impost of the finger replacements to the operating budget gave rival modes of fast transport without such costs a competitive edge.

Mk.2 upgrade early in 1973 and by the summer of that year all three craft skirts were tapered.

August 1973 – SR.N4 Data Collection and Analysis Report

As was originally hoped by the nascent British hovercraft industry, the success of the cross-Channel services attracted the attention of the world at large, not the least of which was the interest shown in the USA. One or two commercial possibilities were trialled but, having no equivalent mass market route like the Channel, the attention across the Atlantic moved, perhaps naturally, into the military sphere.

US SESPO data project 1972. Front row 3rd from left Billy James; David Luff; National Physical Laboratory representative; Ted Ruckert; and Roger Syms (author). Ted Unsted behind David's left. The rest of the group unknown – American, BHC and NPL people.

The concept of the 'Hundred Knot Navy' was formulated and development money poured into budgets of inconceivable size compared with those in post-war, cash-strapped Britain. Nevertheless it was in Britain, in companies like BHC, Hoverlloyd and Seaspeed, where most of the hands-on expertise and experience in handling large hovering machines resided.

In 1972, to tap into this unique body of knowledge, one of the controlling entities of US research, the Surface Effect Ship Programme Office (SESPO), sent a team to Hoverlloyd to gather operational data firsthand over two short periods in summer 1972 and spring 1973.

The data gathered was meticulous and thorough, covering every possible aspect of the operation. The report issued in August 1973, if nothing else, provided a fascinating detailed snapshot of Hoverlloyd at probably one of the most interesting stages of its development. The data itself as an historic record is of course interesting but what are far more relevant, even today, are some of the conclusions in the analysis.

The Biggest Problem

At that time, along with the Mk.2 conversions, the tapered skirt was introduced. *Sir Christopher*, although still a Mk.1 at the time of delivery in 1972, already had the new skirt and *Swift* and *Sure* were fitted with theirs over the following winter period. One of the three craft with a tapered skirt during the 1972 data retrieval period provided an opportune moment to compare the old 'standard', straight line configuration and the new tapered system. To everyone's consternation, the level of finger wear initially appeared to increase with the introduction of the tapered skirt – the very thing that it was supposed to reduce.

On closer inspection of the data today, the major increase appears to be bow finger wear and a general overall increase in finger wear along the whole hemline. This is hardly surprising as in the old straight line system, flying the machine in a 1° bow-up attitude would, as in fact it was intended to, keep the bow fingers clear of the water. Consequently bow finger wear would have been relatively little while the rear end of the craft, being dragged through the water, would have suffered more, with a commensurate increase in wear from forward to aft. The tapered skirt's straight hemline merely redistributed the same amount of wear more evenly.

What this emphasises quite clearly is the ongoing problem of skirt wear as a considerable cost impost to hovercraft operation, which remains to this day a fundamental flaw in the hovercraft concept. Cockerell's aim to achieve high speed over the water at an economical power input only works if there are no additional consumable costs, and skirt wear still remains a considerable one.

This however needs qualifying. The Cockerell concept is watertight, but the technology has not as yet developed sufficiently for it to work.

The American analysis makes no bones about it:

> *So far (in 1973) there has been no real breakthrough in skirt material technology although, perhaps, the real problem here is that few rubber manufacturers really appreciate the tremendous battering that the skirts receive when negotiating a 3-4 foot sea at 55-60 knots. There can be no doubt that if hovercraft speeds were limited to 40 knots skirts would not be the problem they are now.*

One might add – and continue to be so.

It is also interesting to note that the longest running and most successful passenger operation, Hovertravel on the Solent, has comparatively little trouble with skirt wear, and they operate at around 40 knots in sheltered water for most of the time.

It may also be presumed that today's US Navy's Landing Craft Air Cushion (LCAC) assault vessels, which can be assumed to be the ultimate beneficiaries of this early research, have a quoted design speed of 40 knots and, launched from specially equipped assault ships, transit to the beach in comparatively calm inshore waters – again, not quite the same problem.

The cross-Channel operation with such a high intensity schedule, in sea conditions of 3-4 feet, not as a limit but as an average, 'all day, every day', subjected the concept to the ultimate extreme.

Apart from highlighting the need for a breakthrough in a tougher, longer-lasting skirt material the report points out two commensurate requirements in instrumentation and control. There is no possibility of reducing skirt wear if the two basic parameters of measurement and control cannot be improved. In the American opinion the SR.N4, as designed, was lacking:

> 1. *The fact the presentation of craft longitudinal attitude is not sufficiently accurate or readable for the demands of reading trim situations to within ¼°.*
>
> 2. *The facilities for trimming the craft are not sufficient to cater for the extreme C.G. (centre of gravity) to which the craft can be loaded. This is particularly so when the craft is fully loaded.*

There can be no argument with this criticism. The original aircraft 'artificial horizon' style bar indicator was not even capable of clearly showing a change of 1°, let alone fractions. By 1972 it had been replaced by a typically Hoverlloyd 'give it a go' approach – a 'U' shaped piece of plastic tube, filled with coloured water, spread the length of the control cabin. The front vertical ran up the side stanchion of the captain's side window, against which was glued a measurement gauge drawn on a piece of card. This was a crude, but nonetheless, workable improvement and it showed fractions.

As to the method of trimming on the N4, there was nothing that could be done beyond a complete redesign. The existing system made use of the four fuel tanks strategically placed in four corners of the buoyancy tank. Transfer pumps allowed fuel to be moved between the tanks to help trim out any CG discrepancy, which was only evident once the craft lifted off. The hard fact of life in a commercial operation like this was that, although rough calculations were made on 'average' weights of cars onboard – one ton per car – the vast diversity of car design, individual load and the *ad hoc* loading of each on the deck, made the resultant CG position a totally unknown quantity until the final moment when the machine came up on hover.

Using the fuel tanks in this dual purpose manner was a perfectly reasonable idea but the design had not allowed for the possible range of CG movement. In the report's view the tanks were not big enough to allow for sufficient spare space to enable enough weight to be moved around. Furthermore, the transfer pumps were not of sufficient power and capacity to move this fast enough.

Praise for the Proteus

Considering the somewhat out-of-date 1940's technology involved, when it came to the 'Old Prote', the Americans were surprisingly complimentary. The report pointed out that the Rolls Royce Proteus may have been old, but during its lifetime it had built up an enviable pedigree as a 'marinised' engine, one capable of coping with the most extreme of conditions. Furthermore, it would probably take an equally lengthy and expensive development period for any more modern engine to catch up:

> *As far as engines are concerned, there now clearly emerges an engine capable of up to 4,000hp, able to provide the hovercraft power reliably, efficiently and quietly. Any other engine will need a considerable amount of development before it can achieve the same standard. When it is realised that the Proteus already has over 110,000 power hours in hovercraft and considering that the first two years of this experience was in an environment at least ten times worse than the*

manufacturers' requirement, the record of reliability achieved during this period has been excellent.

It must be remembered that the basic design of this engine goes back to the early '50s[6]*. A more modern concept might well bring advantages such as an improved specific fuel consumption or a smaller, quieter engine for the same power, but it is doubtful if the standard of reliability and the period between overhauls presently set by the Proteus could be equalled.*

Praise indeed – in fact the tenor of this American perspective on the British hovercraft industry in general, and Hoverlloyd in particular, is one of admiration for what had been achieved with what they no doubt considered the very minimum of resources.

"Extremely Courageous"

The SR.N4 was an extremely courageous design concept, which is even more remarkable when the very limited hovercraft experience that was available in 1965 is considered. However, it must be emphasised that the SR.N4 was a great leap forward for the hovercraft industry. The experience that has been so far gained clearly indicates problems that must be tackled in future designs. Credit must be given therefore to the design team of the SR.N4 for producing, without previous experience, a large hovercraft that, despite the many problems, has proved a successful commercial vehicle.

Reading between the lines of such comments like 'courageous', 'without previous experience' etc., one can detect a certain amount of headshaking at the mad British for jumping in with both feet without checking the water. At that period when America was still flush with cash they would have spent – and were spending – 'n' billion in development before they put their concept in the water, let alone put the travelling public on it[7].

A Three-Day Snapshot

Hoverlloyd is rightfully proud of its efficiency and professionalism; we strove every day to ensure that flights departed on time and arrived on time.

---

[6] This is incorrect. In fact the Bristol Proteus engine started its development in the mid 1940s.

[7] This was a huge accolade for Ray Wheeler and his team at BHC for making that jump.

Sadly, being able to prove this strict adherence to punctuality is impossible today, as the evidence, contained in our day-to-day operational records, was trashed at the time of the Hoverspeed merger.

Therefore it is a real 'find' to discover in the pages of this unsolicited and unbiased account from a totally independent authority, the following:

> *The Decca Radars (and of course the navigators) on the SR.N4 have given remarkably good service and have enabled the craft to be operated at normal speeds in conditions of fog. As an instance, during August 1972, during a weekend when persistent sea fog was experienced in the Channel, Hoverlloyd operated 21 return crossings per day for the three-day period. All started on time and finished on time. During the same weekend, conventional ship ferries were running up to seven hours late.*

A welcome accolade, and by us, warmly received.

Mk.2 Safety Issues

Although the increase in the passenger load of the Mk.2 was still well within the capacity of the five 30-man life rafts on each side, it was to all intents and purposes a new passenger carrying vehicle. As such, a new Operating Permit was required and the evacuation procedures had to be demonstrated again for the Civil Aviation Authority (the new authority replacing the ARB) and the Board of Trade (BoT).

Robin Paine, having taken over as Safety Captain from Ted Ruckert, was responsible for the success of the exercise. 'Big' John Lloyd was captain and David Ward was the 1st officer for the evacuation. Although it was flat calm, it was also very foggy with visibility restricted to about 200 yards. This turned out to be a blessing in disguise, because the press, including television, had got wind of the event and, as well as hoping to take photographs from the shore, had also chartered boats to picture and film the evacuation at sea. Robin recalls:

> *We decided that this might not be in our best interests as the positioning of the new rafts in a recess was an unknown factor and, based on the previous experience, we knew that things did not always go according to plan. We managed to keep the time of departure under wraps, so together with our advantage of speed, we were able to out manoeuvre the paparazzi. With the two authorities and some BHC observers on board, we proceeded to the evacuation point more or less unobserved, which, as it turned out, was just as well.*

With 140 Royal Marines seated on one side of the modified *Swift*, the craft proceeded to the middle of Pegwell Bay. Having carried out this exercise before with the Mk.1, it would have been reasonable to assume that as the life rafts and procedures were the same it would be a walk in the park. However, there was one distinct difference between the new and the old design. On the Mk.2 the life rafts were neatly recessed into the edge of the walkway, rather than simply placed in a cradle on top of the plenum.

Things did not go according to plan. David Ward and his car deck crewmen struggled to launch the life rafts because the lip on the front of the new recess was a little too large. As a result of the brute force required to lift the heavy rafts over the lip, one turned upside down and another was punctured. David also remembered that by quickly bending over on the narrow walkway (a marginal improvement over the Mk.1), he bumped his backside against the passenger cabin superstructure and nearly launched himself into the water instead of the raft.

With two of the five life rafts out of action David, following procedure, ordered the remaining marines across to the port side and embarked them successfully from that side.

Mercifully the Operating Permit was awarded, but it was quite apparent that there had been no trial by BHC before delivering the craft to Hoverlloyd. Why we received the permit without having to repeat the exercise has always remained a mystery to Robin, but the episode was neatly summed up by one of the BHC observers who said:

> *How is it that the Americans can land a man on the moon and we can't even launch a bloody life raft from the side of a hovercraft?*

Swift as an SR.N4 Mk.2 with nos. 2-5 life rafts now recessed in the plenum chamber, making embarkation easier.

Weekly Drills

Weekly practice drills for each crew were part of the on-going safety awareness and training. They were of course carried out on dry land, quite often in front of passengers waiting to board the hovercraft with their cars in the embarkation car park – a good PR exercise as a bonus. The difference between the practice drills and the real thing was that during the practice drills the life rafts were not launched. Instead, the captain walked around the craft and asked each crew member questions about safety, all of whom by then were standing at their designated stations. The procedure worked well but not without the occasional misunderstanding. George Kennedy recalls:

> *On one particular occasion, a newly appointed captain, Linton Heatley, a New Zealander, who sported a very distinct antipodean accent, was conducting the practice drill and had descended the flight deck ladder to commence questioning the crew. He arrived in one of the passenger cabins and said to a stewardess at her station, "G'day, where are the lifejackets stowed?" to which she correctly replied, "Under the passenger seats." The captain then weighed in with his supplementary – "Where are the 'speers'*[8] *kept? A puzzled look crossed the young lady's face. She thought for a moment and then ventured, "We don't carry spears, captain, but we do have axes!"*

So, although safety was taken seriously, it did have its amusing moments[9].

Safety at Sea

Right from the beginning, running two hovercraft back and forth at closing speeds in excess of 120 knots, the necessity for route separation was recognised as essential for operational safety. On the Decca Navigator roller chart in front of the captain the 140° course from Ramsgate to Calais was drawn in, plus two parallel courses spaced at half a nautical mile on each side. In keeping with nautical anti-collision convention each craft kept to the right-hand (starboard) line, thereby leaving a mile separation from the opposing track, passing port to port.

This generally worked well but these were not railway lines and quite often craft strayed off, usually as a result of alterations to avoid the mass

---

[8] This is the best we could do for the Kiwi phonetic pronunciation of 'spares'. Linton went on to become Operations Manager for Hoverspeed from 1992 to 1993.

[9] In the SR.N4's 33 years of operation the evacuation of passengers only occurred once – the Hoverlloyd incident off Sangatte in 1971. Although transferring passengers in heavy surf and in tidal conditions, which made the craft difficult to control and resulted in circumstances outside the normal safety training scenario, the crew reacted admirably.

of transiting ship traffic. Navigators, as a matter of course, always did their best to make avoidance manoeuvres with conventional vessels off to starboard thereby increasing the hovercraft clearance but on occasion a risky movement to port was unavoidable.

When there was a chance of 'getting on the wrong side of the bed', radio calls were made immediately to warn an approaching craft of the problem and every effort was made to get back to the safe side. It was always known where the opposing craft was likely to be, because another safety procedure was for all craft to report at intervals on the route. Namely, the reporting positions leaving Ramsgate or Calais, were passing the MPC Buoy in mid-Channel and the East Goodwin Light Vessel. With this forewarning the navigator knew where and when to expect a fast-moving target to appear over the edge of the radar screen.

This perfectly logical set of rules should have solved the problem, but in the early years there was a major and dangerous flaw. The maritime VHF in use at sea at that time was limited in the number of channels available and radio discipline among the mass of ships passing through and shore stations using them was appallingly lax. Channel 16, designated as a calling frequency, i.e. one with which you made initial contact and then moved to another channel to send your message, was a farce.

With the hundreds of vessels negotiating the Straits daily the radio traffic on 16 could invariably be continuous and the most irritating point was that most of it was not work-related but conversational. Imagine the worry this caused the hovercraft crews, desperately trying to call their counterparts to relay safety information and being blocked over several minutes by two fishing vessels discussing the weather. Waiting even three minutes for a chance to talk was totally unacceptable. A hovercraft appearing over the edge of the radar six-mile range would be close to the point of collision in that time.

Eventually our safety concerns got to the ears of the BoT communications regulators and a meeting was held at Hoverlloyd, together with representatives from Seaspeed, to try and solve the problem. The meeting lasted little more than an hour – "Quite easy", said the bureaucrats. "Advances in radio technology now allow us to place frequencies closer together so we are just about to introduce a number of new frequencies interleaved between the existing ones, virtually doubling the range of channels available. Although not yet officially available, we will allow you to use one of these frequencies, Channel 72, as of now."

An answer to our prayers it would seem, so we quickly modified our VHF radios to use the new frequency as a result of which we had an exclusive silent channel of our very own. Instant communication whenever we needed it.

Psalm 146 says, "Put not your trust in princes", or any form of government it would seem. It took some three or four years of uninterrupted communication between all fast craft in the Straits before we again started to experience interference from other ships using our safety channel for casual conversation. When we complained to these transgressors we were told with complete assurance on their part, contrary to our belief, they had every right to use it. On further enquiry, disconcertingly, we found that we were in the wrong. What we had thought was our exclusive frequency had been designated right from the outset as a common inter-ship frequency for all maritime interests to use. Our helpful bureaucrats had neglected to tell us that.

Not quite back to square one but almost. Channel 72 never was quite the mayhem of Channel 16, and to be fair, when we explained the use 72 was being put to in the Dover area, the large majority of vessels obligingly moved elsewhere.

This was not the only problem. Later on in the decade, the supposedly quiet frequency was literally blasted out on a daily basis by a French voice at very high volume, "'Allo Marie!" Further investigation discovered this to be a flower seller in Calais talking to his delivery vans. But this was a maritime frequency; what was a shore-based business doing using it? It was then that we learnt another fact of the international communications business. The French maritime frequency range was different and 72 may have been listed as maritime in Britain but it certainly wasn't in France. So "'Allo Marie" was using it quite legitimately. It's a crazy world.

More Traffic into Calais

Seaspeed's original intentions were to run between Dover and Boulogne, because that was the necessary link between British and Continental rail services. As previously explained, the extreme sea conditions off Boulogne Hoverport in bad westerly weather prompted the diversion of several services to Calais. By 1971 this had become formalized and, officially, scheduled flights into Calais were the norm, gradually increasing in number throughout the decade.

This increase in activity at Calais prompted another meeting between the two hovercraft companies to ensure safe movement between the two converging routes. Procedures in and out of the hoverport were agreed and reporting procedures on Channel 72 were modified to suit.

A deterioration in safety communications later in the decade couldn't have come at a worse time, coinciding as it did with the peak of fast craft operations in the area.

By the end of the decade the introduction of the French hovercraft, the SEDAM N500, to the Seaspeed fleet of two Mk3s, Hoverlloyd's fleet of four Mk2s and, although not directly connected with Calais, the Belgian jetfoils on a route from Ostend to Dover, passing close off the approaches, added further complexity of crossing routes.

All fast craft adhered to the Channel 72 protocols, but the possibility of as many as four or five fast craft, in poor visibility, converging on one small area off Calais was a worrying one at times.

Where Do We Go From Here?

Captain David Wise congratulates the proud owner and family of the 'zillionth' car carried by Hoverlloyd.

Throughout the decade the boom in passenger and freight movement between England and France still showed no sign of abating; from a little over two million passengers and 300,000 cars[10] in 1969, to more than three times that number in 1980, with one noticeable exception.

---

[10] These are comparison figures, on all forms of sea transport, for cars only. All other vehicles: trucks, caravans, trailers, motorcycles etc. are not counted.

The figures for 1975 and '76 indicate a plateau, both years showing four million passengers while car figures actually declined. One explanation for the latter is the continued increase in freight carriage, which was becoming the year-round principal business of the conventional ferries. Those particular years show a substantial increase in freight units which would suggest it was not a decline in cars on the ferries but more of a question of them being squeezed out.

The assumption is further supported by the fact that Hoverlloyd's car figures, where freight was not an issue, remained steady like the passenger numbers. The conclusion here is that, in both cases, capacity of the various companies serving the Channel had peaked; demand was still growing but space wasn't.

In reality this would have been no surprise to the operators watching the steady climb in market demand, as plans would have been made some years earlier to expand the fleets, but the new vessels had yet to arrive.

It is worth bearing in mind that these figures are only between Dover and Calais They do not include all the other ports; Folkestone and Harwich on the UK side and Boulogne, Dunkirk, Zeebruge and Ostend on the Continent to name but a few. It must be assumed that expansion was on the cards on those routes too.

To these can also be added the acquisition of the only private operator out of Dover, Townsend Thoresen, by P&O. A company whose aggressive approach to the market was destined to change the game entirely[11].

So what were the hovercraft companies to do? Expansion seemed to be the only route, but how was it to be achieved?

Mk.3 or Mk.2?

Early in 1976 the British Rail Board approved the Seaspeed plan to drastically upgrade the two existing craft from Mk.1 to Mk.3 'Super 4'. Not just widening the craft as per the Mk.2 modification, but extending the length of the machine by slicing it in half amidships and inserting a 55-foot section.

---

[11] Initially Townsend bought P&O Ferries, who had two small ships operating from Dover to Boulogne. The Monopolies & Mergers Commission was asked to investigate the takeover, but P&O told them they either allowed the takeover or the service would just close with the loss of a whole lot of jobs. The takeover went ahead and as soon as that was complete, P&O bought Townsend!

At the time Seaspeed felt they needed to have a better sea-keeping craft to deal with the vagaries of the weather at Dover and Boulogne to improve both the operational reliability and passenger comfort. Peter Yerbury, Chief Engineer, explains:

> *We told Ray Wheeler (BHC's chief designer) we wanted certain criteria for the Mk.3, such as Force 9 operating capability and a much better comfort level than the Mk. 1 or 2 could offer.*
>
> *He went away, did some sums, came back and said this is what they could offer us. We basically had specified the performance of the craft and it was up to them to design something that would achieve those requirements. We couldn't stipulate the actual extra capacity because they were saying from early on that a 55-foot extension would meet the criteria. We couldn't just decree we wanted 100 feet put in. The extra capacity was dictated by the length of the extension as a result of their calculations.*
>
> *We then worked out what we could fit into that space. Initially we calculated that our total passenger capacity would be 418 and then we found a way of putting another six seats in, so the Mk.3's eventual capacity was 424 passengers and 55 cars.*

A drawing showing the concept of converting a Mk.1 to a stretched Mk.3 by inserting a 55-foot section, together with widening it to the same overall width as a Mk.2.

Ray Wheeler said the expensive part would be realigning the engines with the forward gearboxes and the rewiring. Peter Yerbury:

> *Ray was apprehensive too about the rotation of the shafts with nearly 60 feet of extra length, but we had confidence that BHC could solve that problem. My role was to make sure that the thing would work and look at the maintenance aspects of it. Ray was the designer and when he said it would work I said, "Right, put your name down on that piece of paper there", which is what happened.*
>
> *John Lefeaux was involved in it to the extent that he was my boss and I reported to him, but he mainly did the building side, as in the construction of the new terminal (in the Western Docks) – he was project manager for that and I looked after the craft.*

Such a comparatively enormous machine meant that the existing small pad in the Eastern Docks[12] was totally inadequate. Planning had also started in 1976 on a purpose-built hoverport in the Western Docks, the completion of which was lined up to coincide with the introduction of the Mk.3s scheduled for 1978.

In contrast, Hoverlloyd's response was not so radical; instead the easy option was taken to expand the fleet with another Mk.2.

It would appear there were two reasons for this, both concerning money. Firstly BHC had a sixth Mk.2 under construction in the hope that the continued success in the Channel would prompt an overseas buyer. By 1976 such an event had not transpired and in all likelihood never would. In consequence Hoverlloyd's view was that BHC, by now desperate to get rid of it, would be willing to sell it at a bargain basement price.

The second alternative of 'jumboising' their existing fleet, or at least two of them, was simply not an option. It is not clear now if management was fully aware of what the Seaspeed project was costing, but even at half the rumoured outlay today it would have been quite impossible for a privately-financed company to contemplate.

Ray Wheeler records that Emrys Jones, Hoverlloyd's Technical Director, had written a paper for a conference presentation that flatly rejected the idea of the 'Super 4'. Unfortunately the paper is

[12] It was about this time that the Dover Harbour Board required more land for both parking and building extra berths, with modern upper and lower loading ramps for the arrival of the new generation of roll-on roll-off ferries. It was therefore convenient to all concerned to build a new hoverport in the Western Docks.

untraceable now, so we can only guess at Emrys' arguments. It is likely that, with his own background at Vickers, he was well versed in the cost factors involved in airframe manufacture and probably worked out an idea of the eventual cost of such a radical modification – at least near enough to realise that the whole project was well beyond Hoverlloyd's financial resources.

SR.N4 Mk.3 model undergoing trials in the BHC test tank at Cowes.

Dover Western Docks, looking south towards the western entrance. The long pier to the left is Princes Pier and the plan was to reclaim the land between Princes Pier and the end of the shorter pier to the right on which to build the new hoverport.

It is difficult to verify exactly today, but the estimated cost was in the region of £12 million for *The Princess Anne,* the first to be converted, and *The Princess Margaret,* the second, a further £14 million. If the cost of the new hoverport is added (purported to be £14 million), although strictly not a budget expense to British Rail, because the Dover Harbour Board financed its construction, the DHB would have ensured that the new operation would have recompensed them.

If this were a private enterprise like Hoverlloyd the whole thing would have been a financial disaster[13]. However, even with the ability to create much more revenue with the larger craft, not only as a result of its greater capacity, but also its greater sea-keeping ability and therefore higher service reliability, it would not have been nearly enough to cover the amortisation of such a huge sum. When governments are paying with our money, apparently it doesn't matter.

The completed new Dover Hoverport in the Western Docks (looking north) on reclaimed land at a purported cost of £14 million, was opened in July 1978. Note the French N500 to the left and an SR.N4 Mk.3 to the right.

Theoretically, although attempting a similar modification of Hoverlloyd's Mk.2 craft to Mk.3 level would have had the advantage of only needing lengthening, not widening as well, it is still uncertain whether it would have been viable. In fact, the introduction of the low pressure skirt

[13] Financial disaster for Seaspeed or not, it definitely was for BHC and their corporate owners Westland. BHC lost £4 million on the deal which, adding to Westland's considerable existing debt, presaged their demise and subsequent takeover by GKN Aerospace in 1988.

would have required remodelling of the outer skirt hinge line and squaring off the existing 'V' shaped side extension of the Mk.2, which wasn't that simple after all.

If the necessary modifications to the hoverport to cope with the larger size are included, it is debatable whether the cost would have been low enough to be financially managed. The inescapable fact was that, although the SR.N4 was proving itself operationally capable of running a service across the Channel, economically even the 'Super 4' profit margin could not compete with the average ferry, and in fact could not compete with the fast catamarans that eventually replaced it.

Clouds on the horizon

The sad fact was that in 1977, although the hovercraft was reaching the peak of its capabilities, ominous signs were appearing, indicating that the future was not so bright. In a classic boom and bust scenario, all too familiar to the maritime world, earlier decisions by the various shipping interests to build new and bigger ships to cope with the projected end of the decade boom, resulted in all this extra tonnage appearing simultaneously, creating an over-capacity situation on the Channel as a whole. In order to fill their new tonnage and justify their investment, the ferry companies very soon found themselves in a competitive price-war, which the hovercraft companies with their smaller profit margins could not sustain.

Even bigger threats, although comparatively far off, were nevertheless distinctly on the horizon.

The ongoing development of the European Union signalled the eventual complete removal of customs duties across all European borders. That in itself would have been a blow, as the loss of duty-free sales would have wiped out the 10% profit margin of Hoverlloyd in one single stroke.

What was of even greater concern, although at that time put to one side by the British government, was the prospect of the Channel Tunnel, which loomed ever closer with, what seems now, an inevitability of its own.

Chapter 23

The Peak Years 1977 – 1979

1977 brochure showing the new addition to the fleet, *The Prince of Wales.*

The Fourth Craft

The 17th June 1977, the day our fourth addition to the fleet was named by Lady Astor of Hever, heralded a period of great optimism and confidence in the future. Over the previous two years Hoverlloyd had been running at near full capacity in the peak season and there seemed to be no reason why the fleet of four should not achieve full capacity too.

Any thought of competition through a price war, the loss of duty-free sales or the arrival of the ultimate competitor, the Channel Tunnel, were just 'may-be' and not seriously considered as problems, and if they were possibilities at all, they were seen as a long way into the future.

Roger Syms' daughter, Caitlin presents Lady Astor with flowers at the naming ceremony. In the background are James Hodgson and Mrs. Hodgson. 17th June 1977.

To cope with the extra numbers, the engineering block had been enlarged and the standing area was expanded to allow for a second jacking system. A new double-storey block was added to the south-west end of the main building to include a much expanded reservations floor and rest rooms for the burgeoning operations staff. A mezzanine floor was added to the departure lounge, which also included much improved catering facilities for the waiting passengers. To the north-east, embarkation car parks were enlarged and a small kiosk was opened to provide breakfast for early morning passenger arrivals.

Everywhere one looked the company and the hoverport were growing and with it the confidence of the staff. Hoverlloyd's 'can do' attitude and the fierce pride it engendered had thus far achieved spectacular success and there was no reason to believe we would not succeed in the future.

Extracts from the press release issued for the occasion captured the mood of the time:

> *Hoverlloyd's latest hovercraft,* The Prince of Wales, *was named by Lady Astor of Hever at the International Hoverport in Ramsgate on June 17th 1977. This brings to four the number of SR.N4 Mk.2 (Mountbatten Class) hovercraft in Hoverlloyd's fleet, the largest operating between Britain and Continental Europe.*
>
> The Prince of Wales, *which has more than 120 modifications and improvements on previous hovercraft, will go into operation on Hoverlloyd's Ramsgate-Calais service on June 18th 1977,*

bringing the number of daily services in each direction to 27 flights in high summer. The addition of the fourth craft will mean that Hoverlloyd is able to carry up to 15,000 passengers and 2,000 vehicles daily.

Hoverlloyd's policy is to provide frequent services with proven hovercraft which can make the Channel crossing in 40 minutes. The SR.N4 Mk.2 hovercraft disembarks passengers and vehicles and completes the embarkation for the return flight all within 20 minutes of arrival. The 40-minute hovercraft crossing, check-in 30 minutes before flight time and excellent facilities of the custom-built hoverports, mean that it is possible for travellers to be en route the other side of the Channel within 1 hour 20 minutes of check-in.

The Prince of Wales – GH 2054

Of the 120 modifications claimed for *The Prince of Wales*, the principal and most noticeable was the extension of the control cabin in order to accommodate new Decca radars.

New Decca radar installations on *The Prince of Wales.*

Although the original Decca installations were still giving good service to the point where there was never a suggestion that they should be replaced, it had to be admitted that by 1977 they were well out of date.

The replacements on the new craft were desk mounted – hence the need for the extra cabin length – and being matching models allowed for a more efficient back-up in case of failure. In the original configuration the Decca 202, with its tiny 8-inch screen, was barely adequate as an alternative and, mounted laterally, it made for uncomfortable viewing for the navigator who was obliged to twist sideways to operate it. The radar picture on the new sets was clearer and brighter, but still not sufficient enough to dispense with cowls or curtains. At this juncture daylight screens were still being developed.

Both sets could now be viewed simultaneously with most navigators preferring to use the main radar on the usual six-mile range while switching the back-up set to a range higher or lower, depending on whether a look further ahead for traffic patterns or a closer in view if detail was required, such as an approach in fog to either Ramsgate or Calais hoverport.

Compass Problems

So everything was set for a bumper season, but almost immediately the new machine went into service a snag occurred with the compass system, which on the face of it seemed almost trivial, but had the potential to put our expensive new purchase[1] out of action every time the visibility closed in. The compass was not so much inaccurate as erratic and unreliable, which in terms of collision safety was far worse. The prospect of losing a quarter of our earning capability every time fog appeared was not to be contemplated.

In his newly promoted capacity as Flight Manager, Roger had already been following this situation closely:

> *After a few days operation of the new craft, navigators started to report disturbing errors of the compass, but, only it seemed, on the course line to and from Calais. On any other direction the compass reading was perfectly steady, but on the 140°/320° line the compass moved erratically, as much as 10° off course.*
>
> *On such a short route as ours this could be coped with navigationally, but it had serious implications for anti-collision safety, which, after all, was the N4 navigator's principal concern. In any condition of poor visibility, using a faulty compass was not acceptable.*

---

[1] No record remains of the actual price of *The Prince of Wales* but knowing the costs of the first craft and applying inflation it suggests a figure anywhere between £3 million and £4 million.

Murphy's Law being what it is, thick fog rolled in on the very first holiday weekend of the season and threatened to stay for a day or more. The first captain assigned to the Prince of Wales, *quite rightly, declined to depart with a faulty compass in such conditions. As duty captain that day, and therefore desperate not to lose a full to capacity flight because of a technical problem, I asked him if he would be prepared to go if I were to navigate. I hoped my near 2,000 hours navigation experience and my suggestion we do a dog-leg course to Calais, keeping off the problematic normal course, would reassure him enough for him to agree.*

He did, and away we went, but, as soon as we turned on to 140° past our fairway buoy off Ramsgate, the compass waved around erratically. At that point we had little option but to try and stay on the usual course in order to stay clear of the Goodwin wrecks and I was already having some misgivings about the wisdom of it all.

The real worry occurred leaving Calais, with visibility little better than 200 yards and also being initially forced to hold the 320° track, we passed close by a target which, because of the wavering picture, I wasn't at all sure whether it was moving or at anchor. I made a decision there and then that this idea was foolhardy and, like it or not, we would have to cancel any further trips until this problem was sorted out.

However, as luck would have it, by the time we had returned to Pegwell the fog vanished and didn't return for a week or two, which gave us enough time to discover the cause of the error and do something about it.

The only possible explanation for this phenomenon was that there was something in the craft magnetized sufficiently to affect the compass. It was noted that on its arrival from Cowes, *The Prince of Wales* had been placed immediately on the jacks for a check inspection of the skirt, and that night there had been a particularly violent electrical storm. It was plausible that some metal had been struck by lighting and had been magnetized.

This posed the big question, where was there ferrous metal on a craft built almost entirely of aluminium, fibreglass and rubber? The only metal closest to the compass were the pylon hydraulic systems but they seemed too far away to be the possible cause.

After several days of futile investigation the BHC team sent up to solve the problem called in the real experts from the Admiralty Compass

Laboratory. These knowledgeable gentlemen duly turned up with a case full of gadgets and in a very short while identified a small area in the skin of the cabin roof, close to starboard of the front ramp that was highly magnetic. What is more it was within a few feet of the compass receiver unit situated in the forward passenger cabin roof, close by the door to the car deck.

The repeater compass (circled) in front of the captain and the inset shows where the compass unit was moved to in order to cure the problem.

This had to be the culprit, but what in heaven's name was it? It couldn't be the aluminium skin so what was hiding underneath? The BHC team scratched their heads. They had done their best and had worked hard to solve the problem, working well into the night on several occasions, but to be fair they were electronics people not structures. They were now dealing in an area with which they were unfamiliar.

It was in fact a mild steel strap placed between the starboard car deck bulkhead and the aluminium skin. For what purpose the strap was intended is not clear even now. Obviously something needed strengthening and supporting for which the usual duralumin structure was not adequate.

In terms of a solution there was no question of taking the craft structure apart to replace the offending article. The only option was to move the compass unit away from the magnetism. It was moved to a box on the port side of the roof well out of harm's way.

Subsequently, the saying that lightning never strikes twice in the same place proved to be true. It was the first time such a thing occurred on any N4, and as far as is known it didn't happen again – even if it did, from now on we would know what to do about it and rectify it quickly

Unforeseen Costs

Although nothing compared with Seaspeed's budget in moving to the Mk.3 operation, Hoverlloyd's decision to expand its fleet also came with concurrent expenditure, some of which had been unforeseen.

The necessary expansion of the engineering department to cope with the extra maintenance involved, and the unavoidable growth of other areas in traffic and reservations to cope with the increased numbers of passengers, have already been mentioned. However, what was not fully realised was the large increase in operational staff required to handle the half-hourly schedule.

The Prince of Wales in the centre at Calais between *Sure* in the foreground and Seaspeed's *The Princess Anne* in the background.

The smooth transition from two hovercraft to three in 1972 with no need to change or add to the port facilities, and very little increase in staff requirements, may have given the false impression that increasing from three to four would be just as easy.

So much of Hoverlloyd's administrative documentation was destroyed during the merger in 1981 that it is difficult now to ascertain the numbers involved, but if the need for two duty captains per day to manage the operation and the addition of standby crewing to cover for sickness are added, then it is likely that 17 crews were required to run this now highly intensive timetable. This is confirmed by June Cooper:

> *Additional stewardesses were required for seasonal work from Easter to the end of the summer season. The operation expanded by firstly widening the craft and later by the acquisition of two additional craft so that, at its height, there were 36 permanent stewardesses and up to 100 additional temporary girls.*

Extrapolating from there it suggests the likely summer complement was 51 flight crew, 136 cabin staff and over 100 car deck hands. For the cabin crew that must have been an enormous task to train such a large number of seasonal recruits, but probably alleviated to some extent by the immense popularity of the company, ensuring that a high proportion of the previously trained would want to return each year.

When Hoverlloyd started out with two hovercraft in 1969 the schedule was run with five crews, so how did doubling that number of craft eight years later more than triple the number of personnel? Roger has only a 'best guess':

> *When I took on the job of flight manager early in 1977 and with it the responsibility of crewing, I was shocked when starting to map out the summer season requirements, as to how many personnel were needed. I rearranged schedule patterns and recalculated the figures several times, not initially believing them.*
>
> *Now, without the day-to-day statistics to hand, it is impossible to attempt even the simplest analysis. Of course things had changed in the ensuing years. CAA safety rules had become more stringent, leading to slightly less hours worked; the gradual increase in the schedule had demonstrated the need for 'on the ground operational control', and hence confirmed the role of the duty captain, which removed at least three captains from the summer roster, none which could be described as major influences that could have brought about such a drastic increase in crewing.*
>
> *My only educated guess is 'quantum', the idea that this whole process is not a continuum but a series of quantum jumps. Three craft could be operated with more or less the same structure as two, but, add another, then a jump was needed to cope. If my theory is correct we could probably have added a fifth craft with comparatively little adjustment.*

One of the new intakes of cabin crew for the summer season.

Duty Captains

And intensive it certainly was. Craft were departing and arriving on schedule every 20 minutes. Any hiccup or breakdown, however minor, led to bunching up of movements, which needed careful management. Any snag requiring a craft removed from the schedule for a while meant not only the rearranging of passenger loads, but reassignment of crews to cover an adjustment of the timetable. This was further complicated by the necessity to keep within the complex CAA rules concerning flight crew operational safety hours.

Duty captains, who were operational 'supremos' for the day, had to manage all this, and some became good at it. It certainly required a level of quick thinking, a mathematical facility with numbers perhaps, but, above all, the diplomatic skill to handle complaints from crews who were still to do their full schedule, while others might be let off one trip and going home early.

On bad days of weather or breakdown delays – the latter still occurred even at this juncture – the DC would finish his day feeling the need for a good drop of something and a good night's rest.

The most difficult situation to handle was when a craft, for whatever reason, was stranded in Calais for the night. Up to that time the standard procedure was for the crew to spend the night in Calais and bring their craft back the following morning. This arrangement worked well but, under the new, more concentrated schedule, having a machine in the wrong place was headache enough, but a crew out of sync with the next morning's roster was an even bigger one.

As a better option, as far as the schedule was concerned, the standard procedure was changed to return the stranded crew by ferry as soon as possible, so with a night in their own beds they were fresh and available to slot into their roster the next day without interruption to the scheme of things.

The standby crew for the day had already been briefed the previous evening to go over as passengers on the first flight to bring the stranded craft back. If everything went according to plan they would arrive in Calais at 06:40, pick the craft up and be back in Pegwell a little after 07:30 for the rostered crew to pick up that scheduled departure. This meant a late departure of the machine but with every chance of it catching up later in the day. Although, in fact, leaving the crew to bring it back would in fact ensure the 07:30 departing on time, it was felt the delay was a small price to pay for keeping the crew roster in order.

It has to be admitted that the change in procedure was not a popular move, as most of us enjoyed the prospect of a night out on the town in Calais. Stories of crew high-jinks abound, most of which, if related here, would no doubt instigate libel actions.

Set Crews

Two other changes were introduced during this period, which had significant effects on not just the operation of the craft but the life of the operations personnel in general.

The first was the introduction in 1978 of what became known as 'set crews', the combining of all craft staff into one roster. What had previously been organised by each sub-department, due to differing regulatory requirements, resulted in a different work pattern so that at any one time of the day there would be a different mix of personnel. This was not necessarily a problem when things were running smoothly but could present a huge difficulty for the duty captain if the schedule was badly disrupted. The DC had to rearrange three different groups with three different sets of rules, which wasn't easy.

Putting everyone together as one coherent crew could only be done if everyone followed the flight crew regime. They were the only people whose work pattern was regulated by a government department and therefore could not be altered.

Robin Paine (centre), with his set crew of fourteen. 1979.

Initially there was some reluctance to embark on this new approach as the general opinion was that flight crew had it 'cushy' and fitting everyone into their routine would necessitate even more cabin and car deck personnel. On closer examination of the whole scheme this concern turned out to be unfounded as the numbers remained much the same, with in fact a slight decrease in the number of stewardesses.

So the 1978 season saw a total change. Instead of three sets of rosters to consult when things went wrong, only complete teams needed to be juggled, without worrying about conflicting work patterns; which was a definite boon for the duty captain and a clear improvement in efficiency.

This was not the only improvement as the increase in morale was palpable. At all levels the operations staff loved the new set-up. Summer crews not only worked together, but also, because they were on the same days off, spent a lot of leisure time together.

Senior stewardess Cathy Whitnall's comment is typical:

> *Hoverlloyd was more like a family business – everyone knew each other and the different characters made each day enjoyable and interesting. Many good friendships were made, especially when we were rostered in set crews working with the same people for many weeks during the busy summer season. There has been a bond between us all, which is evident when we have met up again in recent years.*

Cathy Whitnall.

Crew outings and evening entertainments were organised on a regular basis. At one point crews sported T-shirts, the logo usually based on the particular captain's name. Brenda King still has hers:

While having a clear out, I recently came across both my 'Biggles Bunch' sweat and T-shirts. You may recall that they were produced to be worn in the bar as opposed to uniform.

Running through the list of crews that ordered the shirts, John Lloyd (Biggles) recalled Captain Ward requesting 'Wards Wonders' before any of his crew made other suggestions that would not have been appropriate!

Can't imagine what the alternative might have been!

But there was one aspect of the change that hadn't occurred to a busy management. Roger, who was the principal architect of set crews, on looking back at his achievement, is not sure whether to be proud or embarrassed.

June Cooper says the crewing system encouraged several summer romances, culminating in marriage. Roger has quite the opposite opinion:

I think most of the liaisons happened before we threw everyone together. In fact I have quite the opposite recollection. As it was solely my idea, which I remember Joan Stroud (Chief Stewardess) at the time being dead against, I carry the full blame for the number of existing marriages that fell apart. Two come immediately to mind, but I am sure there were others. I guess everything has its advantages and disadvantages.

I feel I should have been more aware of the consequences of throwing boys and girls together in such a manner. It was either total naivety, or my mind was on other things.

One wit, on observing the intimacy the set crew situation was creating, was minded to remark:

> *It would appear that the crew that hangs together bangs together!*

Right-hand Seat Change

Concerning the second important change occurring in 1979, Roger has no misgivings:

> *It was long overdue; there had been plenty of previous signals highlighting the problem. Robin's initial uncertainty with a new untried navigator and my recent incident with the faulty compass on* The Prince of Wales, *plus other occasions too many to number, all pointed to one serious flaw in our organisation. Why were we placing the most junior and untried member of the flight crew in the most important seat, the navigator's?*

In poor visibility the most inexperienced person was placed virtually in command, while the man with real expertise who had spent the previous few months training the new recruit sat in the front as flight engineer and watched the dials.

To be fair, we had trained these people to a high degree and those who passed through the rigorous test to become seasonal navigators never let us down. But in the end, however well the navigator was trained and at whatever level of excellence he was considered to be, he was still a vastly inexperienced person occasionally put in charge of the lives of 300 people. This paradox had to be addressed.

In answer to the obvious question – if this anomaly and its implications for operational safety had been appreciated for some time, why was it not acted upon earlier? The answer was the problem of cost. For most of this period the training of 1st officer/flight engineers would have involved sending recruits to the BHC establishment in Cowes for the systems and structures course, followed by another week at Rolls Royce to learn about the Proteus. In comparison to this significant training expense, the cost of training a navigator was nothing as it was handled by our own experts in-house. In our early years of struggle, extra expenditure on training was not to be contemplated.

But times had changed. In the latter half of the decade, as described in the previous chapter, because we could boast our own systems and structure experts in the flight crew, there was no need for an expensive course in the Isle of Wight. Likewise the engineering department now had permission from Rolls Royce to conduct their own Proteus engine overhauls, so the other important training element could also be conducted in-house.

In 1979, with no financial impediments remaining, the step was taken to change the seats. New seasonal recruits were trained at Pegwell as 2nd officer/flight engineers and the 1st officer reverted to the job of navigator.

Not everyone was happy with the change. The right-hand seat with its dual controls was, in airline parlance, the co-pilot's seat and therefore the superior position in terms of rank. On the N4 the fact that the navigator's role was by far the more important did not mitigate for some the feeling that it was a demotion. Apart from which 1st officers liked to sit in front, look out the window and chat to the captain or even steer the craft in suitable conditions during the passage. Being relegated to the back seemed a drop in social status too.

Regardless, like it or not, no one argued that it was the right step to take.

The Joss Bay Incident

Perhaps we might have been entitled to think that after almost ten years of operation we had risen sufficiently up the learning curve to assume that we had most of the ins and outs of hovering weighed off and there was little left to surprise us. As always, when a certain complacency sets in, something occurs that jerks us out of our smug self-congratulation and reminds us that we still don't know it all.

On 15th November 1978, *Sure,* with Ted Ruckert as captain, reported at around 10:30 that he had lost the two starboard engines and was struggling to maintain course, with his speed down to 5 knots. The craft was returning from Calais in rapidly deteriorating weather. Other departures had already been cancelled and passengers diverted to Dover. Ted in fact was not driving, but had allowed his 1st officer, Tony Braddock, having recently checked out as captain, to take control in the left-hand seat to accumulate more experience.

Ted Ruckert:

> *The wind was WSW 30 knots with an 8-foot sea and swell. We had set off from Pegwell at 0845 with Tony in the left-hand seat to get a bit of bad weather experience. In the Kellet Gut we hit a large wave with an almighty 'thump' on the starboard*

bow/side. Apart from the unpleasant bang, nothing seemed untoward mechanically or structurally and we continued via the South Goodwin to Calais, arriving there at about 0945. We fuelled, loaded and departed at about 1010 for Ramsgate (Tony still in the left-hand seat).

Captain Ted Ruckert.

It is likely that by this time the weather had deteriorated well past the 30 knots and was well up to our Force 8 operational limits approaching 40 knots. Although Ted does not mention it, a careful examination whilst in Calais showed no apparent damage and presented no concern for the return journey despite the weather:

At about 1020/1030 (still in very rough and unpleasant conditions) the craft began to struggle and slow down, so I asked the car deck supervisor to check around to see if he could see anything untoward through the windows. After a few minutes he reported back that there appeared to be damage aft on the starboard side by the air filters with large lumps of skirt 'flailing' around. About this time engine temperature indicators on that side forced us to shut down both starboard engines.

We were now about mid channel and our speed was down to about 5 knots. I then changed seats with Tony and decided to let the craft make the best course and speed towards the Kent coast. From our track it appeared that we would be lucky to make landfall at the North Foreland.

Without any doubt Ted made the right decision to carry on towards the Kent coast, as in a fast-deteriorating westerly gale there would have been no shelter along the French side of the Channel and the machine would have been swept up the coast from Calais to who knows where.

Pegwell Bay Hoverport Radio Room, which was the centre of communications and from where operations were co-ordinated in the event of problems and emergencies. (The jack staff in front of the control cabin was put there for the captain to use as a reference point.)

Roger, as duty captain that day, had other concerns:

I had taken a Dover Strait Decca chart into the radio room and asked Ted to give me regular Decca position reports, which I plotted. My principle concern was the worsening weather situation. Sure's *track was moving steadily to the east and if extrapolated further was likely to miss North Foreland altogether. Added to that was the thought that, if another engine failed, progress towards the English coast would finish and the craft would be blown further off track, somewhere out into the North Sea.*

I took the precaution of arranging for one of the Dover tugs to be put on standby, just in case my worst fears were substantiated and we needed assistance.

The port side unreachable towing lug from on board the craft.

It was then, with some disquiet, I realised we were in no way prepared for such an eventuality. We could send the tug to take Sure *under tow, but how were the hovercraft crew to make fast the tow line? It would entail someone walking along the plenum top forward to make fast to one of the anchor rope lugs as highlighted in the photograph above. This was all very well in theory, but without a rail with which to fasten a safety harness – the evacuation handrail stopped by no.1 life raft – the chances in a heavy sea of the crew member being swept away were almost certain.*

Luckily for us all, Ted battled on and, as the craft got near the lee of the land, he managed to pull the track more to the west and south of North Foreland.

There and then Ted made another sensible decision, which was not to try to reach Pegwell, but to land in Joss Bay[2], the first beach to the north of North Foreland. Trying to move around the corner past Ramsgate Harbour into the teeth of an increasing westerly gale was going to be near impossible, on top of which the craft, after its long struggle to return, was running out of fuel.

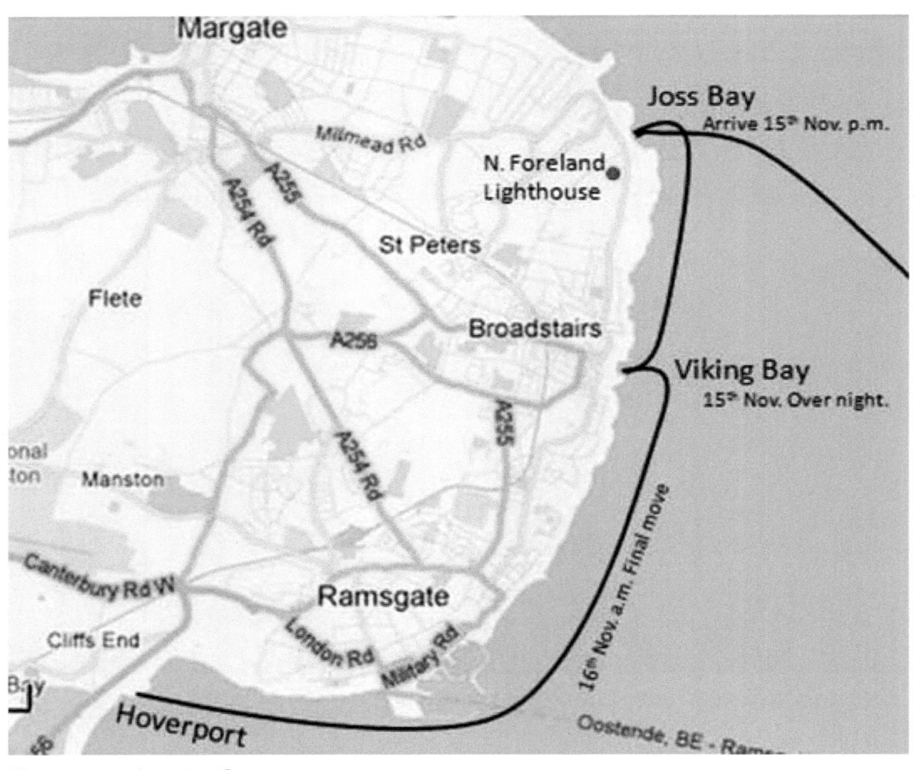

The route taken by *Sure*.

---

[2] Robin, who happened to be living on the North Foreland at the time, was on a day off when he saw *Sure* drifting past his window, as a result of which he was able to take the series of photographs shown on the following pages.

Sure is now about half a mile out to sea opposite the North Foreland lighthouse and well in the lee of the land, enabling her to get closer to the shore.

Sure approaching Joss Bay close in to the shore.

Sure's first contact with the beach at Joss Bay.

Things were immediately organised back at Pegwell and *The Prince of Wales* was rapidly crewed up to arrange a quick transfer of passengers and cars as smoothly as possible so they could continue their journeys with no further mishaps.

Sure arrived at Joss Bay at 1.30pm. Roger was one of the first to arrive at the stricken craft and, after congratulating Ted on his amazing efforts, went into the port cabin, where all the passengers had been transferred, to explain the arrangements:

> *I don't remember exactly what I said now by way of apology for the ordeal, other than, "I can assure you, things like this don't happen every day." I am not at all sure they necessarily believed me.*

Joss Bay – Stage Two

While the transfer of the passenger load was going smoothly the engineers were faced with a difficult repair problem. The damage to the starboard quarter was extreme. Structurally, the whole outer skirt hinge line, and with it the skirt assembly, was missing underneath the starboard no.4 engine space, from forward of the caboose on the side, around to just below the rear of the engine. More seriously, what had disappeared with the accompanying structure was the installation of the four oil coolers, hence the overheating of the two starboard engines.

Ted Ruckert managed to manoeuvre *Sure* to a suitable location further up the beach. The extent of the damage starts to become apparent.

The Prince of Wales comes to Joss Bay to recover *Sure's* passengers and cars for the five-mile 'round the coast' trip back to Pegwell Bay.

If the damaged craft was to be moved at all, the first priority was to reinstall the coolers. The structure and skirt could have also been replaced *in situ* on the beach, but the thinking probably went – "If the craft had got across the Channel in a howling gale on only two engines in its present state, she should get around the corner to Pegwell on four engines without much difficulty."

The engineers, as amazingly efficient as ever, replaced the missing oil coolers in double quick time and a standby crew under the command of Bob Middleton was commissioned to drive *Sure* back to Pegwell.

Captain Bob Middleton. (The Decca flight log shows that this picture was taken in Calais.)

Bob was always considered one of our most skilful captains but in this instance, moving the craft that day proved beyond even his capabilities.

The wild weather had by no means abated and Bob found himself battling straight into a strong headwind which, by this time, probably had increased to Force 10 and daylight was fading. Still dragging one leg, as it were, with the badly damaged starboard quarter, he could make very little progress. He was left with a decision either to return back to the safe shelter of Joss Bay or to find some other haven nearby that was at least a little closer to home.

Bob chose Viking Bay, which is the beach around which the town of Broadstairs is situated. Typical of Bob he had not picked the easiest option. Viking Bay is comparatively small. The beach area is not much bigger than the hard standing at Pegwell Hoverport and is bounded on the north side by a substantial stone breakwater which, unfortunately in this case, was on the downwind side.

Sure leaves Joss Bay in an attempt to return to Pegwell Bay with Bob Middleton in command. Although he has four engines operational, the escape of cushion air means that progress is slow.

An added problem was the very steep slope up to the high tide mark, which was the only secure place to park the craft overnight. Undaunted as ever, Bob gunned the craft in towards the southern end of the beach at full pelt, crested the slope and then with a mere pocket handkerchief of space remaining on the plateau, swung around to leave *Sure* facing north and parallel with the road – an object lesson in SR.N4 control that very few others could emulate.

Overnight the gale blew itself out and left a calm clear day to attempt the final return to Pegwell. Roger assigned himself the job and, with his crew, was able to limp back to Pegwell Bay without incident.

So ended another episode, which might have turned out a lot worse, and another lesson learnt. It was a reminder that for all our experience and success over the preceding decade, the SR.N4 was in many ways still a prototype. There were still many flaws to sort out, not least of which was the expensive skirt system, on top of which it was also a timely reminder that the hovercraft was still 'an aircraft at sea', and at times needed careful handling.

With slower than anticipated progress in still atrocious conditions and fading light with no hope of reaching Pegwell Bay, Bob Middleton decides to attempt a difficult approach into Viking Bay, Broadstairs.

With an ebbing tide, Bob Middleton successfully reaches the beach in Viking Bay, Broadstairs. Bleak House, once the home of Charles Dickens, can be seen in the background.

Sure on the beach in Viking Bay, Broadstairs, where she spent the night before returning to Pegwell Bay.

Engineers assessing the damage to *Sure* in Viking Bay. Note the replaced oil coolers beneath the deck.

Is He Waving?

Another less fraught incident occurred during this period which provided lessons all of their own. The question was – "What do you do with a dead body in the Channel?" We certainly had no standard operating procedure or book of protocol to help, and in the end the answer from both sides of the Channel seemed to be, 'not a lot!'

Roger Syms (author).

Robin was captain and Roger was his 1st officer, neither of who have a record of the time of the year, but both remember it as a pleasant relatively calm day but still a moderate amount of swell running.

Robin estimates that we were approaching mid-Channel when:

> *We spotted a body with a rope around its waist lying face down in the water. This was a new experience for both of us, but we thought we should at least stop (on the basis that there was probably some law that we should) and investigate, which we duly did. As captain, I phoned down to the cabin crew girls on each side of the craft to let them know what was going on. We didn't want to upset the passengers, so I asked the girls to draw the curtains in the forward cabin and made an announcement to the passengers telling them what we had found, but that we were still uncertain as to what to do.*

With the aim of also shielding our passengers from the grisly sight, Roger thought it would be a good idea for him to go over the roof, across to the starboard side and down the ladder by no. 1 life raft, thereby avoiding the necessity to open the side cabin doors:

> *I am full of good ideas, but this one was not the best. Unlike the port side, where there was a rail over the roof to the port side*

ladder, there was just open roof to the ladder on the starboard side[3]. The craft was now settled in the water and, although its movement in the swell was comparatively gentle, walking across the roof without anything to hang onto was worrying to say the least. I had a choice of passing close in front of the yawning hole of no. 2 fan air intake and risk being pitched down into the mincer of no.2 fan, or taking a wider berth forward and risk being projected off the roof altogether and landing in the sea beside the deceased. It didn't take long to decide that both options were not for me.

Robin Paine (author).

Robin was mildly amused to see Roger return:

Roger didn't get very far before he reappeared on the flight deck muttering something like, "Bugger that for a game of soldiers". We then decided that it would be best if he went down the ladder to the car deck and out through the main cabin door with his harness on to investigate.

The initial premise was that we would take a closer look at the body in order to provide a good description to the authorities and possibly actually retrieve the corpse to bring it ashore. Once Roger managed a closer look standing on the plenum by the main cabin door, it soon became clear that any worthwhile description was impossible, and as far as retrieval was concerned we simply didn't have the wherewithal:

He was floating face down on the surface dressed in shirt and trousers with a few feet of rope around his waist, so no description of his features could be ascertained and, without going into detail, enough to say that manhandling a body that had been in the sea, for who knows how long, was not a job for amateurs.

---

3 Another mystery of N4 design that was never explained – why was there no hand rail on the starboard side, as there was on the port side, or a rail around the control cabin to be able to get there?.

Before leaving the scene Robin tried asking for advice:

I called our Ramsgate base on the VHF asking for guidance, but they were at a loss. They then called the UK coast guard, but failed to elicit much interest from them either. The hoverport control room at Calais heard what was going on and a very agitated Jean Louf, the Calais manager, was asking us to be very specific as to which side of the Channel the body was on. In addition to that, he was extremely keen that we did not attempt to pick it up and bring it to Calais, because, if we landed it on French soil, the French authorities would be responsible for dealing with it.

Eventually, with no sense as to what should be done coming from either side of the Channel we got back on our cushion of air and went on our merry way to Calais.

On returning to Pegwell, Robin was decidedly unhappy having wasted time and fuel to no avail:

I was still feeling a bit miffed at having lost about 15 minutes on our schedule and having burnt, unnecessarily, several extra gallons of fuel, thinking we were doing a great job in identifying a body at sea. So I felt inclined to phone the Coast Guard myself. I pointed out that I thought I had complied with whatever law applied to finding a body at sea, but there seemed to be a distinct lack of interest from the very authority that should have shown interest. I then asked, if there was another occurrence, should I even bother to stop. Back came the unhelpful reply – "Only if he waves!"

The international crisis as to who was responsible was obligingly resolved by the corpse two days later, when it was washed up on the beach on the Kent coast. It turned out that the dead man was a sailor from an East German cargo ship.

After such an interesting lesson, perhaps we should have included in our *Operations Manual* and the *Captain's Pocket Book* the following:

If you pass a body at sea:

1. *Wave politely.*

2. *If it waves back, smile and wave again.*

3. *If it doesn't wave, just continue on.*

4. *It is not necessary to inform any authority, they will only get upset.*

Time Moves On

Moving inexorably into Hoverlloyd's second decade of operations it was inevitable that a number of celebratory milestones should appear – the zillionth passenger, car, nautical mile etc. But the occasion that particularly comes to mind is one of our long-serving stewardesses, Betty Dowle, being awarded the Freedom of Calais by the Calais Chamber of Commerce, to commemorate her 10,000th trip across the Channel.

Betty was one of our original stalwarts who joined the company in February 1971. Like so many of Hoverlloyd's young personnel, Betty had an adventurous streak. Apart from travelling extensively around Europe, one of her many adventures prior to joining was working as an extra in the film industry. It was not surprising therefore that her name came most to the fore when someone from the cabin staff was needed for publicity events:

> *It was a very exciting time and I was asked to do many promotional events. Among the highlights of these were newspaper stories involving members of the public. For example, a 100 year old Mrs Hermitage taking her first cross-Channel trip and escorting a Pearly King and Queen from London's East End.*
>
> *Also I filmed with Michael Caine in "The Black Windmill" – Not a lot of people know that.*
>
> *The highlight of my career was being awarded the Freedom of Calais by the Calais Chamber of Commerce on the occasion of my 10,000th crossing. I was presented with a gold watch and a commemorative medal.*
>
> *A photograph of this event appearing in the* Daily Telegraph *prompted a proposal of marriage from a London publisher.*

The proposal must have been flattering, but Betty was already happily married and, besides, shortly after her presentation:

> *I left when I became pregnant after nearly ten very enjoyable years with Hoverlloyd, all served as a stewardess on the hovercraft, and amazing myself by spending that amount of time in one job!*

PART FIVE

The Final Years

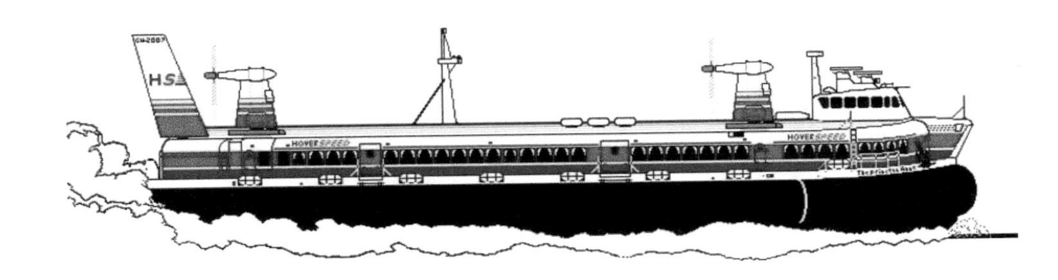

Chapter 24

Bigger and Better?

1978 The Mk.3 Arrives

The Princess Anne as a Mk.3 at speed in the Channel. The trade-off for her large increase in capacity for a relatively small increase in power was a noticeable reduction in speed. Note how the craft now 'sits in', rather than 'sits on', the skirt giving a better suspension and more protection to the structure.

Although Hoverlloyd was rightly proud of its four-craft fleet, the first arrival of the newly converted *The Princess Anne* at Calais was a truly magnificent sight. One could not help a certain twinge of envy.

The extra 55 feet in length gave the machine an air of streamlined elegance which made the Mk.2 by comparison look somewhat short, fat and dumpy. The impression was of a beautiful, leggy, teenage girl, showing herself off to her dumpy middle-aged aunt.

The new 21-foot wide-blade propellers referred to as 'aerodynamic shovels'.

Apart from the length, two other features attracted the eye. Firstly, the remodelled propellers were now 2 feet longer in diameter and wider in the chord. Their appearance very soon brought forth the soubriquet 'aerodynamic shovels' from our ever inventive engineers. But the feature that stood out particularly was what was termed the 'low-pressure skirt'. It was far larger than the Mk.2 construction, adding extra cushion height all round but with a fuller bag design, allowing the hard structure of the Mk.3 to 'sit in' rather than 'sit on' the skirt. The new design provided a softer suspension and also acted as a fender. The combined effect provided a greater degree of passenger comfort and increased the sea-keeping capabilities of the new machine, which was clearly proved in Force 9 conditions on acceptance trials.

It cannot be denied that this was the ultimate vindication for Cockerel's invention; here was a vehicle over double in weight to its original design (150 tons), carrying almost double the numbers of cars and passengers, yet using exactly the same power installation with only a minor increase in output (400 shp) and fuel consumption.

Date	Model	All-up Weight Loaded	Engine Rating
1968	SR.N4 Mk.1	190 tons	3,400 shp
1972	SR.N4 Mk.2	200 tons	3,400 shp
1977	SR.N4 Mk.3	320 tons	3,800 shp

The Princess Anne as a Mk.1 was cut in half, the two sections moved apart and a new 55-foot section inserted.

The great shame was that all this magnificence and efficiency should have cost so much. Although inflation in Britain during the '70s, as in the rest of the world, had increased considerably, it hardly justified the construction costs of the Mk.3. Sharing BHC's loss of £4 million between the two craft, which reputedly cost £12 million and £14 million respectively, brings the total figure for the completion of the second to be converted, *The Princess Margaret,* to a staggering £16 million.

It is said that Hoverlloyd's original Mk.1 craft were purchased in 1969 for £1.2 million, which at the contemporary inflation rate should have been £3.3 million in 1977[1]. In trying to calculate a Mk.3 build from scratch the addition of an extra 55 feet of structure is difficult to estimate now, but on the face of it, it was only structure. All the expensive bits, control systems, propeller assemblies and engines were already there.

The new 55-foot section nearing completion.

Nonetheless, fitting it into the existing structure was not quite so simple at all. Ray Wheeler, BHC's design boss explains the biggest problem:

> *John Lefeaux and Peter Yerbury (Seaspeed's MD and Chief Engineer) approached me to stretch an existing craft, and by then both craft were quite old. There were all sorts of complications in stretching it. For instance, the outer engines had to be remounted because the angle would have been wrong. With regard to the electrics, we had an argument in the design department about whether we were going to cut the*

---

[1] The Bank of England online inflation calculator.

cables and then rejoin them or replace them. I said to John and Peter that I would not recommend stretching the existing craft. It might even be cheaper to build a new one. But Dick Stanton-Jones decided to go ahead and stretch it and, "That we would do a careful estimate." I said we couldn't estimate what our factory was going to do when we didn't know exactly what was involved or what we would find that needed repairing or replacing.

Ray was proved to be right. It most certainly would have been a lot easier and cheaper to build a new hovercraft. So, going back to our best guess at the inflation price above, if an extra 50% is added to the £3.3 million figure to allow for the complexities, it still only brings the total estimate to £4.65 million for a newly built Mk.3. Going back to the Seaspeed conversion, even multiplying by a factor of two, to allow for the added problems of modifying an existing craft, £9.3 million comes nowhere near the actual cost of £16 million.

The new section joined to the two original sections.

It has to be accepted that perhaps these inflation numbers are presumably calculated on an average set of goods and services, which may not necessarily reflect accurately on this type of construction, and it also has to be admitted most of the documents that might support these facts are unavailable now. However, Ray does point to another possible answer:

What you need to understand is that all the aircraft companies up to something like 1965, were funded by the government – on a cost plus basis. So the financial side of an aircraft company was totally isolated and you weren't allowed to know what went on in that department. It was taboo. When I tried to see what costing methods were used to get at the construction costs, I couldn't get the information. And that's how the company was based.

The Princess Anne being rolled out of the Columbine Hangar at Cowes as an SR.N4 Mk.3 before taking to the water on 6th April 1978.

Although this was now long past 1965, obviously the mind-set still lived on in BHC and Westland. The fact was both these companies, although involved to some extent in the commercial world, were first and foremost throughout their history, contactors to the government. The government's rather lax approach, of which the 'cost plus' rules were probably only a part, encouraged a complacency which ruled out any sense of cost control whatever. It simply did not impinge on the company's thinking.

Industrial Problems for Seaspeed

Having made the decision to stretch their Mk.1s to become Mk.3s, Seaspeed's management's attention turned to the operational detail. Peter Barr, as the senior captain, had a semi-managerial role with the title of 'Flight Captain', as a result of which he attended the numerous and lengthy meetings in London with John Lefeaux, Les Thyer, Gordon Harris and Peter Yerbury about the operation of the modified craft. Peter Barr explains:

The Princess Anne as a Mk.3 being shown off to the public with her new, low pressure skirt and its extra height.

The lengthened main cabin with two rows of three abreast seating. (The same configuration as the Mk.2.)

We decided what crew we needed and who was going to do what to whom etc. Then John Lefeaux tried to put this to the unions and they just said, "No". They said the guys on the car deck had to have a special cabin of their own where the inner cabins used to be, and in it they had to have a boiler to make their tea and so on and so forth. They already had an observer's watch cabin (starboard side forward) as part of the safety requirement.

Lefeaux put it to them that we'd had all the meetings with the relevant departments and special cabins for car deck crew with tea boilers did not come into it. My name was quoted and the answer came back, "If you want the craft to operate this will happen and we don't give a bugger what the flight captain says." So they got it and it was at that point that I vacated my 50/50 managerial/flying role and went back into flight crew.

The lengthened car deck, as seen from the bow ramp with the 'watch observers' cabin near left. The 'tea room' required by the unions was half an old inner cabin on the right-hand side of the picture, which took up two car spaces. It was removed when the Hoverlloyd/Seaspeed merger took place. (This picture was taken after the merger.) Note the ladder to the flight deck.

It wasn't just the unions that were presenting a problem to management. The flight crew saw this as an opportunity to improve their own situation as their salaries were below those of Hoverlloyd. The Seaspeed management, however, was unwilling to enter into negotiations despite the 41% increase in passenger capacity (254 to 424), and 46% increase in vehicle capacity (30 to 55) of the Mk.3, with the result that the flight crew threatened strike action. Peter Barr:

I'd been sent to carry out the trials on the stretched craft and I was looking forward to that, as you can imagine, but I was told in no uncertain terms by my colleagues that we weren't going to do it. So the union attitude affected us as well, but strangely it was the only kind of language that the BR system understood. After this ultimatum the management asked if we would go to arbitration, to which we said, "Yes" and got virtually everything we had asked for. Derek Meredith, as management, carried out the pre-service trials.

It was just those initial few days. It was a combination of poor management and strong union power. When you know that this is the language they understand then you tend to launch into it pretty swiftly.

Despite all the modifications and advances in skirt technology, there were still the occasional skirt problems. Note the billowing skirt on the starboard side. It would appear that a long length of the inner skirt has come adrift for a reason not clear in this picture.

The French Join In

In 1976 when Seaspeed submitted its 'Super 4' expansion plans to the British Rail Board Investment Committee for approval, it also included, as part of the application, the intention of SNCF – the French equivalent of British Rail – to operate two SEDAM N500 hovercraft on the Dover/Boulogne route. These two craft, each capable of carrying 400 passengers and 60 cars, would double the throughput of traffic at Dover and thereby help pay off the enormous £14 million cost of building the new hoverport.

Such a large request for funds from a nationalised enterprise, which was currently doing everything it could to minimise expenditure, was bound to be difficult and, as John Lefeaux points out, Seaspeed's financial record according to the investment board was not an impressive one:

> *BRB's Investment Committee... probed it thoroughly and with deep scepticism in view of 'our poor track record with financial results'... In the end we convinced them and our submission was approved on 27th February 1976*[2].

Presumably the additional revenue from the French hovercraft would have been a significant factor in swaying the committee's judgement in Seaspeed's favour. In retrospect, considering the total failure of the SEDAM project, it was a bad decision, but, even without the benefit of hindsight, it is puzzling now to understand why the committee was prepared to provide a substantial amount of money based on the performance of an untried machine.

After all this was not just another BHC hovercraft; another step in a long period of tried and true development. This was a totally different, even radical concept, using different construction, different engines, different control arrangements and, most revolutionary of all, a totally novel approach to skirt design.

The SEDAM[3] N500

Billed by its proud creators as 'the biggest and fastest hovercraft in the world', the N500 was, from the ground up, a completely new concept in every way. It is informative to compare its statistics with the SR.N4 Mk.3 in the table on the following page. Units used are metric.

The claim that it was the largest and fastest was marginal to say the least. It was impressive nonetheless. Observing its towering profile from ground level it cannot be denied, looking at its height and apparent bulk, that it certainly looked bigger. However, considering the similar plan area of the two machines, and the fact that the N500 was double decked, it is remarkable that what should have been twice the space did not translate itself into car and passenger numbers.

One major difference between the two was the method of construction. While the Super 4 was, in the BHC tradition, airframe construction using

---

2 John Lefeaux: *Whatever Happened to the Hovercraft?* Pentland Books, 2001.

3 Société D'Etudes et de Développement des Aéroglisseurs Marins – Aéroglisseurs the French term for hovercraft, roughly translates as 'aero-sliders' or perhaps 'gliders'.

thin duralumin joined by Redux adhesive and rivets, the N500 was marine constructed with thicker sheets of aluminium, which were welded.

	SEDAM N500	**BHC SR.N4 Mk.3**
Length Overall	50 metres	56.38 metres
Beam	23 metres	23.16 metres
Height on hover	17 metres	14 metres
Motive Power	3 Lycoming gas turbines, 3,800shp each.	4 Rolls Royce Proteus gas turbines, 3,800shp each providing motive power and lift.
Lift Power	2 Lycoming gas turbines, 3,200hp each	
Maximum Speed	70 knots	75 knots
Maximum Wave Height	2.5 metres	3 metres
All up weight (AUW)	260 tonnes	325 tonnes
Passengers	400	424
Cars	60	55

It certainly was a stronger structure, not just in the method of construction but in its basic architecture. Looking at the cutaway diagram on the page after next, the two bulkheads running the full length and height of the craft not only provided an extra height lane for freight vehicles and coaches, but had an additional purpose by adding enormously to the longitudinal strength of the machine. Welded firmly to the deck, forming the top of the buoyancy tank, together they formed a very strong inverted T beam. The N500 was never going to break its back.

However, one unfortunate result of such an arrangement was the isolation of one upper-deck passenger cabin from the other. Admittedly the passenger cabins were also separated on the N4, but with the required safety pathway between the cars, it was a quick process to transfer passengers from one side to another in the case of emergency. On the French machine there was a pathway between, but it would entail passengers filing down one set of stairs walking around a coach in the centre laneway, with not a lot space on each side, and back up another steep set of stairs on the other side.

A model of the SEDAM N500 in its original design.

Another aspect of emergency evacuation was that the height of the upper deck made it necessary to use aircraft-style chutes attached to each of the three emergency exits (shown as *issues de secours* in the following French cutaway). The problem here was that these were designed for aircraft, largely to evacuate passengers onto a hard-standing for which they were quite adequate. They were not so effective from high-sided ships onto sea and swell, as there was a tendency for the chute to twist or bend, either preventing evacuees from sliding further or throwing them off altogether. Similar to aircraft procedure the French authority only required a demonstration of the chutes onto dry land for the issue of a safety certificate; it is very doubtful they would have performed satisfactorily at sea.

Luckily the chutes were never called upon in an emergency. In the 30 years since, a great deal of development work has been carried out on ship versions of the chute, which are better designed now to cope with the sea and swell.

N500 Engines and Control

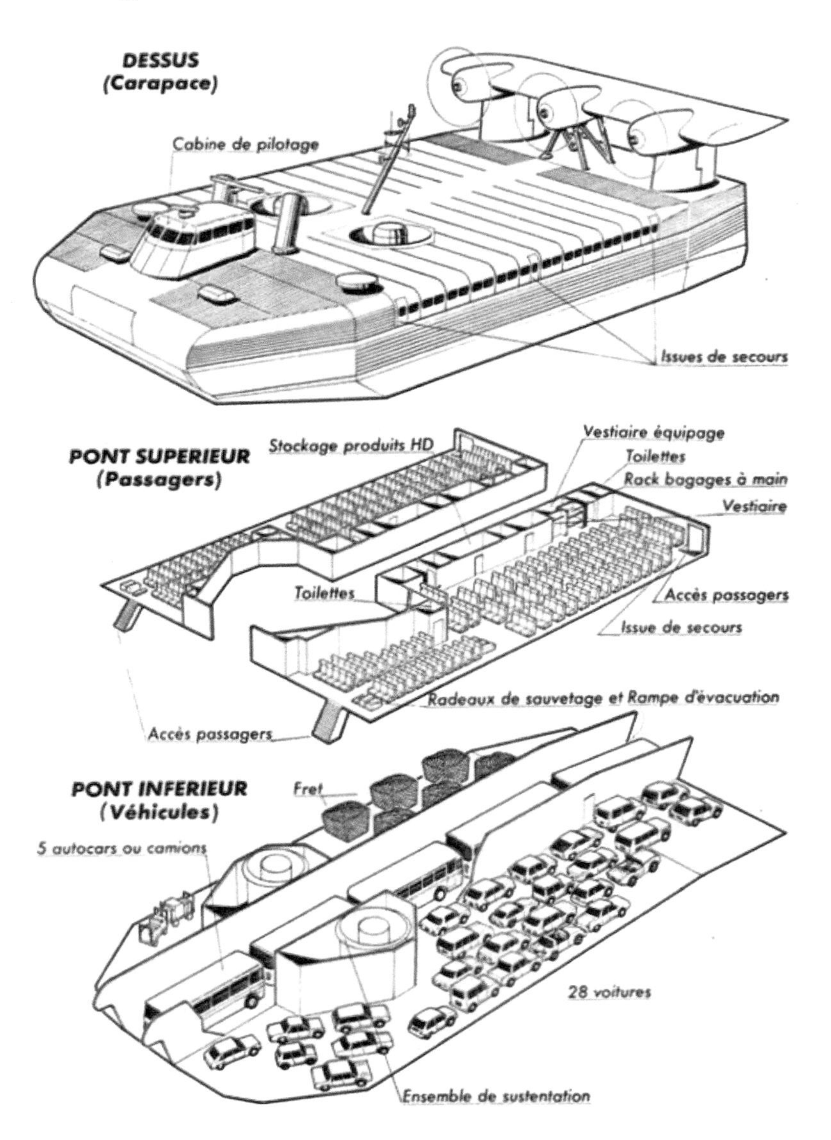

Although the N500, unlike the BHC craft, used separate power units to provide propulsion and lift, there was some economical advantage gained in that the three motive power units and the two lift engines used the same engine, the American AVCO Lycoming TF40. The power output was identical to the Proteus but half its size, with significantly less fuel consumption. A definite plus factor compared with the thirsty SR.N4. The Lycoming was the only other jet engine to be marinised.

What turned out to be a minus was the choice of location for the engines, mounted as they were high up on the rear of the craft.

It can only be assumed in retrospect that the French designers considered that the two outboard engines at the back, winged out as far as possible, would provide an extremely long lever with which to turn the craft; strong enough to cope with all situations. Together with a highly sophisticated airflow control system, which had much the same effect as the original skirt lift arrangement on the SR.N5 and 6[4], the extremely innovative system was claimed to deal with a lateral wind up to 20 knots, as a result of which the manoeuvrability of the French hovercraft should have been equal to the SR.N4.

In practice this was not the case. A typical problem was the steep gradient of the approach ramp to the new Dover hoverport. In calm weather there was no difficulty, but trying to negotiate the ramp in a cross wind meant having to come astern on one engine to keep the craft straight, resulting in less motive power up the slope. In some instances of severe weather it proved more or less impossible and the N500 crew had no choice but to divert to Pegwell Bay where the approach slopes were a lot gentler and more sheltered from the westerly wind.

Skirt under Wraps

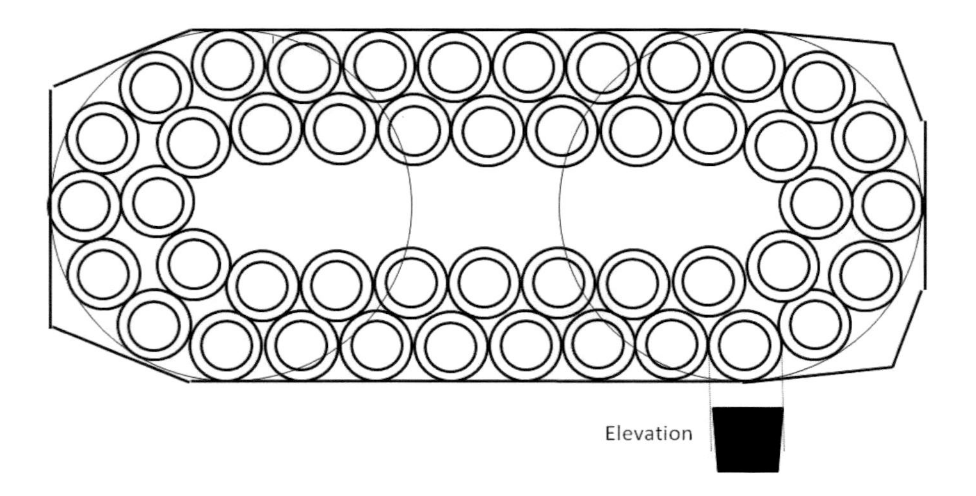

[4] Similarly, BHC's system also proved ineffective in practice. 'Skirt lift' was supplemented latterly by 'puff ports' on the N6.

When reading through the published material of the time, one of the most intriguing aspects of the N500 story is the complete absence of the most basic diagrams showing the configuration of the skirt system. The French were clearly reluctant to give away the secret of their unique design and even written descriptions of the system are brief, ambivalent and tell very little.

The diagram opposite is drawn from information contained in a somewhat poorly translated SEDAM specification for the N500 published in 1974. The extract below is all that is said of the skirt design in a detailed description of the machine numbering 33 pages:

> *The plan shape of the cushion is close to a rectangle prolonged* [extended?] *fore and aft by two half circles about 20 metres in diameter.*
>
> *The air cushion is generated by a number of elementary skirt units arranged around the craft. Each skirt unit is composed of peripheral lobes and inner lobes.*
>
> *The skirt geometry is such as to ensure a good resistance to outward buckle. The skirt conicity* [cone-shaped?] *is calculated to produce the vertical tension, required to balance the internal lobes, which ensure cushion stability.*

The SEDAM N500 with its jupe skirt configuration.

Together with other even less informative sources, this drawing is the best that can be imagined of the system as a whole. The dimensions of the 'jupes'[5] – both inner and outer – can only be guessed. What can be said is that they clearly conform to Cockerell's annular jet concept, but instead of being applied to the whole periphery like the British craft, they compose a multi-element system doing the same job.

One quite logical claim made by the French for their system was, because of its small construction, each jupe could be made of a lighter material and, because of the lighter weight, was less prone to damage. In addition, damage or loss of one or two jupes could not have any effect on their neighbours and, what is more, were easily and quickly replaced.

What about Fenders?

To furnish the whole lift system with a number of individual annular jets appears, on the face of it, to be a perfectly reasonable approach, but it proved to have one major shortcoming. The multi-element construction provided perfectly good lift to the hard structure but afforded zero in the way of protection.

This was particularly so forward, where on more than one occasion the bow door was severely damaged as a result of smashing into waves. No doubt, the individual jet distribution prevented pressure shifts under the craft and thus rendered plough-in unlikely. However, this didn't stop the bow hard structure, and sometimes the stern, from impacting large waves with calamitous results.

In later modifications the peripheral jupes were extended outward all around the hull to provide some sort of fender protection but this proved to have very little effect. Even with the modification, the jupes had no lateral resistance but simply collapsed against oncoming waves, often shutting off airflow altogether.

In contrast, the Mk.3 N4 perfected a feature that had commenced with the addition of the extension of the hinge line higher across the bow, together with the anti-plough bag on the Mk.1 craft in 1969. The advent of the much larger low pressure skirt on the Mk.3 created a resilient air-filled fender all around the craft, not just at the bow.

What is extremely interesting to speculate on now is whether the British, like the French, really appreciated the idea of 'fendering' or whether in their case the whole development was entirely fortuitous, brought about by concerns for pressure control and nothing to do with protection.

---

[5] 'Jupe' is the French word for 'skirt'.

The stern of the N500. The stern door was used as the vehicles ramp. Note the three jupes jutting out from the side. This was an early modification, and in the 1982 final modifications the scalloped effect was continued all round the craft to provide better 'fendering' against the seas.

Fenders on ships are a ubiquitous fact of maritime life, used in a variety of ways and designs, to keep vessels from slamming and chafing against quays or each other. When mariners look at the Mk.3 hovercraft, it is highly likely that it is the fendering propensities that are the first thing they think of. Not so those from an aeronautical background. It is not a feature aircraft manufacturers normally have to take into consideration. As far as can be seen, the term fender does not exist in BHC documentation; it does not appear in the aircraft manufacturer's vocabulary. The British certainly got it right when it came to effective skirt performance, but it is an interesting thought that cannot be ruled out, that it may have occurred by chance rather than by deliberate design.

N500 Performance

Sad to say, the whole N500 story was fraught with calamity right from the very outset. The first craft to be built, named *Côte d 'Argent,* performed her first test flight on April 19th 1977. By all accounts that and ensuing trials went well, up until a mere two weeks later when, on May 3rd, a maintenance accident caused a fire that, within a matter of minutes, totally destroyed the craft.

The interior of the N500 from the bow door, looking at the entrance to the starboard forward cabin, with the high vehicle lane on the right with a height of 3.8m. Below the passenger cabin the starboard car deck can be seen, which had very limited head clearance of 2.2m.

Such a major catastrophe so early in the project appeared to have a deleterious effect on French morale from which it would appear they never recovered. From that point on there seemed to be very little drive behind the enterprise and any enthusiasm to develop the concept or cope with inevitable teething troubles lacked any spirit. One theory suggests that those first trials may have already revealed fundamental flaws in the design, which also dampened the enthusiasm to continue. If all had gone swimmingly one would have thought that the French would have been all too eager to replace it as soon as possible.

The second hovercraft under construction, named after its inventor *Ingénieur Jean Bertin,* was completed and joined Seaspeed's *The Princess Anne*[6] on the Dover routes to Boulogne and Calais for the 1978 summer season. As would be expected with a new and untried machine there were a large number of teething troubles, which reduced its contribution to the service considerably. So what should have been a revenue flow from four large machines was reduced to one and a half and could only at best be three. Not a promising outlook compared with the initial estimates.

Peter Yerbury, Seaspeed's chief engineer, was not impressed:

---

6 *The Princess Margaret* was in Cowes undergoing conversion to Mk.3.

Travelling on the N500 was not at all comfortable. Instead of having a nice cushion of air, you had this air which would lift the craft, and then dissipate, so the craft would come down making the basic motion feel similar to riding on a pogo stick. It was very restricted by its weather limitations. The mechanical reliability was pretty awful too.

In the final round of modifications in 1983, the French craft was fitted with puff ports at the front but they were not as effective as the stretched N4 with two pylons at the front. If you are manoeuvring in tight spaces, which you are at Dover, then you need a great deal of control, which the N500 simply didn't have.

The N500 arriving at Calais Hoverport on a calm day in its first design configuration. It was only able to operate for 58%of the time. June 1979.

During the first three years of operation, according to John Lefeaux, the French hovercraft could do no better than 58% of its scheduled flights, not a viable figure under any circumstances. By the start of the merged Hoverlloyd and Seaspeed operation, to become Hoverspeed at the end of 1981, the N500 had been withdrawn for major modifications.

Ray Wheeler, Chief Designer of BHC recalls:

I got very friendly with the SEDAM sales engineer. When they had problems with their skirt system, we offered to design a system for them. I went to the French Minister of Transport on behalf of BHC to negotiate with him a co-operation between the two companies, but nothing came of it.

Towards the end of the first full Hoverspeed year in 1982, trials on the much modified craft commenced and were scheduled over several weeks. After a very short interval the trials were suspended because of severe damage to the bow structure. Roger, in his new capacity as Flight Manager (Hoverspeed), acted together with a Hoverspeed engineer as the company's representatives reporting back to Peter Yerbury:

My far too short sojourn in Boulogne opened my eyes to differences between the two nations, which went far beyond our language and what we eat. It occurred to me that the French have an opinion of the English as a bunch of fussy prigs, while they have a view of themselves as a cool, laid-back people where style and elegance rank far above prissy detail.

The whole approach seemed lackadaisical in the extreme. Nothing ever seemed to get underway until 10am and then it was usually for coffee and a casual chat. Sometimes the chat would last until noon, which then meant a two hour lunch. Very often the first time you boarded the craft to actually do something was two in the afternoon. Trials usually finished by four. What with the small hotel on the waterfront run by a famous local chef, who had just returned from making his name in Paris, where we embarked on the nightly chore of working our way through his mouth-watering menu, it was the best holiday I had had in years. It was all too good to last!

And of course it didn't.

It was a good westerly blow when we came off the beach straight into one of those famous standing waves caused by the rebound off the southern breakwater of Boulogne Harbour. There was an ear-splitting bang and one of the men below reported that the bow door had buckled inward. It surely had and our return to the pad showed the whole door had been crumpled and twisted as if a giant had screwed it up to put in his wastepaper basket.

Another thing that surprised me was the fact that there was another door in a similar condition lying out on the sand-dunes; no-one had thought to repair it as a spare.

As far as Peter Yerbury was concerned this was further proof that the N500 was simply not up to scratch. However, John Cumberland[7] had other ideas:

---

[7] John Cumberland was Hoverspeed's first Managing Director.

I asked John Cumberland if he really wanted me to accept it. I said it really didn't come up to the standard we required. I didn't want to accept it, but Sedam suddenly rushed around and fulfilled all the requirements that were specified, but it wasn't any more reliable. The weather restrictions remained the same unless they were going to change the whole skirt. Ray Wheeler did have talks with the N500 people about the skirt, but nothing came of that. Cumberland was very keen to accept it and I wasn't. He was a strong personality and said we had to have it – end of story.

The N500 at Calais Hoverport about to discharge. Note the portable ramp at the bow. This was compatible with the SR.N4 stern ramp. The N500's stern door acted as a ramp.

So despite Yerbury's misgivings the saga continued, but only for another year. At the end of 1983 Hoverspeed returned the French machine to SNCF as not coming up to specification. One must therefore conclude that the summer of that year was another series of failures. The *Jean Bertin* remained forlorn and unwanted on the pad at Boulogne until 1985 when it was broken up – a sad end to a sad story.

Chapter 25

1980 – The Beginning of the End

Swift at night. By 1980 darkness had begun to descend on Hoverlloyd and the giant cross-Channel hovercraft.

In 1980 the cross-Channel hovercraft services experienced a change in fortune. The table opposite shows traffic arrival and departure statistics during the Hoverlloyd years from the start of the SR.N6 trial period in 1966 through to 1980, which, although not the company's final year of existence, was the year in which the catastrophic decline in traffic heralded its demise.

Up to that point the Hoverlloyd story had been one of inexorable rise, both in traffic carried and its market share compared with other carriers into Calais. It should be noted that in order to compare apples with apples these figures include passengers and cars only, and therefore ignore the major revenue of the conventional ferries, namely the freight market, which also grew enormously during the period. This goes some way to explain the anomalies such as the steep rise in Hoverlloyd's passenger market share in 1970, occasioned it would seem, not only by the company more than doubling its own passenger numbers but also the steep decline in those of the other carriers. Obviously this was partly due to Hoverlloyd's success, particularly the development of foot passenger services, but the major reason was the significant increase in cross-Channel freight, which was starting to take over as the major component. Between 1969 and 1970 the ferries more than doubled their carriage of freight, so, although they lost foot passengers to Hoverlloyd, they gained more than enough in freight revenue.

CALAIS PASSENGER & CAR STATISTICS 1966 – 1980							
Year	**Fleet**	**Hoverlloyd - Pax Arr/Dep**	**Total Pax all Services**	**HL % Share**	**Hoverlloyd - Vehicles Arr/Dep**	**Total Vehicles all Services**	**HL % Share**
1966	**2 SR.N6s**	9,840	1,922,440	0.51%			
1967		19,699	1,721,961	1.14%			
1968							
1969	**2 SR.N4s**	295,771	2,108,284	14.03%	38,339	307,585	12.46%
1970		710,293	1,618,053	43.90%	57,105	398,284	14.34%
1971		609,954	1,823,763	33.44%	77,832	370,041	21.03%
1972	**3 SR.N4s**	662,459	2,341,388	28.29%	86,809	462,632	18.76%
1973		764,825	3,350,611	22.83%	113,038	667,480	16.94%
1974		827,504	3,541,431	23.37%	119,613	693,122	17.26%
1975		927,980	4,149,606	22.36%	146,476	832,893	17.59%
1976		954,423	4,169,389	22.89%	154,209	618,017	24.95%
1977	**4 SR.N4s**	1,140,283	4,428,536	25.75%	191,139	655,746	29.15%
1978		1,235,069	5,090,643	24.26%	208,746	763,584	27.34%
1979		1,218,913	5,235,511	23.28%	205,585	752,654	27.31%
1980		1,090,773	6,039,971	18.06%	163,380	816,348	20.01%

After a ten year period of expansion where passenger and car numbers steadily climbed and market share remained around a healthy 25%, the decline in 1980, although not entirely unforeseen, still came as a shock. While total traffic statistics in and out of Calais had increased by roughly 150,000 passengers and 60,000 cars by comparison, Hoverlloyd dropped around 128,000 passengers and, even more catastrophically for a company getting by on a 10% profit margin, a drop in cars carried that year of over 42,000.

Merger discussions

For the second time in its short life Hoverlloyd was in dire financial straits, a difficult situation that was not helped this time by the parent company Broström[1], who now owned both Swedish Lloyd and Swedish America Line, also losing money. In fact it was rumoured, prior to 1980, that Hoverlloyd was the only Broström business that was making money, however small its margin.

Of course Hoverlloyd was not alone. After the N500 debacle, Seaspeed was falling badly short of their projected revenue stream and, although the two Mk.3s were working well, their income would not have been sufficient to cover their own amortisation, let alone the port dues for the new Dover Hoverport.

With both hovercraft operations in trouble there was a natural reaction to amalgamate, which they duly did to form a 50/50 company called Hoverspeed.

The new management

Robin Wilkins, who had previously worked for Gillette, joined Hoverspeed at Pegwell Bay as Marketing Manager in 1981:

> *One day I got a phone call from a head hunter who said he had this interesting opportunity and would I like to come up and talk about it. He said they were putting two hovercraft companies together. There was a new managing director coming in and he was looking for somebody out of consumer goods to do the marketing and, by the way, it's going to transform the cross-Channel industry. So I thought I would go along and have a look. I heard in the next door office, because we were more or less open plan, that the head hunter was also calling this friend of mine, who had a 1st from Oxford, and was much more*

1 In 1976 Broström had acquired all the shares of Swedish America Line and Swedish Lloyd that it did not already own.

capable than me. I thought, oh hell, if he goes for the job I won't get it, so I went into his office later and said I'd had this phone call from a head hunter, but said this project didn't sound much to me, so I wouldn't bother, as a result of which I persuaded him not to go for the interview.

I then went for the initial interview. I remember this head hunter saying to me that it was going to be a very dynamic organisation and I'd have lots of freedom to do things. It's a small organisation and they're going to make £5 million in the first year. Well, they did make £5 million in the first year, but unfortunately it was a loss and not a profit. In the end I came down to Pegwell Bay for an interview. It did seem like the middle of nowhere being in the cauliflower infested hinterland of Thanet – very flat. My wife and I went into nearby Ramsgate, which seemed quite cheap – you could get your hair cut for £2, or something like that. My interview was with John Cumberland, the newly appointed MD, and I got the job.

John Cumberland had also appointed a sales manager, but eventually Robin took on both the sales and marketing roles. Robin found working for Hoverspeed very different to Gillette where he had all sorts of people to do things for him and could easily obtain any information required simply by picking up a phone. At Hoverspeed it was all 'do-it-yourself'. He found he got on well with John Cumberland when many others didn't and he found it an exciting place to work.

That first season of the merger did not get off to the best start. In fact Robin describes it 'as a bit of a disaster'. It had been decided to keep operations going at both Dover and Pegwell Bay, resulting in over capacity and too much cost. The difficult task of amalgamating the two reservations systems was given to an ex-Seaspeed reservations manager. He had a huge department at Pegwell Bay with people answering telephones, but unfortunately it disintegrated into a complete shambles, necessitating the reversion to a very basic manual reservation system to try to cope with demand.

It was long before the advent of the Internet and even affordable computer-based reservation systems. At that time tickets were either sent out by mail to travel agents, or passengers, on an individual basis. Alternatively, travel agents held ticket stocks, requiring the filing of a 'return', which had to be checked with all tickets and revenue being accounted for. It was indeed a cumbersome system, but that was how the travel industry operated at the time. Robin Wilkins:

Of course, I was completely new to the travel business and also new to working in a smaller company like this. We did struggle those first few months, but we got plenty of demand as a result of a new TV ad we designed to re-launch the newly merged companies.

I remember going into the final stages of that ad production. John and I had a big argument about the music because John had paid a huge amount of money for it and I couldn't see the point. He just waved me aside and said, "That's what we're doing." It was an Emerson, Lake and Palmer track – I think 'Fanfare for the Common Man' – and it cost tens of thousands of pounds just to borrow the music, let alone make the ad. It was, however, a very good ad about streamlining the Channel. We had moving arrows that went across the Channel and into the craft.

Of course the hovercraft was a fantastic thing to watch. Visually everybody loved it; people who travelled on it loved it even if it was a bit uncomfortable at times – to put it mildly. So off we went, but I think it became quite apparent after a relatively short period of time – probably two or three months – that we shouldn't have kept Pegwell Bay open as well as Dover.

The new management felt that Dover was where *THE* market was because of the Seaspeed rail connection. There were indeed a large number of rail passengers travelling on the Dover-Boulogne route, which was a big source of revenue. The Mk.3s were superior to the Mk.2s, not so much in manoeuvrability, but in ride quality, weather keeping and economics, in addition to which the Mk.3 facilities were at Dover. That was the Seaspeed card, but as Robin explains:

What I think the Hoverlloyd people brought to it was a kind of entrepreneurial spirit and certainly a feeling that they were a cut above everybody else on the Channel, and, yes, I thought so too. I thought the captains were much more RAF than BEA – they would take the craft out in conditions, which some of the Seaspeed captains wouldn't. I thought they were more customer orientated and were very proud of the service.

But the integration went pretty well when you consider you were marrying up the culture of a nationalised industry – part of British Rail on one hand with all the sort of ponderousness that came with that in those days – with the free-thinking pioneering spirit that Hoverlloyd had. The hovercraft was still new, different and dynamic compared to shipping. I think the biggest cultural difference was between Sealink and Seaspeed rather than Seaspeed and Hoverlloyd. Everybody in Hoverspeed regarded the ships as being an alternative to which we transfer people when we have to, but we wouldn't do it willingly.

We really tried to take on the ships – this is a long time pre-Tunnel – and get a premium for our speed product, but unfortunately the loss at the end of the first year of the merged companies was around £5.3 million.

The magnificent, expanded hoverport at Pegwell Bay, which was only open 13 years to traffic. The top of the VT-2 hovercraft, which was purchased from the NHTU in 1981 for spares, can just be seen over the vehicle check-in kiosk to the left.

The decision had been made in the middle of 1981 to move to Dover, although Pegwell Bay operated for the summer season of 1982. Robin Wilkins:

We'd created a lot of demand, but we'd screwed up the reservations system and in fact we were probably turning people away because they couldn't get through on the phone to make bookings. Yet we didn't have enough demand to make the whole thing sensible. We couldn't get the frequency we wanted in Dover because we were trying to run two bases and sharing the craft between the two. We had all the costs of two bases, so it wasn't a great idea.

In the end the decision was taken to move out of Pegwell Bay and amalgamate all the activities down in Dover, but at the end of the second year we had still made a loss. At one stage we were running the two big Mk.3s in addition to two, or possibly three, smaller Mk.2s. It was a lot of capacity, but frequency was very important. We were running every half hour and of course, bizarrely, on the Channel, market frequency and capacity create demand, or at least attracted demand, whereas in a normal business it is the other way round – demand attracts supply, but in this case it was vice versa.

We had far too high a cost base. I can't remember how many people we were employing – five or six hundred. We had two engineering teams, we still had relatively big facilities within the ports – catering and all that sort of thing – and the cost base had to come down. Fuel price was a constant worry, as was the engineering cost. When you think what the fares were – far higher than they are today – we wouldn't have lasted two weeks at today's fares.

Without doubt there is little disagreement anywhere that, from an operational point of view, Pegwell Bay was superior to Dover. First of all, a large portion of the route was protected by the Goodwin Sands as a result of which the craft had a nice long run-in with little traffic about, as nothing else could go up into that area. There was also the protection of the Kent coast from the prevailing westerlies. Robin Wilkins:

If you were designing it from an operational point of view, then Ramsgate-Calais was fine. Unfortunately, the customer and marketing got in the way. The market was in Dover. At that stage the roads into Thanet were very poor; it took a long time to get there; most people didn't know where it was and, particularly when you were running a hovercraft, you knew you were going to have times when you couldn't operate and needed to transfer people. Clearly, if you were transferring people, down the road was easier and a lot more appealing to the customer as an alternative to what you had at Ramsgate.

Not everyone held this view. There was the counterargument that Hoverlloyd's reliability record of 98% was such, that having to re-route cars and passengers on odd occasions was almost irrelevant. The roads were bad to Thanet, but there were plans for improvements and, in any case, Hoverlloyd had proved that they could attract passengers to Ramsgate as their figures were greater than Seaspeed's. As Robin Wilkins acknowledged, operating from Dover negated the need to operate a hovercraft, as in essence one was going from one deep water port to another on the other side of the Channel. In fact it would not be many years before the fast, but slower than a hovercraft, catamaran appeared on the scene operating from a berth within the bounds of Dover Hoverport to a berth in Calais not far from the hoverport terminal where catamaran passengers and cars checked in.

October 1992. Dover Hoverport located in the Western Docks. The remaining five hovercraft on the pad. Clockwise from bottom left: *Swift, Sir Christopher, The Princess Margaret, The Princess Anne and The Prince of Wales. Sure* was left at Pegwell Bay. Note the Dover Dunkirk train ferry lock to the left of the hoverport.

Peter Yerbury, who had been Seaspeed's Chief Engineer at Dover since 1974, was appointed Technical Manager of Hoverspeed. With Emrys Jones and Monty Banks from Hoverlloyd having departed the scene, Peter had the distinct advantage of not having to deal with any of the old Hoverlloyd management or under-management at Ramsgate. Basically. Seaspeed took over the engineering of Hoverspeed:

> *There were some difficulties in merging the two engineering departments, because one of the problems we had was that we had staff in Ramsgate and staff in Dover – we had too many staff, so it's always an unpleasant task when you have to make people redundant.*
>
> *However, we had to make sure that we had enough of the various trades, like electricians, tin ware people etc. It couldn't always be first in last out – we had to have sufficient people of the right expertise. We ended up with just over 200 permanent people in engineering. That's a lot of people. We didn't have many seasonal – just a few for the skirt team.*

Unfortunately, because the hovercraft was built on aircraft lines, we would expect to maintain it with schedules normally applicable to aircraft and, because it was a lightweight structure, it required more maintenance. If you look at what they do on the ferries, they put an engineer on the ship and take him with it, so he's doing the maintenance while the thing is still operating. You can't do that on the hovercraft.

After we moved the craft to Dover we still ran Ramsgate for at least two years. We tried not to take craft up to Ramsgate unless absolutely necessary because of the cost of running them backwards and forwards. So it was generally winter overhauls, engine overhauls (as Hoverlloyd had invested in an engine overhaul facility), which we used initially, and any major work that needed to be done. Day to day work was carried out in Dover.

The Princess Anne on passage between Dover and Calais.

Initially, one factor of the amalgamation structure that might have been better thought out was the, probably unintended, management split between the two merging companies. The operations management was exclusively Hoverlloyd and the engineering management was, equally, exclusively Seaspeed; not an easy situation when putting together two vastly different cultures.

Roger, as flight manager of Hoverspeed, found it extremely difficult:

To be fair, the antagonism between flight crew and engineers, although not helped by the two-culture factor, was more than likely inherited. An animosity nurtured in the British Rail union-dominated era, it was a typical 'oil and water' situation – well known among ships at sea – which in Hoverlloyd we had been fortunate to avoid.

In my old job at Pegwell I had been used to having weekly meetings with senior engineers, where maintenance and day-to-day handling problems were discussed and resolved in an amicable atmosphere. I tried to institute the same regime in Dover with zero results. In contrast, the atmosphere at the first meeting could only be described as poisonous; some topics such as the introduction of a 'snag book'[2] were refused discussion altogether. Nothing could be said to have been achieved, the whole meeting was futile.

A Mk.3, having just entered Dover Harbour through the Western Entrance, needs to clear The Prince of Wales Pier before turning in for the final approach to the hoverport.

The cabin crews were now run by Seaspeed too. Marcelle Connell, a fine Belgian lady, who had received a BEM for services to Seaspeed, had ruled the Seaspeed girls as Chief Purserette with great efficiency since the start of Seaspeed operations in 1968, but retired in 1980

---

[2] A procedure where problems and system failures are recorded in a carbon copy book. The top copy is presented to the engineers daily and the carbon copy remains on the flight deck as a record. It is the only sure way of keeping track of ongoing problems and how the technicians were dealing with them.

before the merger. Barbara Ratcliff, who had been a senior purserette since 1972, was appointed Chief Purserette in 1980 in place of Marcelle. Barbara found the initial weeks of the amalgamation difficult because both the Hoverlloyd and Seaspeed cabin crews, who took on the Seaspeed title of 'purserettes', still had their old company uniforms. To avoid confusion for the passengers, the Hoverlloyd girls manned one side of the craft and the Seaspeed girls the other. Shortly after the merger, and for the first time, male cabin crew were employed.

Marcelle Connell (right) and Barbara Ratcliff (left) at Marcelle's farewell presentation. (Newspaper unknown.)

The merged car deck crews were managed by Alan Holland from Hoverlloyd. The Holland family had provided more members of staff for Hoverlloyd than any other single family. Andy, the father, had arrived in the very early days of Hoverlloyd from the Kent coalmines to be employed as car deck crew and soon went on to become a supervisor. He was closely followed by his three sons.

The move to Dover was the logical decision for the merged company and it would have been a very brave board of directors to decide otherwise. Dover did have a much larger market, the rail passengers could not be ignored and the route length to Calais was shorter. With the Mk.3s consuming 5,000 litres of fuel per hour, the shorter route length was an important consideration. There was also a larger number of 'go-show' cars and passengers without tickets for the taking. Nevertheless, the argument will rumble on as to which base would have been better depending on whether one was a Hoverlloyd or Seaspeed person, but, whatever the view, it is clear with hindsight that the fate of the giant cross-Channel hovercraft would have been the same.

Robin Wilkins, however, remembers asking on arrival at Hoverspeed what the future might hold. When would the company be getting new craft[3] when these current ones wear out? Can they be re-engined to reduce fuel costs? It was all very exciting, but these were valid questions:

> *I arrived in the business when the first phase was over. Hoverlloyd enjoyed tremendous success and public acclaim. It was one of those products that had a discernible differentiation from everything else and you either liked it or didn't like it and the people who loved it really loved it. Not only did they like the whole feeling of modernity, speed and convenience and style, but I think they liked the service and the personality that went with it. They liked the individuality of the people who worked there. I think that did stay with the merged companies all the way through.*

A magnificent sight. A close up of a Mk.3 arriving at Calais.

> *We always positioned ourselves as being a cut above the ferries; the seat-side service on board the craft, somebody greeting you on the car deck. In a way it was much more personal and very 'airline like' in the days when people enjoyed flying and enjoyed being in an airport – days which are long gone. There was definitely that and it was one of the reasons why we were able to keep staff. One of the big motivations was that we were a 'family' – probably not quite in the same way that Hoverlloyd was a family, because that was a real 'start up', whereas we came in in a slightly different era.*

---

[3] BHC's BH88 was on the drawing board, but that was as far as it got.

I had no baggage from before so I didn't have any feelings one way or the other. I came in with my ex-American company, fast-moving consumer goods approach – this is how we're going to sell it. I remember somebody saying to me, "You can't sell this like a tin of beans. It's not a product." I said, "Well, it is a product and you can sell it like a tin of beans, but the only difference is when you buy a tin of beans you know what it is going to taste like and it always tastes the same. The difference in travel is that you don't know what it is going to be like on the day. It could be different."

A change of crew. Purserettes and car deck crew meeting the inbound *The Princess Anne.*

The 'tin of baked beans' analogy could have been in response to a comment made by John Cumberland to staff in the very early days. Peter Barr recalls:

We had a meeting on the evening before the merger occurred at Pegwell Bay and Cumberland introduced himself. He told us something about himself and what he proposed to do. "It was just like selling baked beans", I think he said, "It didn't matter what you were selling in marketing – it was all pretty similar". Right there and then we thought, "I don't know if this is going to work".

And, in the end, for John Cumberland it didn't, but, as Peter Barr explains, he sorted out the unions:

The union people had demanded the car deck crew had their own staff cabin, tea boilers and all the rest of it, but when the merger occurred at the end of 1981 and before the season began in 1982, it was all taken out. There wasn't a murmur from anybody as the company had been de-unionised by then.

*Hoverspeed simply didn't recognise the unions. I think they took the attitude that if you wanted to join the union that's fine by us, but we're not recognising them. Everybody knew that it was s**t or bust so if we didn't do our level best this could all collapse within a year. The fact of putting the two companies together and writing off the debts was enough to convince everybody that it was the last chance.*

Cars loading onto a Mk.3. Note the ladder up to the flight deck and the car deck cabin to the right of the entrance.

Some felt Cumberland was sacrificing maintenance for marketing with the limited funds available, to the detriment of the operation of the hovercraft. You might be able to market the hovercraft like a tin of baked beans, but you couldn't operate them on that basis. His management style provoked serious clashes with both senior and junior managers, particularly on the operational side, resulting in David Wise, the former Hoverlloyd Director of Operations, who was so highly regarded by the pilots, cabin and car deck crews, being moved sideways. David was given the thankless task of attempting to sell off Hoverlloyd's original hovercraft *Swift* and *Sure*. Some interest was shown by a number of overseas buyers, but nothing came to fruition. It is likely that as soon as the interested parties looked into the running and maintenance costs their enthusiasm waned. After a few futile years, David returned to his old job as head of operations in 1987 and remained there until he retired in 1991.

Roger Syms, on the other hand, had a major spat with Cumberland, lost heart and took the first opportunity of redundancy early in 1983[4].

By September 1983 it was clear the 'Cumberland strategy' was not working. The merged companies were still producing losses with no prospect of turning the fortunes of the company around, resulting in the departure of Cumberland in October of that year.

The Prince of Wales in Hoverspeed livery. The Mk.2s were faster than the stretched Mk.3s and were often referred to as 'the greyhounds'.

4 Robin Paine left in 1979 after nine years in command to pursue his own business interests.

Merging of the flight crews

Captain Brian Laverick-Smith with Warwick Jacobs, curator of the Hovercraft Museum at Lee-on-Solent.

From the crewing point of view, two different cultures were thrown together, with some adjusting to their new environment and others still set in their ways. Peter Barr:

> *There were always some who wouldn't settle. Even towards the end there were one or two people who muttered on about Hoverlloyd and Ramsgate and Seaspeed and the good old days. You would never convince Bob Middleton or John (Biggles) Lloyd, for example, that Ramsgate wasn't the only place to run hovercraft, but we had our equivalent on our side – people who could never see their way clear to mix in with the others, but it all largely faded away towards the end.*

We never ran six craft from Dover, but we ran four quite regularly, although it was a bit tight. If you had three craft there and you were coming in and it was poor visibility, you had to get them to shift the others out of the way as best they could. In reasonable weather you could slot in and out, but it was never designed for six hovercraft. But Dover was a damned good hoverport to operate out of. We all thought at one point it would be very 'iffy' because of the approach, whereby you had to round the Prince of Wales Pier and once round it you had to get back onto your line to get on the pad – not easy at times, but it turned out to be generally very manageable.

The merger resulted in the inevitable seniority issues amongst the flight crews with some captains being demoted to 1st officer. Brian Laverick-Smith, a former junior Seaspeed captain was one such unlucky person:

I was bumped down to 1st officer, but at least I had a job. To begin with we had a lot of confrontation with the Hoverlloyd guys. It was difficult. I went on the flight deck the first day of the merger and there was Ian Philip[5]; I'd never met him before. "Hi Ian", I said and I got on with the job. I went to start the engines up and he's sitting there doing his thing. He looked at me, did a double take and just tapped my epaulette and said, "I like that". I had three stripes up and I'd fought like the devil for four stripes, which had now been taken off me.

I was thinking, "Excuse me, I fought for years to become a captain and you like the fact that I am down to three stripes?" I started the engines and my conversation for the trip was, "Put her up, Brian." "Right." "Put her down, Brian." "Right." At Calais I slammed the throttles shut – bang – stopped the engines and said I was going to inspect the skirt. "Righty-oh", said The Laird nonchalantly.

*I went down below, walked across the car deck and one of the old Seaspeed car deck crew came up to me and said, "Ah ha – I like that". I said, "Don't you f***ing talk to me, Sunshine." "No, no, no, no", he said, "you've got four stripes on that shoulder and three stripes on that one." I said, "Oh Jesus", and crept up the flight deck ladder and there was The Laird reading The Times. I said, "Ian", and he said, "Yes?" "I'm sorry," I said, "I thought you were having a go at my demotion." "Oh my*

---

[5] Ian Philip was nicknamed 'The Laird of Patrixbourne' by Roger Syms in the early days of Hoverlloyd because of his aristocratic Scottish voice and mannerisms, together with an authoritarian air, which he generally employed when making succinct pronouncements with dry humour and a straight face.

dear boy, I don't care whether you've got 15 stripes up there so long as you do your bloody job." I said, "Yeah – no problem."

The perceived and real rivalry as a result of the clash of cultures was not an attractive trait on both sides, as Brian notes:

It was difficult if you had been together like us lot had and you lot had for many years when suddenly you get another 'family' forced upon you. It was difficult all round. It was a clash of cultures.

Some Seaspeed people would refer to the 'Black Crew', meaning Hoverlloyd, and I thought it was awful, but there was a better atmosphere if you were on with what they called a 'White Crew' – your own people. You could relax knowing you were on with your old buddies. It was only some of the Seaspeed crew who would refer to the 'Black and White crews', but I don't know whether the Hoverlloyd crews had names for the Seaspeed crews. There was a 'them and us' situation right to the end by a very few. Maybe a quarter felt that way and three-quarters didn't, but after 19 years on from the merger in 1981 the 'Back and Whites' became 'Grey'.

The rivalry also manifested itself in touches of humour:

After the amalgamation, when we were operating from both Dover and Ramsgate, a crew would bring a craft from Ramsgate to Calais for us to pick up and take to Dover and they would take our craft back to Ramsgate. The craft would then run out of Dover and Ramsgate respectively for a while and then they would change again.

One day, with Juian Druce as captain and myself as navigator, we picked up a Hoverlloyd craft in Calais and set off back to Dover with the Decca rolling chart. Julian was very pernickety and liked everything done by the book – it's got to be done properly. It was good visibility and not much traffic about when he said, "What's that, Brian?" I said, "Whereabouts?" "To starboard – out there." I said, "Nothing – what are you talking about?" There on the Decca was this red blob coming down the rolling chart. Somebody had put a red blob on the Decca rolling chart. Then there was another red blob and he said, "What the hell's that?"

As the blobs came down the chart it was apparent that they were not red blobs, but two stiletto high heeled shoes. I left my seat and stood behind Julian in the captain's seat. "I wonder what some bugger from Hoverlloyd's done?" "Gosh, Brian – 10 minutes off – don't forget to tell Dover Port Control we're 10 minutes off", at which point two calves of a pair of ladies legs started to appear, whereupon I went back to my seat to deal with Dover Port Control. "Brian, those bastards – they're awful people at Hoverlloyd." "Roger – three minutes off Western Entrance." "I've got Western Entrance visual, Brian... Jesus Christ, look at the Western Entrance", and there right on the Western entrance on the Decca rolling chart was a fanny. "The baskets at Hoverlloyd have done it again," he said!

At the end of the '82 season when everything moved to Dover there were a considerable number of redundancies, with many people being offered voluntary redundancy, of which several took advantage, including Roger.

Financial difficulties

Earlier in 1983 Robin Wilkins had taken over the job of sales manager, as well as continuing with the marketing, at the request of John Cumberland. A few months later he received a phone call inviting him to see the chairman, B.G. Neilson, a Swedish Broström director, who happened to be in nearby Deal:

I went to see him on a Sunday afternoon thinking, "Blimey, he's going to tell me the whole thing's a shambles and we're all out on our ear and so on." He sat me down, we had a cup of tea and he asked me if I would I like to join the board. Well, I was staggered. It was the last thing I was expecting, so I was very pleased about that.

They appointed Peter Yerbury, (Technical Manager), Tim Redburn and myself to the board. Tim Redburn was brought in as the Finance Director because it was felt that the accounting function wasn't strong enough. He was a very clever guy and he's done very well for himself. Effectively the three of us went on the board, but although the losses were coming down, they were still on-going.

It was at this time that Margaret Thatcher's government was in full swing with the privatisation of nationalised industries. The nationalised industries were also being told that there was no more money in any form and they had to stand on their own two feet, in addition to divesting themselves of non-core activities. In the case of British Rail this meant disposing of, amongst other non-core businesses, ferries and hovercraft. Robin Wilkins:

> *We were approaching the end of 1983 and we needed to re-finance ourselves for the winter. We were running an overdraft of about £4 to £4½ million, from memory, which was real money in those days. So we were dependent on the banks and of course the banks wanted to see that their money was safe with guarantees from both parent companies. Broström was prepared to give a guarantee, but only if British Rail, would give one, but unfortunately British Rail couldn't give a guarantee because the government had said that they couldn't. It had been a 50/50 merger with no mechanism for what would happen if they disagreed, except for the casting vote of the Chairman, which was, I guess, valid except in a financial situation such as this. We all started to get very worried because the banks said, "No guarantee, no money" – that's it.*

With no significant signs of improvement in the company's position, Gerry Draper [6], a former marketing director of British Airways, who now ran his own consultancy business, arrived. He was brought in as the chief executive, but he worked out of an office in London. He was a strong character, making many decisions and ordering a great number of things to be done, but he was not in evidence on a day-to-day basis. The depth and breadth of his reorganisation was such that his management on the ground had to gently say, on occasions, that he was probably on the road to disaster if they did such and such. Robin Wilkins:

> *For example, there were all sorts of things about pricing, because revenue was so critical, and Gerry wanted to have different pricing for each individual day with add-ons here and other bits there. This was well in advance of the low-cost airlines and their pricing strategy and that's where in fact we ended up. But at the time the system just couldn't cope with it – it just wouldn't work.*

Towards the end of 1983 it became clear that the banks were digging their heels in and there was some doubt as to whether they would refinance Hoverspeed. In January 1984, Robin and Tim Reburn decided that just the two of them would make an appointment to visit the NatWest Bank, without involving the others, because they found it hard to believe

---

6 Gerry Draper was Marketing Director of British Airways for some years in the 1970s and early '80s and was heavily involved in the problems of filling the new jumbo jets when IATA regulations prohibited discounting. The joke at the time on the jumbo jets was, "One toilet for every passenger". He was also involved in making Concorde profitable.

that the bank was hearing the whole story as a result of the arguing and 'argy bargy' between Broström and British Rail. Robin Wilkins:

> *We went to NatWest and said, "Look, we know that the parent companies are arguing about the future of the business, but we want to talk to you about the prospects of the business and what can be done." Really we went there with a begging bowl pleading. It was important to tell them that we had hundreds of people employed at Hoverspeed; a very attractive product for the customer; we were improving, but we needed more time and could they please give us another shot.*
>
> *Their reply was that they required the business to be looked at and suggested we approached Bill Mackey, who had carried out the Laker Airways liquidation, at Ernst and Whinny. He came in and looked the company over, after which he sat down with us and we asked him, "What do you think?" He said, "Well, if it's any consolation, I've seen a lot worse."*
>
> *So we enquired as to where that left us and he said, "Well, I think the banks are going to find it very difficult to give you a loan." We said, in that case, why didn't they just pull the plug now? The losses were stacking up and it was hand-to-mouth at that stage, but the question was, "Were they going to pull the plug?" We pointed out that surely September was the time to pull the plug when we'd be in positive cash flow. We wouldn't have had enough to take us through the next winter, but at least we'd have cash in the bank for them to get some, if not all, of their money back.*
>
> *"No," said Bill, "you don't understand how the banks work. This is just a drop in the ocean as far as the banks are concerned. They've just lost billions in Brazil. They don't care about £4 million owing from you because, from a political point of view, it would look better to pull the plug on you now (in winter). They can then argue that you are running an overdraft and have no real prospect of paying it back rather than let you run through the summer when you've paid it back and then pulling the plug. Now's the time when they are at their most mercenary and difficult and they'd rather do it now because you're very minor to them."*

The management buyout

Their luck held in the short term, however, but a requirement of the NatWest was that they be furnished with a monthly report, while discussions were on-going, in which Robin was not fully involved. But he does remember one critical board meeting at which the company lawyer was called in to effectively advise as to whether they could continue trading legally. Robin Wilkins:

From memory, the board meeting was on a Thursday or Friday. The lawyer was a friend of Mike Keeling, who was on the Seaspeed board as part of British Rail. Mike asked him what he thought and he said, "I think you're OK today, but if you haven't got something in place by next Tuesday I really think you should call in the administrators."

Well, this was about four days away, so we were really living on a knife edge because there were rumours starting to circulate in the travel trade and, as you know, confidence is everything, so it was quite a difficult period.

I'm not quite sure where the management buyout idea came from, but I guess it was probably through Mike Keeling and Mike Bosworth[7]*, who worked for British Rail and was on the Hoverspeed board.*

Robin Wilkins, Tim Redburn and Peter Yerbury, who were three of the five directors in the management buyout.

Peter Yerbury recalls:

There were five of us initially in the buyout: Tim Redburn;

---

7 Mike Bosworth was unable to be a member of the buyout team because he worked for British Rail.

Robin Wilkins; myself; Mike Keeling, who agreed to take on the chairmanship of Hoverspeed on the buyout; and Gerry Draper, so we ended up as the 'gang of five'.

The operation was bought for a nominal sum by the directors on 16th February 1984 when the loss stood at £3.4 million. The NatWest required the new owners to put up collateral, which involved putting up their houses as security, but there was a clause in the buyout from Broström that would subsequently come back to bite them. The borrowing requirement was of the order of £6 million, an amount the guarantees didn't cover, but was nevertheless regarded as a sufficient act of faith by the bank, which would severely hurt the directors if things went wrong. Peter Yerbury:

We managed to pay back all the money at the end of the summer of 1984 when the revenue had come in, which pleased them no end, but when we said we wanted them to lend us some more money to get through next winter they said they didn't want to do that.

It was at this stage that realisation dawned that the only solution to their funding problems was to approach a merchant bank. Robin Wilkins:

John Nelson from the merchant bank, Kleinwort Benson, turned up and John said they'd be willing to take a punt on it under certain conditions. They wanted all sorts of options to buy in return for underwriting the cash to operate, but they didn't want to take a shareholding because if it went belly up they weren't prepared to have their name on the paper. So they had a lot of covert share options that they would exercise, which they did when it was sold on.

Some funds from NatWest were now underwritten by Kleinwort Benson and Hoverspeed continued to bank with NatWest, who provided funds up to the Kleinwort Benson guarantee. More money, however, was required by way of bank guarantees. A friend of Gerry Draper's from his British Airways days, who was financially well-endowed, came to the rescue and guaranteed a further £¼ million.

But what was the future for this newly-financed company with a fleet of aging hovercraft? The old technology Proteus engines were not being produced anymore and the hunt was on for sources of supply. The VT-2 had already been sold by the Naval Hovercraft Trails Unit[8] (NHTU) to

[8] The NHTU took over from the Interservices Hovercraft Trials Unit (IHTU) when it was disbanded as a tri-service organisation in 1971.

Hoverspeed when the NHTU was disbanded in 1981, thereby providing two more spare engines, but apart from cannibalising the Mk.2s, Proteus engines were starting to become as rare as hens' teeth. Re-engining was out of the question due to costs. Operational costs were still high mainly due to the cost of fuel and skirt maintenance, but the directors looked at things differently. Robin Wilkins:

> *In 1984 there was a rolling five-year window, as I recall, because the first question I asked when I arrived was, "What's going to happen in five years time?" I was told we wouldn't need to do anything for five years, so I thought, "OK", and then three years later it was still five years because we were making quite dramatic improvements. There was a redesign of the propeller blades – well they put some strip down them to make them last longer; the skirt technology had improved – the skirts had been strengthened in a certain way and we weren't suffering the catastrophic losses when skirts were ripping off. Skirts were, however, still a large part of the operating costs.*
>
> *Even so we were making progress. We had gone a long way under the red financial line, but we were coming up. It was just a question of how long before we got above the surface and I think in 1984, the first year of the management buyout, it washed its face from a cash point of view, but we still lost money. In the second year we were still crawling up, and by the time we sold the business in June 1986 we had made a wafer thin profit – £150,000ish.*
>
> *As Mike Keeling was fond of saying, "We're men of straw", because when people normally go into these management buyouts they have quite a bit of money, but he said, "Well, we're just men of straw going along to the banks asking for this that and the other".*

Sea Containers takes over Hoverspeed

Management buyouts with backing from merchant banks, or venture capitalists, are never designed to be long-term affairs and it had been made clear from the beginning that the management should be on the lookout for a prospective purchaser. Hoverspeed was not an attractive proposition, unless one had more money than sense, or some sentimental multi-millionaire wanted to run it as a hobby, so approaches to the obvious sources such as the ferry companies initially proved fruitless. In fact it would be to their advantage if Hoverspeed went bust.

At the beginning of 1986 the pressure was really on to sell and, as luck would have it, Jim Sherwood of Sea Containers, who had bought Sealink from British Rail in July 1984, indicated he might be interested in acquiring Hoverspeed. Robin Wilkins:

Sherwood at that time was very acquisitive. He was going around buying all sorts of stuff. He was the classic entrepreneur; an extremely bright, very intelligent guy, a very tough guy, but he could spot an opportunity. He was a little bit like Richard Branson – he would swing the bat and occasionally he missed, but every now and again he hit and when he hit he made a lot of money.

Tim Redburn and I went to London to see Sherwood. He is one of those people who doesn't make a lot of small talk. He came in, shook hands, sat down and said, "Right – so what's the deal then?" and then proceeded to ask a lot of questions about the business. Sherwood, in classic style, didn't want to pay anything and said he'd give us a very small amount of money, but if we wanted more than that we'd have to earn it, so we were now involved in an 'earn out'. We asked how an 'earn out' would work, because if they were going to be changing things why wouldn't they wait until the 'earn out' was not there and get rid of us?

So it was all a bit difficult, but in the end a formula was agreed. It was a very tortuous negotiation. We were in Sea Containers House for long periods of the night trying to get everything done. It was very, very last minute – probably a month from when we first started chatting to when it went through.

Finally the deal was signed and we were very pleased. Personally it meant that we would get a little bit of money – it wasn't a huge amount and certainly a lot less than most people thought because the merchant bank had all those various share options which kicked in at the last minute.

We then proceeded to have a long period of difficulty with Broström, because, at the time we bought the company out, there was an agreement in place whereby, if we sold within a certain period of time there would be money owing to them, or we couldn't sell without their authority, or whatever it was. For probably two years after that, if not longer, we had this tit for tat between the lawyers and eventually we had to give up some of the money to Broström, plus pay all the legal costs and other bits and pieces.

There was a time when they had to physically deliver a writ to us individually. We knew the last date on which they could serve the writ, so I sort of absented myself. I went and stayed in a hotel in the middle of nowhere, but in the end it was agreed that we would consider it served because they had served it on some of the others. They were suing us as individuals and we just couldn't afford to go to court. We also thought Broström would just keep going and going. So in the end we had to do a deal, a deal was done and it was enough to buy a house, but that was it.

It certainly seemed a lot of hard work, together with high risk, for relatively little reward, but they were lucky to get out in one piece with some sort of consolation prize.

After the sale on 16th February 1986, Peter Yerbury, Mike Keeling and Gerry Draper departed the scene. To start with, Tim Redburn was made the acting chief executive and stayed for six months while Robin Wilkins continued in his marketing role for the remainder of 1986. Although Mike and Robin had the option to leave, it was felt that in order to protect the interests of the 'earn out', they should stay to keep an eye on things, particularly as due diligence was still progressing after the sale.

The company finances were above the red line when it was sold, but only just, and there were distinct advantages in being part of a large company in order to capitalise on that upward trend. For example, fuel became cheaper to purchase and legal services were provided.

Tim Redburn had made it clear he didn't want to stay on beyond six months as he wanted to go off and do something else, but then the question arose as to what Robin was going to do:

As we were now part of Sea Containers I worked for a guy called Nigel Tatham and I was offered the opportunity of being the Marketing Director of Sealink, and stay in marketing to take over the marketing directorship of the whole of the shipping division – effectively all of Sea Containers – or to go into general management and be the Managing Director of Hoverspeed. As my heart was in the hovercraft company and not in the shipping company that's what I opted to do.

So I worked for Nigel Tatham and became MD of Hoverspeed where I remained until 1991. Nigel Tatham was Senior Vice-President Ships. When you look at the original organisation for Sea Containers you had Sherwood, Mike Stracy, who was the Finance Director, then the Support Directors and then you had the guy who was in charge of containers, a guy who was in charge of properties and hotels and a guy who was in charge of shipping. Nigel Tatham was a long-time Sherwood acolyte coming up for retirement.

I was there when the catamarans (SeaCats) arrived. Sherwood had recognised right from the start when he bought the company that here was a way of making Sealink more profitable because he'd got the Isle of Wight – Sealink Isle of Wight – which was very profitable and he'd acquired us just at the time we'd started to make a profit.

This is slightly off the point, but Sherwood said he needed to re-launch the Isle of Wight services and make it a separate company. So we wrote to him and said it would be a good idea if we ingratiated ourselves with the local population of the Isle of Wight and had a competition for people to submit a name for the new service to be called. We'd choose one of them and that's what it would be called and politically that would be very, very popular.

He wrote back saying, "A good idea – the winning name will be Wightlink." I feared him, but he was a very interesting guy to work for.

The cross-Channel replacement for the hovercraft

Sherwood soon realised that the hovercraft did not have a cross-Channel future. He was wondering what to do when he came across Robert Clifford at Incat in Tasmania. Clifford said he had this wave-piercing high speed catamaran made out of aluminium. He claimed it was cheap to build, it pierced the waves and dispensed with the ride problems. Sherwood told a meeting that Incat had never built a car-carrying wave-piercing catamaran, but he had taken an option on every one to be produced until further notice.

Robin was sent to Australia to see the first one, which was a bit of a revelation – an aluminium cutting yard in the back streets of Hobart. There was this 'thing', which was all bits of aluminium. The whole ship only weighed about 200 tons, of which the engines accounted for 80 tons, so there was no weight in it at all.

By 1985 the Channel Tunnel was a certainty and Sherwood was convinced that if he was going to survive the Tunnel he would have to do something, so he opted for the wave-piercing catamarans. He concluded the costs were too high on the hovercraft, and the SeaCats might be the answer.

In 1990 Captain John 'Biggles' Lloyd was sent to Australia by David Wise to collect the first one, *Great Britain*. Biggles came across Clifford on arrival in Tasmania and it seemed quite clear that he did not like 'Poms'. Biggles recalls:

> *Clifford was adamant that the Pom 'Men from the Ministry' (The Department of Trade and Industry – DTI) would have no right to 'stick their oar in'. When I pinned him to his desk and told him that only the British or the French could grant the Operating Permit, before which they would go through the vessel with a fine tooth comb, I was told (and I quote), "You are nothing more than the rep of a future operator, so you can keep your f*****g opinion to yourself until we ask for it!" From then on I knew we were in trouble.*
>
> *Watching some of the fore and aft metalwork on the car deck flex with the movement in seas was not reassuring, though it was addressed, and nor was the steering behaviour in quarterly swells. Crossing the Atlantic in no more than a two and a half metre sea on the quarter (yes, SR.N4 Mk.2 weather), I was sure I could do better on the helm than the auto-helm, but all I could manage to do was reduce the yaw from 70 degrees to 50. Much later, Bob Middleton had the same problem crossing the Channel.*

SeaCat *Great Britain* berth at The Prince of Wales Pier, Dover, with *The Princess Anne* making the transition from water to the hoverport. 21st October 1992.

The passengers weren't all that keen either. On the cross-Bass Strait route down-under, in typical Australian parlance, it was referred to as the 'vomit comet'. Biggles also had some reservations about the quality of the construction.

Some of the welding wasn't up to much either. A T-bar stiffener on the inner hull near the water-jet snapped off when I stood on it. A visual check found the surveyor's 'OK' written on it, so he wasn't up to much either.

*I saw David Wise and Robin Wilkins on the car deck in New York. When I sought them out Robin did a great job of boosting my ego, until David emerged from the engine room. When asked for an opinion by Robin, David admitted, "With all the stuff John sent back to me, I thought he was joking, but he wasn't." To use Robert Clifford's own words, the SeaCat was a 'crock of s**t'.*

Captain John 'Biggles' Lloyd and his mother, retired actress Rosamund John, celebrate the winning of the Hales Trophy (the Blue Riband) for the fastest trans-Atlantic crossing by passenger ship. 1990.

Biggles' attempt, however, to challenge for the Hales Trophy from New York for the fastest crossing from Ambrose Light to Bishop's Rock, off the Scilly Isles, a distance of 2,938 miles, was successful. The trophy had been held by the American passenger liner *United States* when it completed the crossing in 3 days 10 hours and 40 minutes on her maiden voyage in 1952. The *Great Britain* completed the distance in 3 days 7 hours and 54 minutes. For his efforts, John was rewarded with a message from Margaret Thatcher which read, "Your success is a triumph for Britain".

The 74-metre long *Great Britain* first went into service on 14th August 1990 on the Portsmouth-Cherbourg route, but was withdrawn in January 1991. The first SeaCat service out of Dover to Boulogne commenced on 25th June that year with their second SeaCat, *Hoverspeed France,* followed by a refurbished *Great Britain* on 20th July, providing a Dover-Calais link. The third, *Hoverspeed Boulogne,* operated between Folkestone and Boulogne for a period of time. A total of five SeaCats were ordered by Sea Containers, the later ones being increased in length to 86 metres. These later versions were vastly improved and were also used across the Irish Sea and to the Isle of Man, but SeaCats are another story.

The five SeaCats dominated the Hoverspeed fleet with the end in sight for the two SR.N4 Mk.3s. (Artistic licence employed to produce a Hoverspeed advertisement.)

Jim Sherwood's dream of revolutionising the European ferry industry by using the SeaCats in favour of hovercraft for passengers and cars, and the ships for freight as a means of taking on the Channel Tunnel, came to naught. It does seem ironic that the name *Hoverspeed* should have been used for the SeaCats, the hovercraft's replacement, when in fact the concept of revolutionising cross-Channel transport had swung full circle back to a more or less conventional ferry, albeit faster and with a shallower draft. Nevertheless, *Hoverspeed* was a powerful brand, but people must have either been confused about the product, or just didn't think about it.

Sea Containers sold Sealink to Stena Line in 1991 and in 1996 the Sealink name disappeared when the UK services were re-branded as Stena Line. The agreement with SNCF on the Dover-Calais route also ended at this time and the French-run Sealink services were rebranded as SeaFrance.

Robin Wilkins remained Hoverspeed's MD until 1991, at which point Bill Moses took over and Robin eventually decided to move on from the Sea Containers Group. He went on to become the UK Managing Director of French owned SeaFrance. The Channel Tunnel opened on 6th May 1994 and would have a profound effect, not only on the cross-Channel hovercraft and SeaCat services, but also lead to a complete rationalisation of the conventional ferries.

Chapter 26

The End

The Final Years

Despite the manful efforts of the management buyout team, and a fresh and dynamic input from Jim Sherwood, from 1987 onward the fate of cross-Channel hovercraft was one of slow deterioration and death.

On 29th July 1987 the British Prime Minister, Margaret Thatcher, and French President, François Mitterrand, ratified the Treaty of Canterbury, paving the way for the Channel Tunnel to become a reality. Although it took another seven years to be completed and a further few years of development, nevertheless it meant the end of Dover as a major seaport and conduit between Britain and Europe, and with it the already priced-out hovercraft.

From that point on, the hovercraft fleet was slowly reduced. It was a particularly sad time for Hoverlloyd as their old uneconomical Mk.2s were the first to go; one by one ignominiously plundered for spares to keep the two Mk.3s going, until, stripped of anything of worth, they were finally scrapped.

The Prince of Wales gutted by fire at Dover Hoverport. 2nd April 1993.

Swift, having been towed from Dover to Lee-on-Solent, about to be brought ashore. June 1994.

Swift being winched in the hover from the beach, across the road and into her final resting place at the Hovercraft Museum at *HMS Daedalus*, Lee-on-Solent.

The first casualty, *Sure,* was broken up on the pad at Pegwell in the latter part of 1987.

The start of the slower, but more economical SeaCat catamaran service in 1991 hastened the end of the Mk.2s. On 29th September 1991, *Swift* crossed the Channel for the last time and was laid up pending sale. The sale never materialised, and on 25th June 1994 she was donated to the

Hovercraft Museum and towed from Dover to *HMS Daedalus* at Lee-on-Solent. Due to the deterioration of the structure, *Swift* was broken up in 2004 at the Museum and now only the control cabin and some major components remain.

Swift, having been transformed back into her Hoverlloyd colours, is finally broken up.

On the 10th October 1991, the remaining two Mk.2s, *Sir Christopher* and *The Prince of Wales*, were withdrawn from service and used for spares.

The saddest demise was that of *The Prince of Wales*, once the pride of Hoverlloyd's fleet, destroyed by fire on the pad at Dover in 1993; and so badly damaged she was broken up for scrap immediately.

This rather poignant account (author unknown) of the last days of *Sir Christopher* comes from the Hovercraft Museum's archive:

> *On a trip to Dover on 21st February, it was apparent that* Sir Christopher *was entering her final few months on the Dover pad. She had been stripped of all her useful fittings, including cockpit equipment, doors and even windows etc., which could usefully be used on the Mk.3 craft, and the remainder of the structure was being progressively cut up and binned. She was still painted in the current Hoverspeed livery on her port side which faced the pad, but her starboard side was looking very*

> *sorry indeed. Her skirt had been cut off about 12 inches down from its attachment to the hull, just leaving the fixings and a strip of rubber to show where it was. This was particularly saddening for me, as she was the first of the craft on which I travelled back in the late '70s.*
>
> *On a visit on 14th April,* Sir Christopher *was noticeable by her absence. All that remained were the bow ramp and rear doors, as well as a number of propeller pylons and lift fans stacked up at the side of the pad. At the side of the maintenance building, there was a heap of small items salvaged from the craft, including doors, hatches, seats, propeller shafts etc.; basically anything that might be of some use on the Mk.3 craft sometime in the future. I am told by Warwick Jacobs of the Hovercraft Society that some sections of the craft have been saved and donated to various individuals and museums. For example, apparently the control cabin is now someone's garden shed and I am told the keel skirt is in use by a light hovercraft club on their track!*

It would be nice to know where the garden shed is now.

Soldiering On

Martine Watson (front) and Helen Hutchinson (back) preparing in the galley for the new 'Blue Riband' service on the Mk.3s, designed to compete with the 'Club Class' on the conventional ferries. 1993.

Serving passengers in a 'Blue Riband' cabin. (A publicity picture).

The remaining two Mk.3 craft, despite being the oldest structures, sustained by transplants from the Hoverlloyd organ bank, soldiered on.

In 1993, further refurbishment of the Mk.3s hailed the introduction of the 'Blue Riband' service, which was essentially an on-board premium class towards the rear of the craft, providing better seats with complimentary coffee and tea, to compete with the premium lounges now available on the giant ferries. In the same year, the hovercraft service to Boulogne was curtailed and the two craft only operated the Dover-Calais route, leaving another hoverport at Le Portel to an abandoned and unwanted future.

Later in the decade, early in 1998, with surprising optimism for the future, *The Princess Margaret,* followed by *The Princess Anne* were each taken out of service for their annual refit, the alternate craft maintaining services with six return flights each day. During the refit, each craft was stripped right down to basics for deep structural maintenance, which included stripping paint off the structure, removal of elements such as rudders, pylons and doors, engine and propeller overhaul and re-fitting, control cabin equipment upgrades and a complete re-paint. This was one of the most intensive overhauls the craft had had since being stretched to Mk.3 status in the late '70s. The craft were now proclaimed to be 'as good as new'.

The optimism was short lived. Like all new enterprises, the Channel Tunnel took time to be fully accepted by the public so, for a while, the remaining hovercraft, together with the SeaCats, still managed to take what would become an ever diminishing share of the market; albeit at considerably reduced fares due to the fierce competition above and below the Channel. By the end of the millennium, Jim Sherwood's vision of taking on the Tunnel with his SeaCats must have begun to fade.

Certainly it had to be accepted that hovercraft could compete no longer. The last SR.N4 flight left Dover on the evening of 1st October 2000, piloted by Captain Nick Dunn; thus closing 32 years of hovercraft on the Dover Strait, almost certainly never to be seen again.

Subsequently, both Mk.3s made their way, at least in a more dignified manner under their own steam, to join what was left of *Swift* at the Hovercraft Museum at Lee on Solent, and there they remain to this day.

A refurbished *The Princess Margaret* leaving Dover Hoverport in 1998.

1st Officer Simon Brown, Captain Nic Rose and 2nd Officer Claire Galgy, the second of two female flight crew. The other was Kathy Spain.

Long Service Medals

Not just the craft but people went too. Few who were there at the beginning saw the saga through to the end. Peter Barr who joined Seaspeed in 1966 as one of the very first captains stayed almost to the end:

> *I retired in January 1996. They asked me to go back for the summer season that year and I went back for a few weeks the following year. They reduced the hovercraft timetable as a result of having the SeaCats. There were two or three of us who were employed on a seasonal basis, but after that they thanked us warmly for our services and sent us on our way rejoicing.*

Likewise, Brian Laverick-Smith, who sadly had to resign as captain in 1996 as a result of hearing problems, decided he wanted to commemorate the end of the giant SR.N4 in some style:

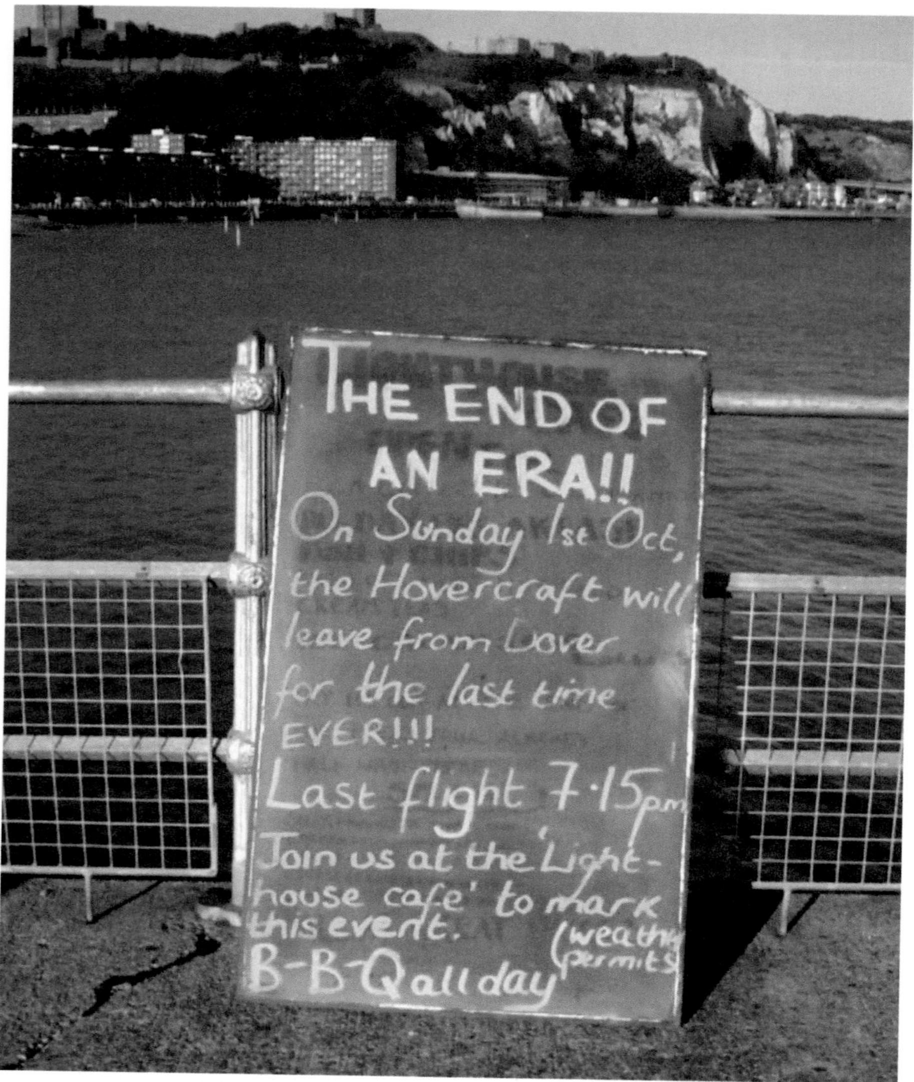

Announcing the farewell party of the SR.N4 cross-Channel hovercraft.

I block booked 40 seats out on the last but one flight for my two wives (one ex), my friends, my family – the whole lot – and I hired out the Churchill Hotel on the Dover seafront. They all came there – 200 ex-Seaspeed, Hoverlloyd and Hoverspeed people.

I then arranged with Captain Nick Dunn, who commanded the last flight, to bring the craft to the front of the hotel, with the permission of the Port Control. They killed the disco and instead played 'The Ride of the Valkyries'. The hovercraft lights were flashing, they dropped the bow ramp, people were crying, and now it's all dead.

The only captain who started at the commencement of cross-Channel operations and saw it through to the very end was Bob Middleton, who began with Hoverlloyd in 1969[1].

Captain Nick Dunn, who commanded the last Channel crossing by SR.N4 hovercraft.

Other Departures

Almost, it seems, on cue, two significant figures in the industry departed, or in nautical parlance 'went aloft', at much the same time as the cross-Channel hovercraft. Sir Christopher Cockerell and Peter 'Sheepy' Lamb.

---

1 Sad to say there is no opportunity of any comment from Bob, the acknowledged ace driver. He passed away in 2007, on the golf course; there are worse ways of departing.

Christopher Cockerell died on 1st June 1999, not a bitter man, according to his daughter, but a disappointed one. Disappointed by the way he had been treated and the fact that his great invention had never attained what he felt might have been its true potential. Even so, he was held in great esteem by so many, including those at Hoverlloyd who flew with pride the only hovercraft ever named after him. According to Ray Wheeler, the day before he passed away at the age of 89, this remarkable man was still working on a project for generating power using waves.

Looking at the faded broken white line of the northern approach ramp of what remains of Pegwell Bay Hoverport and out across the mud flats, where up to 54 SR.N4 departures and arrivals to and from Calais took place each day.

Sheepy Lamb, who did so much for the industry during its early development, through his consummate skills as a test pilot, died on 1st July 2000. One of his greatest hovercraft achievements was when, 41 years earlier, he took a huge risk and made the first Channel hovercraft crossing from Calais to Dover in the SR.N1, which was still in the very early stages of development as an experimental machine. The success of the mission, and subsequent publicity, brought the hovercraft to the attention of the public and with it the enthusiasm to continue the research necessary to kick-start the industry.

By 2000, a large part of the industry had long since departed. In 1984 the British Hovercraft Corporation was renamed Westland Aerospace, as the nature of the business had changed, with hovercraft being only a very small part. In any case, BHC's costs were far too high. There had been no quantum leap in skirt technology and the cost of 'marinising' an economical jet engine was prohibitive for the few hovercraft orders that might have been placed. As BHC's parent company, Westland's financial status had been bleak for years, of which the £4 million loss on the conversion of Seaspeed's Mk.1 to Mk.3 SR.N4s played a small part, and by the autumn of 1985 Westland[2] was facing bankruptcy, with debts purported to be £100 million.

A satellite picture showing the ghosts of yesteryear.

After it was closed as a maintenance base, the hoverport at Pegwell Bay languished neglected, gradually falling into disrepair. With no alternative use being found for it, the hoverport buildings were finally demolished in 1995. Today, among the overgrown concrete foundations, all that remains to remind us of its original purpose are the white lines still showing on the approach ramps. But wait a moment, if one looks at the satellite image of Pegwell today the outlines of the jacking areas can clearly be seen at the south-west end. Shaped like the plan of an SR.N4, the two concrete slabs provide a ghostly image, a reminder of

[2] A rescue package was eventually put together with Sikorsky and Fiat, but before then two of Margaret Thatcher's ministers had resigned in what became known as 'The Westland Affair'. GKN bought into Westland in 1988 and soon acquired the shares owned by Fiat, which gave them absolute control. In 1994, Westland became a wholly-owned subsidiary of GKN. It was merged with Finmeccanica's Agusta helicopter division in 2001. In 2004, Finmeccanica S.p.A. acquired GKN's share in the joint venture.

the huge machines that once called the place home, perhaps waiting for the ghosts of old crews to take them back to sea again.

Questions

John Chaplin, who was there at the beginning, accompanying Sheepy Lamb and Christopher Cockerell on the memorable SR.N1 Channel crossing, left BHC to work in the USA on the far more viable military landing craft projects. He puts forward a very succinct theory as to the chances of success for commercial hovercraft operations:

> *In the few places where there is a river or a bay crossing, or something like that, for which the hovercraft was suited, some bastard would come along and build a bridge or a tunnel.*

Very true, but no one has built a bridge or tunnel between Portsmouth and the Isle of Wight, so Hovertravel still flourishes.

Still the question remains: Would Hoverspeed have survived if 'some bastard' hadn't built the Channel Tunnel?

The final resting place – The Hovercraft Museum at Lee-on-Solent. The picture of *The Princess Anne* was taken from in front of the control cabin of *The Princess Margaret*.

It all boils down to running costs. The gas turbine, for example, is still a necessarily light power unit for a very large machine that has to lift many hundreds of tonnes off the ground. Small hovercraft are able to use diesels today. The gas turbine is, however, not only more expensive in fuel compared with the common or garden marine diesel engine, but it is expensive to 'marinise' and also more expensive to maintain. Whether we will ever see a gas turbine that can be serviced by 'a man and a boy' on any street corner is a matter for conjecture.

To view the hovercraft as an 'aircraft at sea', albeit amazingly robust in practice, was not really proven and most small hovercraft today, although constructed of aluminium, are marine in structure, not so vulnerable and less costly to repair.

But without doubt the biggest single consumable cost on any type of hovercraft, which conventional vessels of whatever size do not have to contend with, is the skirt system. Although tremendous strides were made in skirt development in the SR.N4's lifetime, the cost of replacing rubber on a daily basis was crippling, mainly because of the often hostile environment in which the SR.N4s were operating and the speeds at which they were travelling. The smaller hovercraft of today, by the nature of their size, operate in much friendlier environments and at speeds which rarely exceed 45 knots.

BHC deserves an accolade for producing an excellent solution to the basic problem of interfacing a hard plenum structure with a moving and unpredictable surface like the sea. What let them down was the available material. Until a virtually indestructible and inexpensive skirt material can be found, large and fast hovercraft will always be at a huge disadvantage compared with their more conventional competitors.

The verdict on the English Channel experiment was that large hovercraft had proved themselves operationally, but failed economically. It was a fascinating experience for those of us involved, and one we wouldn't exchange for the world.

The Future

Commercial prospects for large hovercraft may have been all doom and gloom, but Cockerell must have taken some comfort in the development of military applications.

Some countries now have hovercraft amongst their naval fleets, with the largest promoter being the USA. The ultimate result of the 'hundred knot navy' project, in which Hoverlloyd played a small but significant part back in 1972, is not far short of a hundred Landing Craft Air Cushions (LCAC), which, when launched from specially designed mother ships, can carry a fully equipped battle tank up the beach at 45 knots.

These craft are soon to be replaced by Ship to Shore Connectors (SSC), an upgrade with more versatility and carrying capacity. The project, which is still in the design stage, includes Britain's Griffon Hoverwork as a participating company.

Hoverlloyd – a Eulogy

What Made Hoverlloyd Special?

The best answer to the above question comes from Nick Smith, who as a young student spent his vacations working for Hoverlloyd as a car marshaller:

> *Since those days I have worked in other very large organisations, mainly in the IT field, and I look back to my early working years at Hoverlloyd and realise that I 'peaked early' in terms of the quality and cohesion we had. I have smiled many times over the years when I have listened to managers talking about team spirit and loyalty and remembered our company where, if things were going wrong, far from having to cajole people into helping out, we would start getting spontaneous phone calls from off-duty people at home wanting to know what they could do!*

That just about says it all. It cannot be denied that any company operating something as new and exciting as the giant SR.N4 hovercraft was bound to be considered unique, but there was something more about Hoverlloyd. The difficulty now is trying to understand why.

As Nick describes so well, there was an *esprit de corps* about the company, which few of us have experienced before or since. Yes, it was an exciting thing to do, but was that sufficient to engender the dedication and loyalty that it did?

Was it perhaps simply the exuberance and flexibility of youth? In 1969 a large majority of the operations and traffic staff would have been in their twenties. In the whole company few were beyond their mid-thirties. This was the late '60s and early '70s, a time of flower power and general youth rebellion and, even if one was not part of the revolution, it was still party time for all. Everyone on the right side of 40 was having a good time. Being involved in something as exhilarating as the SR.N4 just added to the fun.

It could also be said that it was not just the age factor that was a major influence, but also the type of personnel Hoverlloyd attracted. Without doubt the adventure of something so new attracted the adventurous; people who were prepared for anything and determined to enjoy themselves doing it. In Captain Dennis Ford's opinion:

> *We were doing something that no one else had done, virtually making it up as we went along, and doing a job that was*

interesting and fun in its own right. Add in the fact that the majority of us were not so dissimilar in age, came from a whole variety of interesting backgrounds and were all 'new' to the company with no 'old lags', who had 'been there, done it', thereby putting a drag on progress and enthusiasm. And as a bachelor, surrounded by well turned out attractive women, what else could one ask for?

The youthfulness was exemplified by the fact that both 'Big' John Lloyd and Robin Paine had their commands by the age of 28 and some of the senior stewardesses had barely reached their 21st birthdays. Dennis touches on another interesting factor that must have added to the atmosphere of fun and excitement. So many of the staff, being young and in their twenties, were inevitably single, so attraction between the sexes added an extra *frisson* all of its own.

Who can argue with Dennis' final comment?

Good Management

There is something of the chicken and the egg conundrum here. Did the type of personnel we started out with produce the good management, or did good management produce the people?

> *"Most people in big companies today are administered, not led. They are treated as personnel, not people."*
>
> Robert Townsend – CEO of AVIS and author of *Up the Organization* in the 1960s.

Or, in ghastly modern parlance, 'a human resource' – possibly one up from a cockroach.

Robert Townsend's book – forty something years later – still remains one of the most read, revered and quoted treatises on good management. It is certainly widely read but whether the author's sage advice is followed is quite another question. The very essence of Townsend's approach was to look after the people at the bottom and work up from there.

It is by no means certain that any of the senior management of Hoverlloyd had ever read or even heard of Townsend and his book, but our company nonetheless followed his precepts to the letter. It really doesn't matter whether this was simply a matter of chance; the fact remains that Hoverlloyd's management style was one of concern and encouragement for each and every employee regardless of their place in the hierarchy.

The full credit for that philosophy, which helped so much to foster the feeling of pride and wellbeing, which pervaded all aspects of working life during the company's brief existence, can be placed exclusively at the feet of one man -- Les Colquhoun. Like Robert Townsend, he didn't just believe that everyone in the workforce was equally important, he worked tirelessly to ensure that everyone knew it. Robin Paine's recollection is typical:

At the end of the first season, Les Colquhoun held the first of a series of meetings with the flight crew. From his own background it was probably natural that he identified with operational personnel probably more than other departments.

Even so, as MD he never allowed himself to lose the overall picture. At the meeting in question he gleaned the impression that the flight crew thought they were rather special and their views held sway over those of others. I never knew Les to get excited or raise his voice. Although he obviously took exception to this attitude, Les quietly explained that, although the flight crew had special skills that were recognised, so did everyone else in the company and, in particular, the cleaners, for no better reason than the cleanliness of the craft and terminal played an important part in fare-paying customers wanting to return a second time.

Asking Hoverlloyd people today what it was they particularly liked about the company and the word most used is 'valued'. Regardless of their position, high or low, everyone felt they were worthwhile to the organisation, not just a code number in a personnel list.

It is clear that Les Colquhoun's message of fundamental respect for people, not just 'personnel', flowed laterally as well as vertically down the chain of command. Appreciation of the value of others existed between departments and lasting friendships were made, engendering a family feeling within the company, which was so apparent at the 40-year reunion, held in 2009. Dennis Ford again:

I think that most people had a genuine respect for the work done by other departments. No one was going anywhere without reservations, all done by hand, since computers were not in common use. The engineers worked wonders when we brought the craft back in tatters, sharing their expertise with each other to keep the machines working. And our amazing traffic staff who, with mathematical dexterity, shuffled passenger numbers, adjusted load factors and, with calm diplomacy, placated frustrated travellers during weather delays, ensured the company actually made the money that kept us all in employment.

Esprit de Corps

> Esprit de corps*, a French phrase much used by English writers to denote the common spirit pervading the members of a body or association of persons. It implies sympathy, enthusiasm, devotion, and jealous regard for the honour of the body as a whole.*
>
> Webster's Revised Unabridged Dictionary.

When Hoverlloyd finally amalgamated with Seaspeed to form Hoverspeed, the new senior management team were fond of saying, "We must ensure that Hoverlloyd's *esprit de corps* is transferred to the new company".

However, the sad fact is that just parroting the wish does not necessarily make it happen. It is easy to say what it is, but how does one create it?

So the question remains, why did it occur in Hoverlloyd? The good fortune to have a leader of the calibre of Les certainly fertilised the ground, but were there any other aspects that encouraged the garden to grow? It is a measure of its ethereal nature that those of us who experienced it, who know without doubt it was there, would be hard put to explain why.

The exuberance of youth, the air of adventurousness, the excitement of changing times, all played their part, but were there any other contributing factors?

Looking through the Internet, today's fount of all knowledge and information, apart from the Webster's definition of *esprit de corps* quoted above, it is surprising how little explanation or advice there is on how to achieve it.

Wikipedia, for example, lists factors influencing morale[1] within the workplace as:

1. Job security.
2. Management style.
3. Staff feeling that their contribution is valued by their employer.
4. Realistic opportunities for merit based promotion.
5. A perceived social or economic value of the work being done by the organisation as a whole.
6. The perceived status of the work being done by the organisation as a whole.
7. Team composition.
8. The work culture.

---

[1] In this instance Wikipedia equates *esprit de corps* with morale.

Items 2 and 3 have already been discussed, but how relevant are the others to the Hoverlloyd experience? For instance, 'job security' and 'reasonable opportunities for merit based promotion' were non-existent in the first three years of the N4 operation, when the large majority of us joined the company. It was a time of struggle and uncertainty with genuine worries as to the company's survival. Yet, undeniably, pride in the enterprise and the camaraderie engendered was as strong then as it ever was. In fact it could be claimed that those first difficult years laid the foundation for the strong feelings of pride and affection which followed.

Proceeding further down the list, it's not at all evident that anyone in Hoverlloyd felt any sense of providing something of value to the public other than helping them to cross the water and go on holiday with the minimum of fuss. This was hardly unique, as plenty of other ferry companies out of Dover, Folkestone and Harwich were achieving the same thing and in larger numbers.

However, the perceived status of the organisation was definitely a factor, but perhaps not in the sense implied. It wasn't so much a sense of the importance of the company's contribution, as the feeling of the sheer individuality of the method of transport used. We were different and proud of it.

Team Spirit

The list refers to 'team composition', and this is clearly a factor to be considered when examining Hoverlloyd.

The shift work nature of any transport operation demands a split of the company personnel into teams. How those teams work and how well they get on together is a matter of individual personalities and also how the system is managed.

An important factor in Hoverlloyd's case was the amount of trust and support given to the supervisors. The 'A Girls' in charge of each group of cabin staff, and the car deck supervisors were left to themselves to manage their teams in every respect; check those reporting for duty, admonish the latecomers, make sure they were turned out satisfactorily; this not only applying to the cabin staff, but the car deck personnel too.

A great pride was taken in the company uniform, from the captain down. Everyone, regardless of rank, wanted to present a good impression to each and every passenger. Flight crew never left the control cabin unless they were in full uniform complete with cap.

Of course, these separate flight crew, cabin and car deck teams teamed up as one crew, a total of 13 or 17, depending on the seasonal demand. Although there was a rank hierarchy, from the start there was a very easy relationship and mutual respect between the sections. Everyone was on a first name basis but managed to be highly professional at the same time; a difficult balance to achieve.

A Ticking Off for Roger Syms

In Hoverlloyd there was never any feeling of rank or hierarchy between the flight crew and the rest. I am still embarrassed to remember on one occasion one of the cabin staff giving me a right royal ticking off.

During the turn round at Calais I dashed down to the port cabin to pick something up, just minutes before departure. Because time was running out, I just jumped out of my seat as I was, sleeves rolled up, collar unbuttoned and tie loosened. I think I had also unplugged my headset and had it hanging around my neck.

At the sight of me the girl, whose name I sadly no longer remember, said something to the effect of, "How dare you turn up here looking like that, what will the passengers think?" I was suitably chastened and never went down ever again unless properly dressed.

Interesting to reflect that airlines today set great emphasis in their Crew Resource Management courses on the right of junior members of a flight team to 'challenge' – the right to question or correct those of a higher rank. If the captain makes a mistake the co-pilot has the duty and the right to query or correct him.

The idea of CRM didn't appear until well after Hoverlloyd commenced operations, but it would appear we were already following its precepts, more or less naturally, as a matter of company culture. I may have felt a total twit, but I recognised even then that the ability to challenge, and for the challenge to be accepted, was a healthy sign for us all.

Theory and Reality

Looking back over the Wikipedia list, anyone like ourselves, who has experienced the atmosphere of a model organisation like Hoverlloyd has to ask: Do the people presenting this list of necessary criteria, have any real first-hand experience of the phenomenon, or is it all just theory?

The doubt arises, not so much out of what is listed, but what has been left out.

What is not mentioned for example is the self-enhancing aspect of the team spirit, which once recognised, like 'Topsy' just grows. For anyone working in a well-coordinated and efficient team, dedicated to the highest degree of professionalism and service, there develops an enormous feel-good factor, which instils a desire in each individual to work even harder to perfect the team, simply to prolong the pleasure.

The strange thing is that, to an outsider, it may appear altruistic. Possibly it starts out that way, a subjugation of self for the greater good of the team, but paradoxically in the end, it is probably perpetuated by downright selfishness in the pursuit of happiness. Team spirit is strongly addictive.

The Importance of Trivia

Let us be clear, trivia is important and therefore, in fact, is not trivial at all. It only appears so to bad managers.

The whole essence of Robert Townsend's approach to good management was to observe the little things, and if they needed changing or fixing, make them a priority.

In Hoverlloyd there were many examples of this ethos but these two below will suffice.

For example, how important are reserved parking spaces for top executives? From their lofty perspective it only seems right to provide a 'perk' to hardworking people, making them feel valued and important. The problem is by making these people feel valued and important there is clear message to all those minions below, "sorry mate you aren't".

Townsend's response to reserved parking was typical, "If you're that bloody important get to work first, besides you meet a nice type of person in the staff car park." Needless to say, Hoverlloyd only ever had a staff car park.

The practice of giving everyone a Christmas hamper each year was much appreciated. In itself it was a magnificent gesture but what was really trivial about it and therefore really, really important was the simple fact that regardless of how far up or down the hierarchy you were, everyone got the same one. There was no such thing as the 'deluxe executive hamper' and the 'other one' – they were all the same. In any organisation these things get noticed. The real message, in addition to the staff's appreciation, was that there was no hint of class distinction and that was much appreciated too.

In Memoriam

It is inevitable that, to a lot of readers, particularly those who haven't had the fortune to work in such an atmosphere, all this may sound overly sentimental and nostalgic – an idyllic fantasy, woven around what was no more than a common or garden organisation, with nothing really to recommend it. After all, it failed after 15 years didn't it?

To those who know anything of British history in the 1970s, what also will add to the feelings of doubt is the fact that this story goes totally against the grain of industrial relations in the country during that period.

Hoverlloyd wasn't just special, it completely bucked the trend.

All that we who experienced it can say is, "We were there, it really happened." We carry Hoverlloyd in our hearts to this day. In the hope that history does not forget us we had planned to install a memorial on the cliff above what is left of the Pegwell Bay Hoverport. It was to be a refurbished SR.N4 propeller on a marble plinth standing near another reminder of bygone glories, the Viking Ship.

Unfortunately due to local opposition and funding difficulties this was not to be. As a substitute remembrance this book has taken its place.

Read it and remember us.

Appendix 1

Extracts from Bob Strath's Log Book

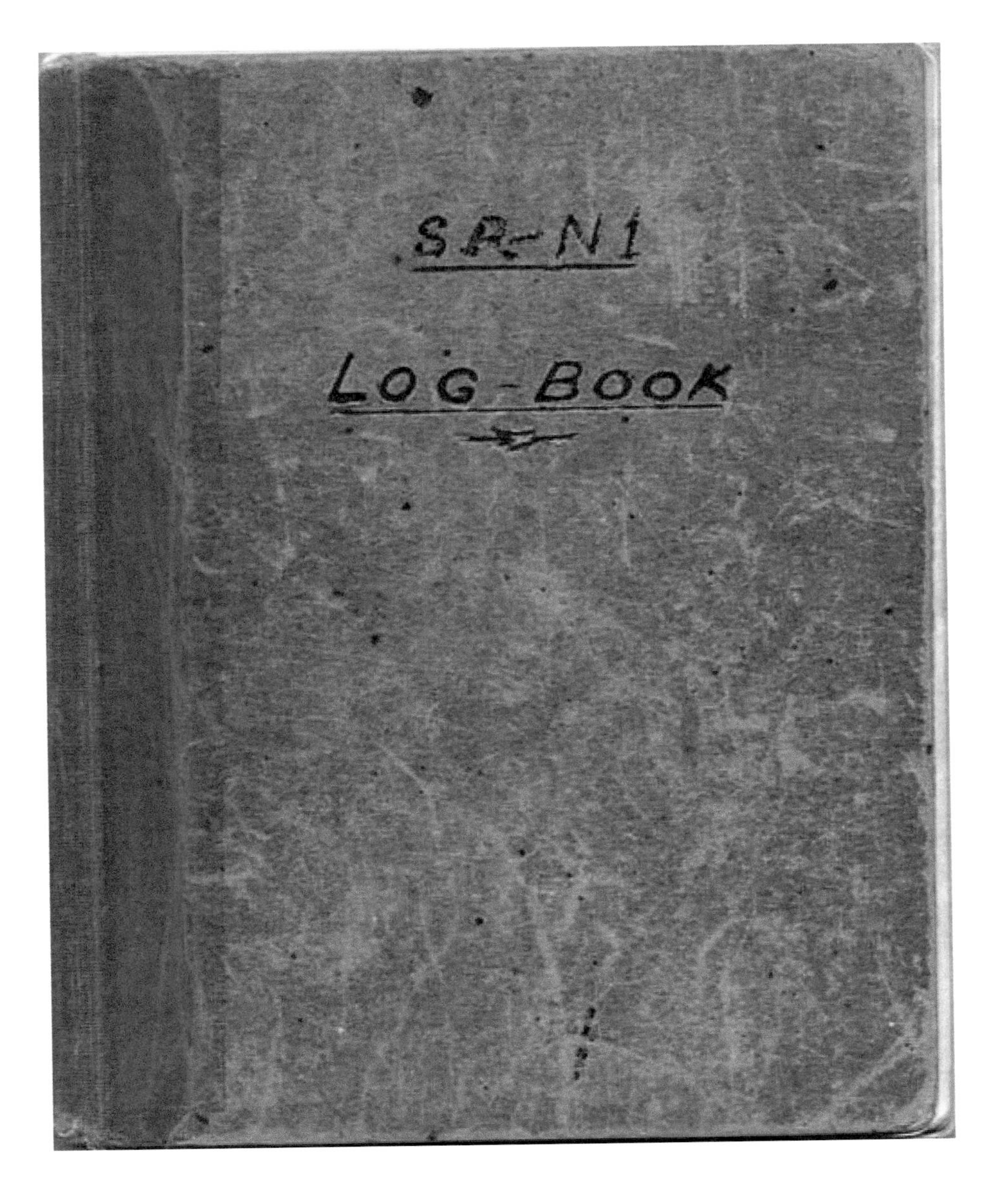

DATE	LEONIDES				SEA	TIME AT SEA				REMARKS.
	DAILY		TOTALS		SORTIE	DAILY		TOTAL		
	HRS	MINS	HRS	MINS	No.	HRS	MINS	HRS	MINS	
29/30·5·59	-	53	-	53	-	-	-	-	-	1st Tethered engine run. High Air Intake Temps & Low Oil Inlet Pressure
30/31·5·59	-	55	1	48	-	-	-	-	-	2nd Tethered engine run. Cushion pressure measured - High A.I. Temp. [illegible]
7·6·59	2	56	4	44	-	-	-	-	-	Engine & electrics D.O.S. Cushion pressure measured from 2000 to 2950. Free Hover. [illegible]
8·6·59	1	57	6	41	-	-	-	-	-	Six free Hovers - Inner curtain removed to ascertain cushion pressure effect.
9·6·59	-	43	7	24	-	-	-	-	-	Three free hovers. Forward lateral aft aileron disconnected.
10·6·59	1	13	8	37	-	-	-	-	-	Forward & aft ailerons interconnected - 3 free hovers.
11·6·59	1	15	9	52	1	-	35	-	35	Press Day - First trial at Sea, craft winched down slipway on wheels.
11·6·59	-	15	10	7	-	-	-	-	35	Engine washing in fresh water.
12·6·59	1	23	11	30	2	1	13	1	48	2nd Sea trial. Ditchings up to 20 kts. - full power run. [illegible]
12·6·59	-	10	11	40	-	-	-	1	48	Engine washing in fresh water [illegible]
21·6·59	2	55	14	35	3	2	25	4	13	[illegible]
22·6·59	2	43	17	18	4	2	23	6	36	First sea voyage [illegible]
24·6·59	-	28	17	46	5	-	23	6	59	First failure. Demonstration to NRDC. Trial abandoned because of broken exhaust pipe.
30·6·59	2	31	20	17	6	2	01	9	00	Redesigned exhaust system. First handling by Harry Phillips, beaching at Osborne [illegible]
3·7·59	2	38	22	55	7	2	30	11	30	Demonstrations for Admirals [illegible]
8·7·59	1	55	24	50	8	1	43	13	13	Beachings at Osborne Bay. Turning manoeuvres filmed from 'copter.
10·7·59	-	30	25	20	-	-	-	13	13	[illegible]
11·7·59	2	00	27	20	9	1	53	15	06	Demonstration for Duncan Sandys [illegible] Under-water photography [illegible]
12·7·59	1	35	28	55	10	1	15	16	21	Demonstration for [illegible]

11th June 1959. Press day when the SR.N1 made its first demonstration on water.

Date	Ops. No.	Leonides Daily Hrs	Leonides Daily Mins	Leonides Totals Hrs	Leonides Totals Mins	Sea Sortie No.	Time at Sea Daily Hrs	Time at Sea Daily Mins	Time at Sea Total Hrs	Time at Sea Total Mins	Remarks
.	.	B/F		38	09	.	B/F		23	55	
24.7.59	34	.	40	38	49	15	.	30	24	25	Demonstration in Calais Harbour & outside on the beach.
25.7.59	35	2	45	41	34	16	2	30	26	55	The first Channel crossing by Hovercraft
26.7.59	~	-	20	41	54	.	-	-	26	55	
31.7.59	-	-	15	42	09	.	-	-	26	55	Engine inhibiting runs only. Factory "close-down" period
3.8.59	.	-	20	42	29	.	-	-	26	55	
7.8.59	.	-	15	42	44	.	-	-	26	55	
13.8.59	-	-	15	42	59	.	-	-	26	55	
14.8.59	36	.	25	43	24				26	55	
15.8.59	37		39	44	03				26	55	Handling assessment on apron after mods to control.
16.8.59	38		52	44	55				26	55	Measurement of hover height with increased payload.
19.8.59	39	1	37	46	32		.		26	55	(upward of 20 bodies being put aboard)
20.8.59	40	-	33	47	05				26	55	~~[illegible]~~) (Working up for SBAC)
24.8.59	41	1	19	48	24				26	55	
26.8.59	42	1	16	49	40				26	55	Operating at R.N.A.S. Lee-on-Solent. First handling assessment over land.
27.8.59	43	1	10	50	50				26	55	Demonstrations before numerous V.I.P.s for N.R.D.C. & H.D.L.
3.9.59	44	.	20	51	10				26	55	Over 20 persons were carried at a time for these demonstrations.
4.9.59	45	1	18	52	28				26	55	Rehearsal runs for SBAC display - at Farnborough
7.9.59	46	-	35	53	03				26	55	1st Day with Commando aboard.
8.9.59	47	.	25	53	28				26	55	Demonstrations at S.B.A.C. display, Farnborough

24th July 1959. A late afternoon demonstration in Calais Harbour.
25th July 1959. First Channel crossing from Calais to Dover, which Bob Strath missed.

Date	Ops. No.	Leonides Daily Hrs	Leonides Daily Mins	Leonides Total Hrs	Leonides Total Mins	Sea Sortie No.	Time at Sea Daily Hrs	Time at Sea Daily Mins	Time at Sea Total Hrs	Time at Sea Total Mins	Remarks
		B/F		55	08	-	B/F		26	55	
13.9.59	52	-	30	55	38	-	-	-	26	55	Last demonstration of the week at S.B.A.C. display Farnboro
18.9.59	53	-	36	56	14	-	-	-	26	55	Demonstration at Lee-on-Solent
21.9.59	54	-	38	56	52	-	-	-	26	55	Demonstration for Lord Louis Mountbatten at Lee-on-Solent. my first hover over land
23.9.59	55	2	10	59	02	17	1	00	27	55	Towing drag tests at Lee-on-Solent. Return to Cowes first beaching at Columbine under own steam.
25.9.59	56	1	03	60	05	-	-	-	27	55	Bow adjustments. Stability measurements on slipway
28.9.59	57	1	18	61	23	.	.	.	27	55	Thrust measurements on slipway
1.10.59	58	1	42	63	05	18	-	20	28	15	Stability tests on apron. Demonstration for [illegible] of Readers Digest
2.10.59	59	1	44	64	49	.			28	15	Hover height measurements at various C.G. positions & constant power
19.10.59	60	-	30	65	19	.			28	15	Stability tests on apron
21.10.59	61	-	30	65	49	.			28	15	Demonstration for Col. Nelson on apron
22.10.59	62	1	40	67	29	19	-	20	28	35	Handling at sea with inner jet blanked
23.10.59	63	1	25	68	54	-			28	35	Stability tests on apron
30.10.59	–	-	10	69	04	.			28	35	Engine inhibiting run.
2.11.59	64	1	00	70	04	.			28	35	Pressure plotting of fan distribution
3.11.59	65	1	30	71	34	20	1	10	29	45	demonstration for [illegible] visitors [illegible] beaching at Folly Inn. Wheels on for beaching at Columbine Slip.
4.11.59	66	1	25	72	59	21	-	40	30	25	Demonstration for Col. Nelson
5.11.59	67	1	05	74	04				30	25	Lateral stability tests & pressure plotting
10.11.59	68	2	45	76	49				30	25	Fan total head measurement

13th September 1959. Last demonstration of the week at Farnborough Air Show.

21st September 1959. Demonstration for Lord Louis Mountbatten at Lee-on-Solent and Bob Strath's first hover as pilot over land.

DATE	OPS No	LEONIDES DAILY Hrs	Mins	TOTALS Hrs	Mins	SEA SORTIE No.	TIME AT SEA DAILY Hrs	Mins	TOTALS Hrs	Mins	REMARKS
		B/F		81	09	·	B/F		31	05	
20.11.59	74	-	10	81	19	·	·	-	31	05	Engine run on quayside at Portland. Trials party left for weekend at home.
23.11.59	75	2	00	83	19	23	1	50	32	55	Trials on the Race Pressure & G9 Range at Portland.
24.11.59	76	1	40	84	59	24	1	40	34	35	Trials on Acoustic Range at Portland
25.11.59	77	-	10	85	09	-			34	35	Acoustic measurements on quayside at Portland
26.11.59	78	1	40	86	49	25	1	40	36	15	Trials on Pressure Range at Portland. Craft loaded on lighter [illegible] for return.
1.12.59	79	-	51	87	40	26	-	50	37	05	Demonstration at sea for H.D.L.'s visitors
2.12.59	80	1	15	88	55	27	1	15	38	20	ditto
12.12.59	81		20	89	15	-	-	-	38	20	Engine run after changing rotating unit & priming nozzle
15.12.59	82	1	25	90	40	28	1	25	39	45	First sea trial with doors fitted. The luxury of it
16.12.59	83	-	30	91	10	·	-	·	39	45	Handling trials on apron
17.12.59	84	-	30	91	40	·	-		39	45	ditto
18.12.59	85	1	25	93	05	29	-	45	40	30	Demonstration for & handling by H.R.H. Prince Phillip Duke of Edin.
29.1.60	86	·	45	93	50				40	30	Engine running on apron
1.2.60	87	1	20	95	10	30	-	45	41	15	Spray measurements and water speed indicator calibration
2.2.60	88	1	30	96	40	31	-	30	41	45	As above + demonstration on apron for visitors
4.2.60	89	1	30	98	10	32	-	30	42	15	Further trials with waterspeed indicator & spray measurements.
9.2.60	90	·	40	98	50	-	-	-	42	15	Demonstration on apron for P.L. Cowes visitors
15.2.60	91	1	20	100	10	33	1	20	43	35	Recorded acceleration in ditching case. Spray measurements made.
17.2.60	92	1	45	101	55	34	-	40	44	15	Roll stability tests with single jet on apron. Spray measurements + [illegible]

18th December 1959. His Royal Highness The Duke of Edinburgh takes control of the SR.N1 at Lee-on-Solent with Sheepy Lamb and Bob Strath.

1st October 2000. The final day of SR.N4 operations

SUNRISE : 0700
SUNSET : 1838
TWILIGHT : 1910

SUNDAY
DATE: 01.10.00

HIGH WATER : 0155, 1412
LOW WATER : 0922, 2141

DEST	STD	ATD	CPT	CAPTAIN	CREW NO	ETA	ATA	B/S	CROSSING TIME		REMARKS
Calais	0700	0700	07	SOUTH		0835					
Ostende	0730		RAP	~~Scenadeen~~ Pottier		1250					
Calais	0800	0800	06	ROSE		0935					
~~OST~~ Dov	0830	/	D/A	Ramakers		0935					
Calais	0900	0900	07	SOUTH		1035					
Calais	1000		06	ROSE		35					
Ostende	1045		D/A	Ramakers		1605					
Calais	1100		07	ELSAM		1235					
											FRI/SAT/SUN
Calais	1200		06	FRY		1335					
Calais	1300		07	ELSAM		35					FRI/SAT/SUN
Calais	1400		06	FRY		1535					
Ostende	1400		RAP	Pottier		1920					
Calais	1500		07	MACFARLAN		1635					
Calais	1600		06	HAWKINS		1735					FRI/SAT/SUN
Calais	1700		07	MACFARLAN		1835					
Ostende	1715		D/A	~~Ramakers~~ Mille		2235					layover Dover
Calais	1800		06	HAWKINS		1935		W			
Ostende	2030		RAP	Pottier		X	/	/	/	/	layover Ostende
Calais	1900		07	DUNN							Staff Flight

Captain John Hawkins commanded the last commercial flight on *The Princess Margaret*.

Captain Nick Dunn commanded the very last Channel crossing made by an SR.N4 as a 'staff trip' on *The Princess Anne*.

Appendix 2

Hovercraft Licensing

Is it a Ship, an Aircraft or a Motor Vehicle?

By the end of 1965, hovercraft had been operating commercially on the rivers and coastal waters of the United Kingdom in a limited form for three years. When these operations had been for the purpose of public transport, they had been carried out without the burden of legislation, which covered the operation for a similar purpose of ships, aircraft and motor vehicles.

With Clyde Hover Ferries and Hovertravel making a serious attempt to run passenger-carrying hovercraft services, and with an ever-growing awareness of the potential for these remarkable machines by other interested parties, it was inevitable that the lawmakers would have to put a handle on the legislation regarding their operation and all that followed. One of the real problems in deciding what law, if any, might apply to Air Cushion Vehicles[1] (ACVs), as hovercraft were also known, was to decide precisely what a hovercraft was. Hovercraft proponents argued that it was an entirely new class of vehicle. To the lawyer, however, the question at the time was: Was it a ship, an aircraft or a motor vehicle?

The Merchant Shipping Act 1894 provided that a passenger steamer which carried more than 12 passengers must not proceed on a voyage with passengers on board unless the owner or Master had a survey certificate from the Minister of Transport, which was in force and applicable to the voyage on which it was about to proceed. It also stated that the provisions in the Act applicable to steamers applied to ships propelled by electricity or other mechanical power, with such modifications as the Minister of Transport may prescribe. So if a hovercraft was a ship it must, *prima facie*, have a survey certificate. But was it a ship? The Act defined a 'ship' as one that includes, "every description of vessel used in navigation not propelled by oars." So in that case what was a 'vessel'? Parliament had already asked itself that question and given an unlikely answer. The Act defines a 'vessel' as "any ship or boat or any description of vessel used in navigation". This was the argument in a circle with a vengeance.

---

[1] An 'ACV' was the term used to cover every type of vehicle that 'hovered', from a warehouse platform to passenger-carrying or military hovercraft. In this Appendix we have simply used the term 'hovercraft' where ACV might have been more legally correct.

Textbooks were not helpful either since they suggested that whether or not a craft came within the statutory definition of a 'ship' depended upon the facts of each case.

The next question was to enquire whether or not a hovercraft might be an aircraft. The power to require Certificates of Airworthiness derived from the Civil Aviation Act 1949, which empowered the Minister to make Orders in Council prohibiting aircraft from flying without valid Certificates of Airworthiness being in force. But, curiously enough, the Civil Aviation Act 1949 and the Air Navigation Order 1960 contained no definition of an 'aircraft'. The Chicago Convention 1944, to which the United Kingdom was a ratified party, used the term 'aircraft' but did not define it, although, in various annexes to it, 'aircraft' was defined for certain purposes only as meaning "any machine that can derive support in the atmosphere from the reaction of the air".

This definition gave rise to a great deal of argument. There were those who considered that a hovercraft was an 'aircraft' within the meaning of the definition because it could not move without the presence of the cushion of air which it created, and that this satisfied the term 'reaction of the air'. On the other hand there were those who said that a hovercraft could not be an 'aircraft' within this definition because, although the hovercraft could not move without the cushion of air, it could not equally move without the existence of the ground or water against which that cushion of air was created and sustained. This, they claimed, made it a surface vehicle.

But perhaps a hovercraft was a motor vehicle? A 'hover vehicle' is defined in the Road Traffic Act 1962 as being a motor vehicle, whether or not it is adapted or intended for use on roads, but at the same time was not to be treated as a vehicle of any of these classes such as 'motor cars', 'motor cycles', 'light locomotives', 'heavy locomotives' defined in the Road Traffic Act 1960. The Road Traffic Act 1962, allowed the Minister of Transport to regulate that any provision of the Road Traffic Act 1960, or the Road Transport Lighting Act 1957, which would otherwise have applied to 'hover vehicles', would not apply to them, or would apply to them subject to such modifications as may be specified and vice versa. As of the end of 1965 no such regulations had been made by the Minister.

The situation appeared to be most unlikely that a hovercraft was a ship, more likely that it was some sort of aircraft, and quite certain that, for the purposes of its use on roads, it was a motor vehicle.

Until July, 1972, the control over hovercraft in the United Kingdom was based on Air Navigation legislation. The method of construction of hovercraft in many ways was similar to the method of construction of aircraft, and the early manufacturers of hovercraft had been mainly aircraft manufacturers. It was for these reasons that the certification of the design and construction of hovercraft and the supervision of maintenance had been entrusted to the Air Registration Board. The Air Registration Board, with powers handed down by the Minister of Aviation given to him by the Air Navigation Act 1960, exercised control by means of 'Permits to Fly' and 'Certificates of Construction and Performance'. The 'Permit' was subject to conditions designed for operational safety. A 'Certificate of Construction and Performance' was issued in respect of an identified Air Cushion Vehicle which had been examined by the Air Registration Board on the basis of the British Air Cushion Vehicle Safety Requirements. The safety of the vehicle as examined was considered satisfactory for the uses listed, provided that the limitations were observed, and that the vehicle was suitably maintained and operated. Lloyd's Register of Shipping issued its own Certificate of Construction and Performance.

It could, however, not be assumed that the grant of a 'Permit to Fly' made a hovercraft an aircraft. All it did was to attempt to cover the use of a vehicle, in circumstances where its use might otherwise be found to be illegal, with a coat of respectability. What was clearly required was a proper definition and proper regulation of ACVs for the future.

Since the hovercraft had some of the characteristics of a ship, some of an aircraft and some of a land vehicle, the easiest solution seemed to be to rationalise the situation by regarding hovercraft as a novel means of transport, different in kind from ships, aircraft and motor vehicles. An Enabling Act in simple terms would enable an appropriate minister by regulation to provide for the use of hovercraft. Ideally, such legislation and regulation would cover the major topics of 'Construction and Fitness for Use', 'Legal Liability to Passengers', 'Legal Liability to Third Parties', 'Safety', and 'Crew Licensing'.

The Hovercraft Act

When Parliament finally took notice of the existence of this new type of craft it was found convenient to do so largely by applying to hovercraft, with modification, provisions of existing statutes such as the Merchant Shipping Acts, the Carriage of Goods by Sea Act, Civil Aviation Acts and one section of a Road Traffic Act. So, while hovercraft have an Act of their own, the Hovercraft Act 1968, which brought hovercraft within the Admiralty jurisdiction, its effect was to give powers for Orders in Council to be made to apply such existing statutes to hovercraft.

In the Hovercraft Act 1968 it gave the legal definition of a hovercraft as:

A vehicle which is designed to be supported when in motion wholly or partly by air expelled from the vehicle to form a cushion of which the boundaries include the ground, water or the surface beneath the vehicle.

This followed fairly closely the technical definition of the Air Cushion Vehicle which was:

Capable of being supported by a continuously generated cushion of air that is dependent for its effectiveness on the proximity of the surface over which the vehicle operates, whether at rest or in motion.

The name 'hovercraft', thought up by Cockerell and his wife, came within the meaning of the Air Cushion Vehicles manufactured by, or under licence from, Hovercraft Development Limited. The term 'hovercraft' was more restricted in meaning than 'Air Cushion Vehicle', but in the absence of the copyright protections, and having regard to the legal definition, the terms are synonymous. The virtual acceptance of the technical definition for the purpose of the law was unfortunate, for no differentiation, apart from the existence of government departmental discretion, was made between the safety requirements of hovercraft transporting people or goods for reward and, say, a hover platform used in a warehouse to carry boxes three inches above the floor.

The definition given by *Lloyd's Register of Shipping* is, perhaps, the most comprehensive and also the longest: An Air Cushion Vehicle is:

A self propelled vehicle which operates amphibiously or only on water, and is capable of supporting at least 75 per cent of its fully loaded weight both whilst stationary and whilst underway on one or more cushions of air. These cushions are to be continuously generated within the vehicle and are to be dependent for their effectiveness on the proximity of the surface over which the vehicle operates.

The Hovercraft Act 1968 was passed on 26th July 1968. It was an enabling Act giving powers to make Orders in Council with respect to hovercraft, providing for the application of certain enactments to hovercraft and providing that certain enactments which relate to Admiralty jurisdiction and the rules of law which relate to maritime liens should have effect as they apply to ships. The subordinate legislation was thus based on two different corpus of legal rules, although the actual method of control does not differ greatly from the previous one. The 'Control of Construction and Safety of Hovercraft' was the task of the Air Registration Board (from 1972 the Civil Aviation Authority), and the 'Certificate of Fitness' took the form of a 'Safety Certificate'. This included supervision of maintenance and corresponded with the formerly issued 'Certificate of Construction and Performance'. The 'Permit to Fly' was replaced by the 'Operating Permit' granted by the Secretary of State, i.e. Department of Trade, Marine Division, and contained many of the provisions that were included in the earlier permit. Other requirements previously included now formed part of the various Orders in Council made under this new Act.

From 12th July 1972, a hovercraft was not treated as a ship, aircraft or motor vehicle for the purposes of the law and, since that date, stands as a vehicle in its own right. Fitted and machined into niches in dozens of Acts of Parliament and hundreds of statutory rules and orders concerned with ships, aircraft or motor vehicles, the legal system of the United Kingdom achieved near perfection in providing a workable legislative framework for a novel form of transportation. In some respects the near perfection is illusory but the British achievement was without parallel, and would provide a starting point for any eventual internationally agreed control over hovercraft operations.

The Experiences of a CAA Hovercraft Test Pilot

The Department of Trade and Industry, Marine Division, was largely responsible for issuing the 'Operating Permit' and monitoring the safe operation of hovercraft operators, while the CAA was more concerned with the operational performance of new and modified hovercraft.

John Syring, left Seaspeed as a Senior Captain in 1974 to join the CAA as a test pilot:

> *I relieved Geoff Howitt, a former Battle of Britain pilot, who became a test pilot for the CAA and, as his years advanced, he stopped flying aeroplanes and became their first hovercraft test pilot. He was looking forward to retirement and so his job came up. It was offered to David Wise of Hoverlloyd and he was 'wise' not to take it, so I did. But in some ways I shouldn't have taken it because Geoff Howitt, who was a very quiet understated sort of man, said to me, "John, you will find the job pretty slow". He didn't want to let the CAA down, but he did want to give me a tip because he said I'd spend ten days sitting in the office for half a day on a hovercraft at best. I stayed until 1977 and by then the industry was perceptively declining. The hovercraft department within the CAA was declining too. We started off with a hydraulics specialist, a structures specialist, a propeller specialist, Deputy Head of Department, a Head of Department, a senior clerk and two secretaries. It was set up as a fully fledged*

department for a big industry, but I could see it would eventually disappear. I needed to move on before it was too late to easily get back into seafaring and so I had this glorious two weeks when I was working for both the CAA and P&O.

During my time there I didn't need to know very much about the Hovercraft Act. I had to prepare trials programmes and list all the things the CAA would like to investigate on a particular craft. Each time I had a trial to do – obviously the information about the craft was flowing through the office – I knew how it was supposed to perform so I made out a trials schedule accordingly, but the interesting point was that the way that the hovercraft was administered, or supervised, put the responsibility on the developers and manufacturers for certifying that their craft was safe. The CAA only came in to look at various aspects of it just to confirm that things were in fact as they were supposed to be. We didn't do a full flight test programme. We might do the lightweight or heavyweight trials, but might not do the heavy weather trials, or we might do any two of three. We were only cherry picking because the responsibility rested with the manufacturer and his chief designer.

The hovercraft had profound disadvantages and I got more and more into this. Essentially they can really only operate in relatively sheltered waters not too far away from a port of refuge. I left the CAA in 1977, but right up to 1997 I stayed on as the test pilot for the CAA as a sub-contractor, so on my days off from P&O I carried on being the man. You see, they didn't have anybody and they only needed somebody very occasionally, so I did all sorts of jobs for 20 years – HM2s, VT-1 and VT-2, BH-7s and all sorts of funny little craft. It was a very interesting time. What happened was I retired from P&O and it was just about the time that the CAA was going to give up its interest in hovercraft and hand it all over to the Marine Division, which they did eventually, because the hovercraft was really treated as a ship.

You've got to remember that over those 30 years the Marine Division became far more competent in dealing with exotic craft – jetfoils, catamarans, oil rigs, lifting barge vessels. Suddenly their expertise grew exponentially, so they just took the hovercraft in their stride in the end. The CAA bailed out and I bailed out at the same time.

So the end result is that it can safely be concluded that the hovercraft is more of a marine vehicle than an aircraft. In the early days the hovercraft was essentially a marine vehicle built by aircraft manufacturers with an amphibious capability, using aircraft engines and systems. Today the modern Griffon hovercraft has done away with the biggest cost items in the older aviation-orientated hovercraft. They now have welded aluminium type structures, or composites, which are simpler and cheaper to produce; lighter and less complicated main skirts and diesel engines. A propeller and rudder pedal are all that are left of the world of aviation.

Even so this unique concept did require its own rules and regulations.

Appendix 3

Measuring Speed

In the early decades there was no totally accurate method of measuring the hovercraft's speed. The deficiency remained for most of the cross-Channel era; right up until full coverage Global Positioning (GPS) became available in the early '90s.

The SR.N4 was fitted with two aero speed measuring devices, one measuring speed over the ground, the other to measure speed through the air. Although perfectly reliable in the air, both were next to useless down at sea level, particularly the Doppler groundspeed indication, which at times could be grossly in error.

The Air Speed Indicator (ASI) was marginally less prone to error and tended to be the instrument the captain kept an eye on, but its indication was never treated as gospel.

The pitot head for the ASI above the control cabin.

Speed through the air is important information for an aircraft, which relies on wind speed over the aerofoil to provide lift, something the hovercraft doesn't need. At the comparatively low speeds of hovercraft the device in, say, a 20 knot headwind would show 20 knots more speed than the actual speed over the ground. Perhaps not a problem, subtracting to compare with the ground speed indication, but with no guaranteed knowledge of the actual wind speed the captain was reduced to guesswork. He was left trying to juggle one guessed at number with another that could be badly in error.

If this wasn't problem enough, the sensing device for the Air Speed Indicator was simply a forward facing tube[1], which undoubtedly would have been another victim of the 'dirty' environment. Interestingly, the technicians say that although the tube was constantly in need of cleaning, snags concerning the unreliability of the ASI were rarely reported, most probably an indication that the flight crew had low expectations of its usefulness from the start.

Consequently, as the craft slowed on the approaches, more often than not the captain fell back on his visual sense of movement and speed. Of course this was fine until the visibility clamped down. In those circumstances the captain simply had to trust the navigator and hope that his speed was OK.

Ground speed measurement, a Doppler system typically used on aircraft, didn't translate at all. The transceiver head, which transmitted and received a radio signal off the sea surface, was placed at the rear of the craft. One only has to look at photographs of the SR.N4 at speed to note the massive amount of spray caused by the air escaping from under the skirt; where it was situated the Doppler head was in the thick of it. Most of the time the Doppler couldn't distinguish between spray and sea surface, so the consequent ground speed was erratic at best and most of the time downright misleading.

The Civil Aviation Authority, who regulated such things, sanctioned our permit to operate based on an upper service speed limit of 55 knots. All very well, but because most of the time we had no real idea what speed we were actually achieving, many times we overstepped the mark, only finding out the actual speed once we had arrived and worked it out, but even then it was an average speed and not a speed at any particular time.

Later down the line we experimented with placing the Doppler head just in front of the control cabin and facing forward. Initially this worked like a dream and for the first time we were getting consistent and even accurate ground speed. Why hadn't anyone thought of it before? We soon found the answer to that question the first day we experienced flat calm – contrary to popular belief it does happen occasionally in the Straits. As soon as the craft moved over the glassy smooth surface,

---

[1] Known technically as a Pitot Head.

probably increasing speed into the 70 knot region, the Doppler indication dropped firmly to zero.

Initial Doppler head position.

The Doppler signal was obviously hitting the mirror surface and bouncing off into infinity, so back to the drawing board.

Eventually the most efficient place for the transceiver was found, under the craft, prompting as always, the inevitable question – why wasn't it thought of before?

On the car deck there were five manhole size plates, which, when uncovered, revealed vertical tubes passing right through the basic raft structure. Their purpose was to provide the mounting for a set of portable jacks that could be used in the event of the craft needing to be lifted whilst away from the usual jacking system at our base in Pegwell Bay. On a few occasions the portables were used in Calais, when more extensive repairs to the skirt system were needed to return the craft safely. One case already mentioned was the incident when *Swift* lost her skirt at Sangatte in November 1971. (Chapter 21: Hoverlloyd's SR.N4 - The First Years.)

Experimental forward position.

The Doppler head was placed at the bottom end of one of the apertures, slightly protruding up into it but not sufficient to block the signal. In this position it worked well and, seeing that this was probably accomplished in 1980, showed once more that for most of Hoverlloyd's brief life we were still 'prototype testing with passengers'.

Bibliography

The Beginning – The SR-N1 Hovercraft by R.L. Wheeler and J.B. Chaplin. Published 2007 by Cross Publishing.

From River to Sea, The Marine Heritage of Sam Saunders by Raymond L. Wheeler. Published 1993 by Cross Publishing.

Whatever Happened to the Hovercraft? by John Lefeaux. Published 2001 by Pentland Books.

The Hoverspeed Story by Miles Cowsill and John Hendy. Published 1993 by Ferry Publications.

The Law of Hovercraft by Laszlo Joseph Kovats

Goodwin Sands Shipwrecks by Richard Larn. Published 1977 by David and Charles.

Air-Cushion Vehicles (Flight International supplement)

Hovercraft World

Hovering Craft & Hydrofoil

Kent Life

The Institute of Marine Engineers

Glossary of Nautical and Hovercraft Terms

Auxiliary Power Unit (APU) — The APU provides electrical power. In the case of the SR.N4, the Rover Gas Turbine was originally developed as a car engine in the 1950s.

Air Cushion Vehicle (ACV) — The official generic term for a hovercraft.

All Up Weight (AUW) — The total weight of the hovercraft when fully laden. Includes passengers, cars, baggage, crew and fuel.

Annular Jet — Air directed downward around the outer edge of the hovercraft.

Anti-Plough Bag — A plough-in safeguard. An extra section of the bow skirt seals off automatically as the bow drops, providing buoyancy and a certain element of fendering.

Automatic Radar Plotting Aid (ARPA) — A computerised system that calculates target movement for the navigator. Not available on the N4 until the last decade of operation.

Bag — The major part of the skirt system attached to the bottom of the plenum.

Buoyancy Tank — The flat watertight raft structure of a hovercraft, around which everything else is built.

Cable — One tenth of a nautical mile.

Caboose — The box-like structure at the after corners of the SR.N4, housing the extensive air filtration systems for the Proteus engines.

Carbo Blasting — A method of cleaning salt from the gas turbines. Entails pouring a bag of finely ground walnut shells into a running engine.

Decca 202
Part of the original SR.N4 radar fit. A smaller back-up radar to the principal radar, the Decca 629.

Decca 629
The original main radar, fitted on the SR.N4 in 1968.

Decca Navigator
An electronic navigation system now superseded by GPS. The model used on the SR.N4 was an aircraft version with a roller chart mounted in front of the captain.

Displacement
Total weight at any time.

Duralumin
A type of aluminium alloy commonly used in aircraft construction.

Elephant's Feet
Colloquial term for the five landing pads supporting the SR.N4 when 'off hover'.

Elevons
Combines the function of ailerons and elevators into a single set of control surfaces.

Fingers
Attached to the bottom of the skirt 'bag', the fingers move up and down individually and thus provide a flexible interface with the surface of the sea.

Global Positioning Systems (GPS)
A satellite-based navigational system, with world-wide coverage since the mid-nineties. Provides positional accuracy, generally within a few metres.

Hinge Line
The connection between skirt segments and the plenum hard structure. A series of metal connections which look and act like a hinge.

Hump
The depression in the water formed by the downward pressure of air under the hovercraft. As the craft gains speed (hump speed) it leaves the hump behind and can accelerate to maximum speed.

Jupes

The skirt system on the French N500. A set of (estimated) 46 individual annular jets instead of one single one as in the British designs.

Keel Bag

A single 'bag' which formed a 'T' with the stability trunks running aft to the rear skirt bags, effectively dividing the rear chamber into two lateral air chambers.

Knot

Nautical measure of speed in nautical miles per hour. 1 knot = 1.15 mph.

LCAC

Landing Craft Air Cushion. US Navy hovercraft designed to deploy from specially designed assault ships.

Low Pressure Skirt

A much modified skirt configuration for the SR.N4 Mk.3.

Mountbatten

The 'class' name given to the SR.N4 hovercraft, in honour of Lord Louis Mountbatten who was largely responsible for drawing attention to Cockerell's discovery.

Nautical Mile (nm)

Longer than a statute mile. (Statute mile = 5,280 feet. Nautical mile = 6,080 feet)

North-Up

A radar presentation which shows north at the top of the screen, similar to a map.

Plenum Chamber

Part of a hovercraft's hard structure which directs the lift air under the machine via the skirt system.

Plough-In

Plough-in occurs in conjunction with a sudden increase in drag at the forward end of the skirt system. The result is a tipping moment between the drag at the sea surface and the unimpeded thrust of the propeller(s), resulting in the craft increasingly being dragged further into the water. Once plough-in commences it is invariably impossible to stop and it can result in severe damage and even capsize small craft.

Proteus

The Rolls Royce Proteus Gas Turbine, originally the Bristol Proteus, designed at the end of World War Two. The only engine 'marinised' sufficiently to cope with large amounts of salt water thrown up by the hovercraft skirts.

Puff Ports

Devices that direct plenum air laterally as a means of directional control.

Pylon

A swivelling structure supporting the propeller gearbox and propeller.

Relative Motion

Targets on the radar screen move relatively to the centre of the screen, where the hovercraft is situated. They do not show their geographical movement.

Segment

One section of the skirt configuration. Separate segments allow easier and cheaper repair of skirt damage.

Significant Wave Height

Defined as the mean wave height (trough to crest) of the highest third of the waves, but not the highest wave that can be encountered.

Sperry G4B Compass

Standard fit on the SR.N4, the G4B is in effect a gyro-stabilised magnetic compass.

Spring

A ship's wire mooring rope.

Stability Trunks

Two laterally fitted 'bags' which divided the air under the hovercraft into two separate fore and aft chambers.

Swell

The underlying wave pattern which may have been engendered a considerable distance away. As distinct from 'sea', which are waves caused by local wind movement.

Tail Planes	Used on some hovercraft as a trimming device.
Target	A common term for a moving echo on the radar screen.
Ticket	Seafaring parlance for 'certificate'.
Torque	The twisting or rotating force on a mechanism – the load the engine has to cope with. Also the amount of strain applied to a screw or bolt.
Tracking	Refers to the actual direction in which a vessel at sea is moving. As distinct from 'course', the direction in which the vessel is pointing.
True Motion	Targets on the radar screen show their geographical 'true' movement.
Variable Pitch Propeller	A propeller whose blades can be turned to give either forward or reverse thrust.
Vector	A line drawn on the radar screen by the ARPA. Shows a target's course and speed either as true or relative movement.
Wave Rider Buoy	A hydrographic surveying device, which measures and transmits data on wave movement.
Yoke	A 'W' shaped control device, which on the SR.N4 controls lateral movement of the pylons.

INDEX

PRINTED AND BOUND BY

Copytech (UK) Limited trading as Printondemand-worldwide,
9 Culley Court, Bakewell Road, Orton Southgate. Peterborough, PE2
6XD, United Kingdom.